Nonlinear and Convex Analysis

PURE AND APPLIED MATHEMATICS

A Program of Monographs, Textbooks, and Lecture Notes

EXECUTIVE EDITORS

Earl J. Taft
Rutgers University
New Brunswick, New Jersey

Zuhair Nashed
University of Delaware
Newark, Delaware

CHAIRMEN OF THE EDITORIAL BOARD

S. Kobayashi
University of California, Berkeley
Berkeley, California

Edwin Hewitt
University of Washington
Seattle, Washington

EDITORIAL BOARD

M. S. Baouendi
Purdue University

Jack K. Hale
Brown University

Marvin Marcus
University of California, Santa Barbara

W. S. Massey
Yale University

Leopoldo Nachbin
Centro Brasileiro de Pesquisas Físicas and University of Rochester

Anil Nerode
Cornell University

Donald Passman
University of Wisconsin-Madison

Fred S. Roberts
Rutgers University

Gian-Carlo Rota
Massachusetts Institute of Technology

David Russell
University of Wisconsin-Madison

Jane Cronin Scanlon
Rutgers University

Walter Schempp
Universität Siegen

Mark Teply
University of Wisconsin-Milwaukee

LECTURE NOTES

IN PURE AND APPLIED MATHEMATICS

1. *N. Jacobson,* Exceptional Lie Algebras
2. *L.-Å. Lindahl and F. Poulsen,* Thin Sets in Harmonic Analysis
3. *I. Satake,* Classification Theory of Semi-Simple Algebraic Groups
4. *F. Hirzebruch, W. D. Newmann, and S. S. Koh,* Differentiable Manifolds and Quadratic Forms (out of print)
5. *I. Chavel,* Riemannian Symmetric Spaces of Rank One (out of print)
6. *R. B. Burckel,* Characterization of C(X) Among Its Subalgebras
7. *B. R. McDonald, A. R. Magid, and K. C. Smith,* Ring Theory: Proceedings of the Oklahoma Conference
8. *Y.-T. Siu,* Techniques of Extension on Analytic Objects
9. *S. R. Caradus, W. E. Pfaffenberger, and B. Yood,* Calkin Algebras and Algebras of Operators on Banach Spaces
10. *E. O. Roxin, P.-T. Liu, and R. L. Sternberg,* Differential Games and Control Theory
11. *M. Orzech and C. Small,* The Brauer Group of Commutative Rings
12. *S. Thomeier,* Topology and Its Applications
13. *J. M. Lopez and K. A. Ross,* Sidon Sets
14. *W. W. Comfort and S. Negrepontis,* Continuous Pseudometrics
15. *K. McKennon and J. M. Robertson,* Locally Convex Spaces
16. *M. Carmeli and S. Malin,* Representations of the Rotation and Lorentz Groups: An Introduction
17. *G. B. Seligman,* Rational Methods in Lie Algebras
18. *D. G. de Figueiredo,* Functional Analysis: Proceedings of the Brazilian Mathematical Society Symposium
19. *L. Cesari, R. Kannan, and J. D. Schuur,* Nonlinear Functional Analysis and Differential Equations: Proceedings of the Michigan State University Conference
20. *J. J. Schäffer,* Geometry of Spheres in Normed Spaces
21. *K. Yano and M. Kon,* Anti-Invariant Submanifolds
22. *W. V. Vasconcelos,* The Rings of Dimension Two
23. *R. E. Chandler,* Hausdorff Compactifications
24. *S. P. Franklin and B. V. S. Thomas,* Topology: Proceedings of the Memphis State University Conference
25. *S. K. Jain,* Ring Theory: Proceedings of the Ohio University Conference
26. *B. R. McDonald and R. A. Morris,* Ring Theory II: Proceedings of the Second Oklahoma Conference
27. *R. B. Mura and A. Rhemtulla,* Orderable Groups
28. *J. R. Graef,* Stability of Dynamical Systems: Theory and Applications
29. *H.-C. Wang,* Homogeneous Branch Algebras
30. *E. O. Roxin, P.-T. Liu, and R. L. Sternberg,* Differential Games and Control Theory II
31. *R. D. Porter,* Introduction to Fibre Bundles
32. *M. Altman,* Contractors and Contractor Directions Theory and Applications
33. *J. S. Golan,* Decomposition and Dimension in Module Categories
34. *G. Fairweather,* Finite Element Galerkin Methods for Differential Equations
35. *J. D. Sally,* Numbers of Generators of Ideals in Local Rings
36. *S. S. Miller,* Complex Analysis: Proceedings of the S.U.N.Y. Brockport Conference
37. *R. Gordon,* Representation Theory of Algebras: Proceedings of the Philadelphia Conference
38. *M. Goto and F. D. Grosshans,* Semisimple Lie Algebras
39. *A. I. Arruda, N. C. A. da Costa, and R. Chuaqui,* Mathematical Logic: Proceedings of the First Brazilian Conference

40. *F. Van Oystaeyen,* Ring Theory: Proceedings of the 1977 Antwerp Conference
41. *F. Van Oystaeyen and A. Verschoren,* Reflectors and Localization: Application to Sheaf Theory
42. *M. Satyanarayana,* Positively Ordered Semigroups
43. *D. L. Russell,* Mathematics of Finite-Dimensional Control Systems
44. *P.-T. Liu and E. Roxin,* Differential Games and Control Theory III: Proceedings of the Third Kingston Conference, Part A
45. *A. Geramita and J. Seberry,* Orthogonal Designs: Quadratic Forms and Hadamard Matrices
46. *J. Cigler, V. Losert, and P. Michor,* Banach Modules and Functors on Categories of Banach Spaces
47. *P.-T. Liu and J. G. Sutinen,* Control Theory in Mathematical Economics: Proceedings of the Third Kingston Conference, Part B
48. *C. Byrnes,* Partial Differential Equations and Geometry
49. *G. Klambauer,* Problems and Propositions in Analysis
50. *J. Knopfmacher,* Analytic Arithmetic of Algebraic Function Fields
51. *F. Van Oystaeyen,* Ring Theory: Proceedings of the 1978 Antwerp Conference
52. *B. Kedem,* Binary Time Series
53. *J. Barros-Neto and R. A. Artino,* Hypoelliptic Boundary-Value Problems
54. *R. L. Sternberg, A. J. Kalinowski, and J. S. Papadakis,* Nonlinear Partial Differential Equations in Engineering and Applied Science
55. *B. R. McDonald,* Ring Theory and Algebra III: Proceedings of the Third Oklahoma Conference
56. *J. S. Golan,* Structure Sheaves over a Noncommutative Ring
57. *T. V. Narayana, J. G. Williams, and R. M. Mathsen,* Combinatorics, Representation Theory and Statistical Methods in Groups: YOUNG DAY Proceedings
58. *T. A. Burton,* Modeling and Differential Equations in Biology
59. *K. H. Kim and F. W. Roush,* Introduction to Mathematical Consensus Theory
60. *J. Banas and K. Goebel,* Measures of Noncompactness in Banach Spaces
61. *O. A. Nielson,* Direct Integral Theory
62. *J. E. Smith, G. O. Kenny, and R. N. Ball,* Ordered Groups: Proceedings of the Boise State Conference
63. *J. Cronin,* Mathematics of Cell Electrophysiology
64. *J. W. Brewer,* Power Series Over Commutative Rings
65. *P. K. Kamthan and M. Gupta,* Sequence Spaces and Series
66. *T. G. McLaughlin,* Regressive Sets and the Theory of Isols
67. *T. L. Herdman, S. M. Rankin, III, and H. W. Stech,* Integral and Functional Differential Equations
68. *R. Draper,* Commutative Algebra: Analytic Methods
69. *W. G. McKay and J. Patera,* Tables of Dimensions, Indices, and Branching Rules for Representations of Simple Lie Algebras
70. *R. L. Devaney and Z. H. Nitecki,* Classical Mechanics and Dynamical Systems
71. *J. Van Geel,* Places and Valuations in Noncommutative Ring Theory
72. *C. Faith,* Injective Modules and Injective Quotient Rings
73. *A. Fiacco,* Mathematical Programming with Data Perturbations I
74. *P. Schultz, C. Praeger, and R. Sullivan,* Algebraic Structures and Applications Proceedings of the First Western Australian Conference on Algebra
75. *L. Bican, T. Kepka, and P. Nemec,* Rings, Modules, and Preradicals
76. *D. C. Kay and M. Breen,* Convexity and Related Combinatorial Geometry: Proceedings of the Second University of Oklahoma Conference
77. *P. Fletcher and W. F. Lindgren,* Quasi-Uniform Spaces
78. *C.-C Yang,* Factorization Theory of Meromorphic Functions
79. *O. Taussky,* Ternary Quadratic Forms and Norms
80. *S. P. Singh and J. H. Burry,* Nonlinear Analysis and Applications
81. *K. B. Hannsgen, T. L. Herdman, H. W. Stech, and R. L. Wheeler,* Volterra and Functional Differential Equations

82. *N. L. Johnson, M. J. Kallaher, and C. T. Long,* Finite Geometries: Proceedings of a Conference in Honor of T. G. Ostrom
83. *G. I. Zapata,* Functional Analysis, Holomorphy, and Approximation Theory
84. *S. Greco and G. Valla,* Commutative Algebra: Proceedings of the Trento Conference
85. *A. V. Fiacco,* Mathematical Programming with Data Perturbations II
86. *J.-B. Hiriart-Urruty, W. Oettli, and J. Stoer,* Optimization: Theory and Algorithms
87. *A. Figa Talamanca and M. A. Picardello,* Harmonic Analysis on Free Groups
88. *M. Harada,* Factor Categories with Applications to Direct Decomposition of Modules
89. *V. I. Istrăţescu,* Strict Convexity and Complex Strict Convexity: Theory and Applications
90. *V. Lakshmikantham,* Trends in Theory and Practice of Nonlinear Differential Equations
91. *H. L. Manocha and J. B. Srivastava,* Algebra and Its Applications
92. *D. V. Chudnovsky and G. V. Chudnovsky,* Classical and Quantum Models and Arithmetic Problems
93. *J. W. Longley,* Least Squares Computations Using Orthogonalization Methods
94. *L. P. de Alcantara,* Mathematical Logic and Formal Systems
95. *C. E. Aull,* Rings of Continuous Functions
96. *R. Chuaqui,* Analysis, Geometry, and Probability
97. *L. Fuchs and L. Salce,* Modules Over Valuation Domains
98. *P. Fischer and W. R. Smith,* Chaos, Fractals, and Dynamics
99. *W. B. Powell and C. Tsinakis,* Ordered Algebraic Structures
100. *G. M. Rassias and T. M. Rassias,* Differential Geometry, Calculus of Variations, and Their Applications
101. *R.-E. Hoffmann and K. H. Hofmann,* Continuous Lattices and Their Applications
102. *J. H. Lightbourne, III, and S. M. Rankin, III,* Physical Mathematics and Nonlinear Partial Differential Equations
103. *C. A. Baker and L. M. Batten,* Finite Geometries
104. *J. W. Brewer, J. W. Bunce, and F. S. Van Vleck,* Linear Systems Over Commutative Rings
105. *C. McCrory and T. Shifrin*, Geometry and Topology: Manifolds, Varieties, and Knots
106. *D. W. Kueker, E. G. K. Lopez-Escobar, and C. H. Smith*, Mathematical Logic and Theoretical Computer Science
107. *B.-L. Lin and S. Simons*, Nonlinear and Convex Analysis: Proceedings in Honor of Ky Fan

Other Volumes in Preparation

KY FAN

Nonlinear and Convex Analysis

PROCEEDINGS IN HONOR OF KY FAN

Edited by

BOR-LUH LIN
The University of Iowa
Iowa City, Iowa

STEPHEN SIMONS
University of California
Santa Barbara, California

MARCEL DEKKER, INC. New York and Basel

ISBN: 0-8247-7777-8

COPYRIGHT © 1987 by MARCEL DEKKER, INC. ALL RIGHTS RESERVED

Neither this book nor any part may be reproduced or transmitted in any form or by any means, electronic or mechanical, including photocopying, microfilming, and recording, or by any information storage and retrieval system, without permission in writing from the publisher.

MARCEL DEKKER, INC.
270 Madison Avenue, New York, New York 10016

Current printing (last digit):
10 9 8 7 6 5 4 3 2 1

PRINTED IN THE UNITED STATES OF AMERICA

Preface

Professor Ky Fan retired at the end of the 1984-85 academic year, and the Department of Mathematics of the University of California at Santa Barbara held a conference in his honor June 23-26, 1985.

This volume contains expanded versions of the talks given at the conference, as well as papers contributed by others, many of whom would have liked to attend the conference but were unable to come.

Before writing about the conference, I will give a brief history of Professor Fan's career, although it is not really possible to do it justice in a few lines.

Professor Fan obtained a B.S. from the National Peking University in 1936 and a D.Sc. Math from the University of Paris in 1941. He was a member of the Institute for Advanced Study at Princeton from 1945 to 1947, and held professorial positions at the University of Notre Dame, Wayne State University, and Northwestern University before coming to the University of California at Santa Barbara in 1965. He has been a visiting professor at the University of Texas at Austin, Hamburg University in Germany, and the University of Paris IX in France.

Elected a member of the Academica Sinica in 1964, he served as director of the Institute of Mathematics at the Academica Sinica for two terms from 1978 to 1984. He has been on many editorial boards, including that of the Journal of Mathematical Analysis and Applications since its founding in 1960.

A student and collaborator of M. Fréchet, Professor Fan was also influenced by J. von Neumann and H. Weyl. He made fundamental contributions to operator and matrix theory, convex analysis and inequalities, linear and nonlinear programming, topology and fixed-point theory, and topological groups. Many of his results have become part of the basic literature in these fields. Furthermore, his work in fixed-point theory has found wide application in mathematical economics and game theory, potential theory, calculus of variations, and differential equations, and has had many consequences in nonlinear functional analysis.

Professor Fan has trained twenty-two Ph.D. students, many of whom hold positions at prestigious universities both in the United States and abroad. Seven of them were at the conference: John Cantwell, Michael Geraghty, Donald Hartig, Charles Himmelberg, Ronald Knill, Bor-Luh Lin, Michael Powell, and Raimond Struble.

In all, sixty-three mathematicians from three continents came to the University of California at Santa Barbara for the conference. The one hour talks were given by Professors Jean-Pierre Aubin of the University of Paris, Felix Browder of the University of Chicago, Zhang Gong-Qing (K. C. Chang) of Peking University, Andrzej Granas of the University of Montreal, Paul Halmos of the University of Santa Clara, and Shizuo Kakutani of Yale University. Professor Samuel Karlin of Stanford University was also scheduled to give a one hour talk, but was prevented from coming by illness. Twenty-seven shorter talks were also given, many in areas pioneered by Professor Fan.

On the evening of June 25, those attending the conference joined members of the Department of Mathematics for a banquet in honor of Professor Fan. The banqueteers were rather startled when one of the speakers—a student of Professor Fan who prefers to remain nameless—started to undress. To the relief (and amusement) of everyone it was merely to demonstrate that he was wearing a Ky Fan T-shirt.

I spoke to many of the participants in the conference, trying to find out more about Professor Fan's career before he came to Santa Barbara. They confirmed unanimously my own feelings; they spoke of his high standards in mathematics, not only for himself, but also for his colleagues and students. But they also spoke of his fine qualities as a human being and his generosity.

Although Professor Fan has now retired from his teaching duties, he remains mathematically active, and will undoubtedly continue to do so for many years to come.

Thanks are due to Marvin Marcus, Associate Vice Chancellor of Research and Academic Development, and David Sprecher, Provost of the College of Letters and Science at the University of California at Santa Barbara, whose offices provided financial support for the conference; to James Robertson, Chairman of the Department of Mathematics at the University of California at Santa Barbara and all his staff, who worked so hard to make the conference a success; to all those who participated - in many cases coming a great distance and braving an unpleasant airline strike to get to Santa Barbara; to all those who could not participate in the conference but sent their good wishes; and to Professor Earl Taft and Professor Zuhair Nashed and Marcel Dekker, Inc., for including the Proceedings in their Lecture Notes in Pure and Applied Mathematics Series.

Stephen Simons

Contents

Contributors

JEAN-PIERRE AUBIN Mathematical Research Center, University of Paris-Dauphine, Paris, France

CHANG KUNG-CHING (ZHANG GONG-QING) Department of Mathematics, Peking University, Beijing, People's Republic of China

GONG-NING CHEN Department of Mathematics, Beijing Normal University, Beijing, People's Republic of China

CHARLES K. CHUI Department of Mathematics, Texas A&M University, College Station, Texas

CARL L. DeVITO* Department of Mathematics, University of Arizona, Tucson, Arizona

HALINA FRANKOWSKA Mathematical Research Center, University of Paris-Dauphine, Paris, France

ANDRZEJ GRANAS Department of Mathematics, University of Montreal, Montreal, Quebec, Canada

CHUNG-WEI HA Department of Mathematics, National Tsing Hua University, Hsinchu, Taiwan, Republic of China

NORIMICHI HIRANO Department of Mathematics, Yokohama National University, Yokohama, Japan

CHARLES D. HORVATH Champlain Regional College, St. Lambert, Quebec, Canada

TATSURO ICHIISHI** Department of Economics, The University of Iowa, Iowa City, Iowa

GEORGE ISAC Department of Mathematics, Royal Military College, St.-Jean, Quebec, Canada

---

*Current affiliation: Naval Postgraduate School, Monterey, California

**Current affiliation: Ohio State University, Columbus, Ohio

HIDETOSHI KOMIYA College of Commerce, Nihon University, Tokyo, Japan

HANG-CHIN LAI Institute of Mathematics, National Tsing Hua University, Hsinchu, Taiwan, Republic of China

M. J. LAI Department of Mathematics, Texas A&M University, College Station, Texas

FON-CHE LIU Institute of Mathematics, Academia Sinica, Taipei, Taiwan, Republic of China

ROBERT H. LOHMAN Department of Mathematics and Sciences, Kent State University, Kent, Ohio

JONG YEOUL PARK Department of Mathematics, Busan National University, Busan, Republic of Korea

W. V. PETRYSHYN Department of Mathematics, Rutgers University, New Brunswick, New Jersey

SIMEON REICH* Department of Mathematics, The Technion-Israel Institute of Technology, Haifa, Israel

I. SHAFRIR Department of Mathematics, The Technion-Israel Institute of Technology, Haifa, Israel

VICTOR L. SHAPIRO, Department of Mathematics and Computer Science, University of California, Riverside, Riverside, California

SHI SHU-ZHONG** Mathematical Research Center, University of Paris-Dauphine, Paris, France

MAU-HSIANG SHIH Department of Mathematics, Chung Yuan University, Chung-Li, Taiwan, Republic of China

THADDEUS J. SHURA Department of Mathematics and Sciences, Kent State University at Salem, Salem, Ohio

IVAN SINGER Department of Mathematics, National Institute for Scientific and Technical Creation, Bucharest, Romania

WATARU TAKAHASHI Department of Information Science, Tokyo Institute of Technology, Tokyo, Japan

KOK-KEONG TAN Department of Mathematics, Statistics and Computing Science, Dalhousie University, Halifax, Nova Scotia, Canada

KENSUKE TANAKA Department of Mathematics, Niigata University, Niigata, Japan

---

*Current affiliation: The University of Southern California, Los Angeles, California

**Current affiliation: Nankai University, Tianjin, People's Republic of China

M. THERA Department of Mathematics, University of Limoges, Limoges, France

YAU-CHUEN WONG Department of Mathematics, The Chinese University of Hong Kong, Hong Kong

DAOXING XIA Department of Mathematics, Vanderbilt University, Nashville, Tennessee

Conference Participants*

Archbold, R. (Aberdeen, Scotland)
Aubin, J.-P. (Paris, France)
Browder, F.E. (Chicago, IL)
Beer, G.A. (Los Angeles, CA)
Bruck, R.E. (Los Angeles, CA)

Border, K.C. (Pasadena, CA)
Cantwell, J.C. (St. Louis, MO)
Chang, K.C. (Beijing, China)
Chu, H. (College Park, MD)
Chuan, J.-C. (Taiwan, China)

Chui, C.K. (College Station, TX)
Crandall, M.G. (Madison, WI)
Deprima, C.R. (Pasadena, CA)
Donoghue, W.F. (Irvine, CA)
Fournier, G. (Sherbrooke, Quebec)

Frankowska, H. (Paris, France)
Geraghty, M.A. (Iowa City, IA)
Granas, A. (Montreal, Quebec)
Greim, P. (Charleston, SC)
Gretsky, N. (Riverside, CA)

Halmos, P.R. (Santa Clara, CA)
Hartig, D.G. (San Luis Obispo, CA)
Hilario, R. (Chino, CA)
Himmelberg, C.J. (Lawrence, KS)
Hirano, N. (Yokahama, Japan)

Hoffman, H. (Los Angeles, CA)
Horvath, C. (Montreal, Quebec)
Hu, T. (Taiwan, China)
Ichiishi, T. (Iowa City, IA)
Kakutani, S. (New Haven, CT)

Kalisch, G.K. (Irvine, CA)
Knill, R.J. (New Orleans, LA)
Komiya, H. (Tokyo, Japan)
Komuro, N. (Hokkaido, Japan)

Kottman, C.A. (Springfield, VA)
Laetsch, T. (Tucson, AZ)
Lai, H.-C. (Taiwan, China)
Lai, M. (College Station, TX)
Lange, L.H. (San Jose, CA)

Lapidus, M. (Iowa City, IA)
Lassonde, M. (Clermont, France)
Lau, A.T. (Edmonton, Alberta)
Lee, T.Y. (University, AL)
Lin, B.-L. (Iowa City, IA)

Liu, F.-C. (Taiwan, China)
Liu, S.-C. (Taiwan, China)
McLinden, L. (Urbana, IL)
Mikusinksi, P. (Orlando, FL)
Namioka, I. (Seattle, WA)

Ng, K.F. (Hong Kong)
Pak, H. (Detroit, MI)
Petryshyn, W.V. (New Brunswick, NJ)
Phelps, R. (Seattle, WA)
Powell, M.H. (Santa Cruz, CA)

Reich, S. (Los Angeles, CA)
Schechter, M. (Irvine, CA)
Shapiro, V.L. (Riverside, CA)
Shekhtman, B. (Riverside, CA)
Shih, H.-H. (Taiwan, China)

Struble, R. (Rayleigh, NC)
Swaminathan, S. (Halifax, Nova Scotia)
Takahashi, W. (Tokyo, Japan)
Tan, K.-K. (Halifax, Nova Scotia)
Thera, M. (Limoges, France)

Wang, J.L.-M. (University, AL)
Wong, Y.C. (Hong Kong)
Wu, S.Y. (Iowa City, IA)
Xia, D. (Nashville, TN)

*Other than those from The University of California at Santa Barbara

List of Publications of Ky Fan

1. Sur une représentation des fonctions abstraites continues, C.R. Acad. Sci. Paris 210(1940), 429-431.
2. Sur les types homogènes de dimensions, C.R. Acad. Sci. Paris 211 (1940), 175-177.
3. Espaces quasi-réguliers, quasi-normaux et quasi-distanciés, C.R. Acad. Sci. Paris 211(1940), 348-351.
4. Caractérisation topologique des arcs simples dans les espaces accessibles de M. Fréchet, C.R. Acad. Sci. Paris 212(1949), 1024-1026.
5. Sur les ensembles possédant la propriété des quatre points, C.R. Acad. Sci. Paris 213(1941), 518-520.
6. Sur les ensembles monotones-connexes, les ensembles filiformes et les ensembles possédant la propriété des quatre points, Bull. Soc. Royale Sci. Liege 10(1941), 625-642.
7. Sur le théorème d'existence des équations différentielles dans l'analyse générale, Bull. Sci. Math. 65(1941), 253-264.
8. Sur quelques notions fondamentales de l'analyse générale, J. Math. Pures et Appl. 21(1942), 289-368.
9. Exposé sur le calcul symbolique de Heaviside, Revue Scientifique 80(1942), 147-163.
10. Sur le comportement asymptotique des solutions d'équations linéaires aux différences finies du second ordre, Bull. Soc. Math. France 70(1942), 76-96.
11. Les fonctions asymptotiquement presque-périodiques d'une variable entière et leur application à l'étude de l'itération des transformations continues, Math. Zeitschr. 48(1943), 685-711.
12. Une propriété asymptotique des solutions de certaines équations Linéaires aux differences finies, C.R. Acad. Sci. Paris 216(1943), 169-171.
13. Quelques propriétés caractéristiques des ensembles possédant la propriété des quatre points et des ensembles filiformes, C.R. Acad. Sci. Paris 216(1943), 553-555.
14. Nouvelles définitions des ensembles possédant la propriété des quatre points et des ensembles filiformes, Bull. Sci. Math. 67 (1943), 187-202.
15. Entfernung zweier zufälligen Grössen und die Konvergenz nach Wahrscheinlichkeit, Math. Zeitschr. 49(1943/44), 681-683.
16. Sur l'extension de la formule générale d'interpolation de M. Borel aux fonctions aléatoires, C.R. Acad. Sci. Paris 218(1944), 260-262.
17. A propos de la définition de connexion de Cantor, Bull. Sci. Math. 68(1944), 111-116.
18. Un théorème général sur les probabilités associées à un système d'événements dépendants, C.R. Acad. Sci. Paris 218(1944), 380-382.
19. Une définition descriptive de l'intégrale stochastique, C.R. Acad. Sci. Paris 218(1944), 953-955.
20. Sur l'approximation et l'intégration des fonctions aléatoires, Bull. Soc. Math. France 72(1944), 97-117.
21. Le prolongement des fonctionnelles continues sur un espace semi-

ordonné, Revue Scientifique 82(1944), 131-139.

22. Conditions d'existence de suites illimitées d'événements correspondant à certaines probabilités données, Revue Scientifique 82 (1944), 235-240.

23. Généralisations du théorème de M. Khintchine sur la validité de la loi des grands nombres pour les suites stationnaires de variables aléatoires, C.R. Acad. Sci. Paris 220(1945), 102-104.

24. Remarques sur un théorème de M. Khintchine, Bull. Sci. Math. 69 (1945), 81-92.

25. Two mean theorems in Hilbert space, Proc. Nat. Acad. Sci. U.S.A. 31(1945), 417-421.

26. (with M. Fréchet) Introduction à la topologie combinatoire, I. Initiation, Vuibert, Paris (1946).

Spanish translation: Introducción a la topología combinatoria (traducida por D.A.H. Nogués), Editorial Universitaria de Buenos Aires (la. edición, 1959; 2a. edición, 1961; 3a edición, 1967).

English translation: Initiation to combinatorial topology (translated by H.W. Eves), Prindle, Weber & Schmidt, Boston, London-Sydney (1967).

27. On positive definite sequences, Ann. of Math. 47(1946), 593-607.

28. (with S. Bochner) Distributive order-preserving operations in partially ordered vector sets, Ann. of Math. 48(1947), 168-179.

29. On a theorem of Weyl concerning eigenvalues of linear transformations, I, Proc. Nat. Acad. Sci. U.S.A. 35(1949), 652-655.

30. On a theorem of Weyl concerning eigenvalues of linear transformations, II, Proc. Nat. Acad. Sci. U.S.A. 36(1950), 31-35.

31. Partially ordered additive groups of continuous functions, Ann. of Math. 51(1950), 409-427.

32. Les fonctions définies-positives et les fonctions complètement monotones (Mémorial des Sci. Math., Fasc. 114), Gauthier-Villars, Paris (1950).

33. (with A. Appert) Espaces topologiques intermédiaires (Actualités Sci. et Industr., Fasc. 1121), Hermann, Paris (1951).

34. Maximum properties and inequalities for the eigenvalues of completely continuous operators, Proc. Nat. Acad. Sci. U.S.A. 37 (1951), 760-766.

35. Fixed-point and minimax theorems in locally convex topological linear spaces, Proc. Nat. Acad. Sci. U.S.A. 38(1952), 121-126.

36. Note on a theorem of Banach, Math. Zeitschr. 55(1952), 308-309.

37. A generalization of Tucker's combinatorial lemma with topological applications, Ann. of Math. 56(1952), 431-437.

38. (with N. Gottesman) On compactifications of Freudenthal and Wallman, Proc. Kon. Nederl. Akad. Wetensch. Amsterdam, Ser. A, 55 (1952), 504-510. Also in: Indag. Math. 14(1952), 504-510.

39. Minimax theorems, Proc. Nat. Acad. Sci. U.S.A. 39(1953), 42-47.

40. Some remarks on commutators of matrices, Archiv. der Math. 5 (1954), 102-107.

41. (with R.A. Struble) Continuity in terms of connectedness, Proc. Kon. Nederl. Akad. Wetensch. Amsterdam, Ser. A, 57(1954), 161-164. Also in: Indag. Math. 16(1954), 161-164.

42. (with A.J. Hoffman) Lower bounds for the rank and location of the eigenvalues of a matrix, Contributions to the Solution of Systems of Linear Equations and the Determination of Eigenvalues, 117-130. (National Bureau of Standards Applied Math. Series, vol. 39, Washington 1954.)

43. Inequalities for eigenvalues of Hermitian matrices, Contributions to the Solution of Systems of Linear Equations and the Determination of Eigenvalues, 131-139. (National Bureau of Standards Applied Math. Series, vol. 39, Washington 1954.)

44. (with G.G. Lorentz) An integral inequality, Amer. Math. Monthly 61(1954), 626-631.

45. (with J. Todd) A determinantal inequality, J. London Math. Soc. 30(1955), 58-64.

46. (with A.J. Hoffman) Some metric inequalities in the space of matrices, Proc. Amer. Math. Soc. 6(1955), 111-116.

47. (with O. Taussky and J. Todd) Discrete analogs of inequalities of Wortinger, Monatsh. Math. 59(1955), 73-90.

48. Some inequalities concerning positive-definite Hermitian matrices, Proc. Cambridge Philos. Soc. 51(1955), 414-421.

49. (with O. Taussky and J. Todd) An algebraic proof of the isoperimetric inequality for polygons, J. Washington Acad. Sci. 45(1955), 339-342.

50. (with I. Glicksberg) Fully convex normed linear spaces, Proc. Nat. Acad. Sci. U.S.A. 41(1955), 947-953.

51. A comparison theorem for eigenvalues of normal matrices, Pacific J. Math. 5(1955), 911-913.

52. On systems of linear inequalities, Linear Inequalities and Related Systems, 99-156 (Annals of Math. Studies, No. 38, Princeton Univ. Press, 1956).

Also in Russian translation, Moscow, 1959.

53. (with G. Pall) Imbedding conditions for Hermitian and normal matrices, Canad. J. Math. 9(1957), 298-304.

54. (with P. Davis) Complete sequences and approximations in normed linear spaces, Duke Math. J. 24(1957), 183-192.

55. (with I. Glicksberg and A.J. Hoffman) Systems of inequalities involving convex functions, Proc. Amer. Math. Soc. 8(1957), 617-622.

56. Existence theorems and extreme solutions for inequalities concerning convex functions or linear transformations, Math. Zeitschr. 68(1957), 205-216.

57. Topological proofs for certain theorems on matrices with non-negative elements, Monatsh. Math. 62(1958), 219-237.

58. Note on circular disks containing the eigenvalues of a matrix, Duke Math. J. 25(1958), 441-445.

59. Linear inequalities and closure properties in normed linear spaces, Seminar on Analytic Functions, Vol. 2, 202-212 (Institute for Advanced Study, Princeton 1958).

60. (with I. Glicksburg) Some geometric properties of the spheres in a normed linear space, Duke Math. J. 25(1958), 553-568.

61. On the equilibrium value of a system of convex and concave functions, Math. Zeitschr. 70(1958), 271-280.

62. Convex sets and their applications, Argonne National Laboratory, Argonne (1959).

63. (with A.S. Householder) A note concerning positive matrices and M-matrices, Monatsh. Math. 63(1959), 265-270.

64. Note on M-matrices, Quarterly J. Math. (Oxford Second Ser.) 11 (1960), 43-49.

65. Combinatorial properties of certain simplicial and cubical vertex maps, Archiv der Math. 11(1960), 368-377.

66. A generalization of Tychonoff's fixed point theorem, Math. Ann. 142(1961), 305-310.

67. On the Krein-Milman theorem, Proceedings of Symposia in Pure Mathematics, Vol. 7, Convexity, 211-219 (Amer. Math. Soc., Providence 1963).

68. (with R. Bellman) On systems of linear inequalities in Hermitian matrix variables, Proceedings of Symposia in Pure Mathematics, Vol. 7, Convexity, 1-11 (Amer. Math. Soc., Providence 1963).

69. Invariant subspaces of certain linear operators, Bull. Amer. Math. Soc. 69(1963), 773-777.

70. Invariant cross-sections and invariant linear subspaces, Israel J. Math. 2(1964), 19-26.

    Russian Translation: Matematika: Period. Sb. Perevodov Inostran. Statei. (Mathematics: Periodical Collection of Translations of Foreign Articles) 13 (6), Moscow, 1969. MR 40 #7057.

71. Inequalities for M-matrices, Proc. Kon. Nederl. Akad. Wetensch. Amsterdam, Ser. A, 67(1964), 602-610.

    Also in: Indag. Math. 26(1964), 602-610.

72. Sur un théorème minimax, C.R. Acad. Sci. Paris 259(1964), 3925-3928.

73. A generalization of the Alaoglu-Bourbaki theorem and its applications, Math. Zeitschr. 88(1965), 48-60.

74. Invariant subspaces for a semigroup of linear operators, Proc. Kon. Nederl. Akad. Wetensch. Amsterdam, Ser A. 68(1965), 447-451.

    Also in: Indag. Math. 27(1965), 447-451.

    Russian Translation: Matematika: Period. Sb. Perevodov Inostran. Statei. (Mathematics: Periodical Collection of Translations of Foreign Articles) 13 (6), Moscow, 1969. MR 40 #7057.

75. Applications of a theorem concerning sets with convex sections, Math. Ann. 163(1966), 189-203.

    Also in: Contributions to Functional Analysis (Springer, Berlin-Heidelberg-New York, 1966).

76. Some matrix inequalities, Abhandl. Math. Seminar Univ. Hamburg 29 (1966), 185-196.

77. Sets with convex sections, Proceedings of the Colloquium on Convexity, Copenhagen, 1965, 72-77 (Københavns Univ. Mat. Inst., 1967.

78. Subadditive functions on a distributive lattice and an extension of Szász's inequality, J. Math. Anal. Appl. 18(1967), 262-268.

79. Inequalities for the sum of two M-matrices, Inequalities, Proceedings of a Symposium, 105-117 (Academic Press 1967).

80. Simplicial maps from an orientable n-pseudomanifold onto S^m with the octahedral triangulation, J. Combinatorial Theory 2(1967), 588-602.

81. An inequality for subadditive functions on a distributive lattice, with application to determinantal inequalities, Linear Algebra and Appl. 1(1968), 33-38.

82. On infinite systems of linear inequalities, J. Math. Anal. Appl. 21(1968), 475-478.

83. A covering property of simplexes, Math. Scandinavica 22(1968), 17-20.

84. Asymptotic cones and duality of linear relations, J. Approximation Theory 2(1969), 152-159.

Also in: Inequalities II, Proceedings of the Second Symposium on Inequalities, 179-186 (Academic Press, 1970).

85. Extensions of two fixed point theorems of F. E. Browder, Math. Zeitschr. 112(1969), 234-240.

Abstract in: Set-valued mappings, selections and topological properties of 2^X. Proceedings of the Conference held at SUNY at Buffalo, May 1969, edited by W. M. Fleischman (Lecture Notes in Mathematics, Vol. 171, Springer, 1970), 12-16.

86. On a theorem of Pontryagin, Studies and Essays presented to Yu-Why Chen, Mathematics Research Center, National Taiwan Univ. (1970), 197-200.

87. A combinatorial property of pseudomanifolds and convering properties of simplexes, J. Math. Anal. Appl. 31(1970), 68-80.

88. On local connectedness of locally compact Abelian groups, Math. Ann. 187(1970), 114-116.

89. Simplicial maps of pseudomanifolds, Annals of the New York Academy of Sciences 175 (Art. 1), (1970), International Conference on Combinatorial Mathematics, 125-130.

90. Combinatorial properties of simplicial maps and convex sets, Proceedings of the Twelfth Biennial Seminar of the Canadian Mathematical Congress: Time Series and Stochastic Processes, Convexity and Combinatorics (Canadian Math. Congress, 1970) 231-241.

91. On the singular values of compact operators, J. London Math. Soc. (2) 3(1971), 187-189.

92. A minimax inequality and applications, Inequalities III, Proceedings of the Third Symposium on Inequalities, 103-113 (Academic Press, 1972).

93. Covering properties of convex sets and fixed point theorems in topological vector spaces, Symposium on Infinite Dimensional Topology (Annals of Math. Studies, No. 69, Princeton Univ. Press, 1972), 79-92.

94. Generalized Cayley transforms and strictly dissipative matrices, Linear Algebre and Appl. 5(1972), 155-172.

95. Fixed-point theorems in functional analysis (Supplementary manual for recording of a lecture), Audio Recordings of Mathematical Lectures, No. 46, Amer. Math. Soc. (1972), 7 pages.

96. On Dilworth's coding theorem, Math. Zeitschr. 127(1972), 92-94.

97. On real matrices with positive definite symmetric component, Linear and Multilinear Algebra 1(1973), 1-4.

98. Sums of eigenvalues of strictly J-positive compact operators, J. Math. Anal. Appl. 42(1973), 431-437.

99. On similarity of operators, Advances in Math. 10(1973), 395-400.

100. On strictly dissipative matrices, Linear Algebra and Appl. 9 (1974), 223-241.

101. Two applications of a consistency theorem for systems of linear inequalities, Linear Algebra and Appl. 11(1975), 171-180.

102. Orbits of semi-groups of contractions and groups of isometries, Abh. Math. Sem. Univ. Hamburg 45(1976), 245-250.

103. Extension of invariant linear functionals, Proc. Amer. Math. Soc. 66(1977), 23-29.

104. Analytic functions of a proper contraction, Math. Zeitschr. 160 (1978), 275-290.

105. Distortion of univalent functions, J. Math. Anal. Appl. 66(1978), 626-631.

106. Julia's lemma for operators, Math. Annalen 239(1979), 241-245.

107. (with T. Ando) Pick-Julia theorems for operators, Math. Zeitschr. 168(1979), 23-34.

108. Fixed-point and related theorems for non-compact convex sets, Game Theory and Related Topics (Proceedings of the Seminar on Game Theory and Related Topics, Bonn/Hagen, 26-29 September, 1978; Managing Editors: O. Moeschlin and D. Pallaschke), North-Holland, 1979, 151-156.

109. Schwarz's lemma for operators on Hilbert space (lecture presented at the Romanian-American Seminar on Operator Theory and Applications, 20-24 March 1978), Analele stiintifice ale Universitatii Al. I. Cuza Iasi, Supliment la tomul 25, s. I a 1979, 103-106.

110. Harnack's inequalities for operators, General Inequalities, Oberwolfach; edited by E.F. Beckenbach), Birkhäuser Verlag, Basel-Boston-Stuttgart, 1980; 333-339.

111. A further generalization of Shapley's generalization of the Knaster-Kuratowski-Mazurkiewicz theorem, Game Theory and Mathematical Economics (edited by O. Moeschlin and D. Pallaschke), North-Holland, 1981, 275-279.

112. Evenly distributed subsets of S^n and a combinatorial application, Pacific J. Math. 98(1982), 323-325.

113. Iteration of analytic functions of operators, Math. Zeitschr. 179(1982), 293-298.

114. Iteration of analytic functions of operators. II, Linear and Multilinear Algebra 12(1983), 295-304.

115. Normalizable operators, Linear Algebra and Appl. 52/53(1983), 253-263.

116. Some properties of convex sets related to fixed point theorems, Math. Ann. 266(1984), 519-537.

117. An identity for symmetric bilinear forms, Linear Algebra and Appl. 65(1985), 273-279.

118. The angular derivative of an operator-valued analytic function, Pacific J. Math. 121(1986), 67-72.

Nonlinear and Convex Analysis

Smooth and Heavy Viable Solutions to Control Problems

JEAN-PIERRE AUBIN Mathematical Research Center, University of Paris-Dauphine, Paris, France

ABSTRACT

We introduce the concept of viability domain of a set-valued map, which we study and use for providing the existence of smooth solutions to differential inclusions.

We then define and study the concept of heavy viable trajectories of a controlled system with feedbacks. Viable trajectories are trajectories satisfying at each instant given constraints on the state. The controls regulating viable trajectories evolve according a set-valued feedback map. Heavy viable trajectories are the ones which are associated to the controls in the feedback map whose velocity has at each instant the minimal norm. We construct the differential equation governing the evolution of the controls associated to heavy viable trajectories and we state their existence.

DEDICATION

I would have liked to find an original way to dedicate this lecture to Professor Ky Fan, but I did not see any better solution than to simply confess that it is both an honor and a pleasure to have been invited to this conference held in his honor.

I have been deeply influenced by the theorems discovered and proved by Professor Ky Fan, and, in particular, by his 1968 famous inequality. Let me just repeat what I tell my students when I begin to teach the Ky Fan inequality. I tell them a lot of stories, how the young Ky Fan came to Paris in 1939 for one year with only a metro map, how he had to survive during the darkest years of the history of my country, how he met Fréchet and worked with him, etc. But most important, I choose the Ky Fan inequality as the best illustration of the concept of "labor value" of a theorem.

Indeed, most of the theorems of nonlinear functional analysis are equivalent to the Brouwer fixed point theorem. But when we prove that statement (A) is equivalent to statement (B), there is always one implication, say "A implies B", that is more difficult to prove than the other one. We then can say that statement (B) "incorporates" more labor value than statement (A). An empirical law shows that the more labor value a theorem incorporates, the more useful it is. And my point is that among all the theorems equivalent to the Brouwer fixed point theorem I know, the Ky Fan inequality is one which is the most valuable.

1. VIABLE SOLUTIONS TO A CONTROL PROBLEM

Let $X = \mathbb{R}^n$, $U: X \rightrightarrows X$ be a set-valued map with closed graph and $f: \text{Graph } U \to X$ be a continuous map. We consider the control problem with feedbacks

$$(1.1) \quad \begin{cases} \text{i)} & x'(t) = f(x(t), u(t)) \quad , \\ \text{ii)} & \text{for almost all } t \geq 0, \quad u(t) \in U(x(t)) \\ \text{iii)} & x(0) = x_0 \text{ given in Dom } U \quad . \end{cases}$$

Instead of selecting a solution $x(\cdot)$ to (1) which minimizes a given functional, as in optimal control theory[1], we are only selecting solutions which are <u>viable</u> in the sense that, given a closed subset $K \subset X$

$$(1.2) \qquad \forall t \geq 0 \ , \quad x(t) \in K \quad .$$

A first issue is to provide necessary and sufficient conditions linking the dynamics of the system (described by f and U) and the constraints bearing on the system (described by the closed subset K) such that the viability property

(1.3) $\forall x_0 \in K$, there exists a solution to (1) viable in K

holds true. This allows us to describe the evolution of the <u>viable controls</u> u(•), (the controls which govern viable solutions).

A second issue is to provide conditions for having smooth viable solutions to a control problem, in the sense that the <u>viable control</u> function is absolutely continuous instead of being simply measurable.

A third issue is to give a mathematical description of the "heavy viable solutions" of the control system which we observe in the evolution of large systems arising in biology and economic and social sciences. Such large systems keep the same control whenever they can and change them only when the viability is at stakes, and do that as slowly as possible. In other words, heavy viable solutions are governed by those controls who minimize at each instant the norm of the velocity of the viable controls. In the case when $f(x,u) = u$, system (1) reduces to the differential inclusion $x'(t) \in U(x(t))$, $x(0) = x_0$: heavy (viable) solutions to this system minimize at each instant the norm of the acceleration of viable solutions; in other words, they evolve with maximal inertia. Hence the name <u>heavy</u> viable solutions (or <u>inert</u> viable solutions).[2]

For solving this problem, we need to introduce and study two concepts: viability domains of differential inclusions and derivatives of set-valued maps.

Let me mention that these results were obtained in collaboration with Halina Frankowska and Georges Haddad.

2. VIABILITY DOMAINS AND INVARIANT SETS OF A SET-VALUED MAP

In this section, we consider a set-valued map F from X to X satisfying once and for all

$$(2.1)\quad \begin{cases} \text{i)} & \text{the graph of } F \text{ is nonempty and closed} \\ \text{ii)} & \forall x \in \mathrm{Dom}(F)\ , \quad \sup_{v\in F(x)} \|v\| =: \|F(x)\| \le a\|x\| + b \end{cases}$$

(This implies that F is upper semicontinuous with compact images.) We propose to extend the concept of invariant subspace K by a single-valued map f, defined by

$$(2.2)\qquad \forall x \in K\ ,\quad f(x) \in K \quad \text{or} \quad f(K) \subset K\ .$$

When we think about the extension of the concept of invariant subset K, we have the choice of using either the property $f(K) \subset K$ or the property

$$(2.3)\qquad \forall x \in K\ ,\quad f(x) \in T_K(x)$$

because a vector subspace K is always the tangent space to every points of K. ($T_K(x) = K$ for all $x \in K$.)

When K is any subset, there are many ways to introduce "tangent cones" $T_K(x)$ to K at x which coincide with the tangent space when K is a smooth manifold and to the tangent cone of convex analysis when K is convex. The 1943 Nagumo theorem shows that we have to choose the contingent cone introduced by Bouligand in the thirties. The contingent cone $T_K(x)$ to K at x is defined by

$$(2.4)\qquad T_K(x) := \left\{ v \in X \;\middle|\; \liminf_{h\to 0+} \frac{d(x+hv,K)}{h} = 0 \right\}$$

Nagumo's theorem states that if a continuous map f satisfies property (2.3), then for any $x_0 \in K$, there exists a viable solution to the differential equation $x'(t) = f(x(t))$, $x(0) = x_0$.

When we consider the differential inclusion

$$(2.5)\qquad x'(t) \in F(x(t)),\quad x(0) = x_0$$

there are two ways of adapting property (2.3).

Definition 2.1

We shall say that a subset $K \subset \text{Dom } F$ is a viability domain of F if

(2.6) $\forall x \in K \ , \quad F(x) \cap T_K(x) \neq \phi$

and is invariant by F if

(2.7) $\forall x \in K \ , \quad F(x) \subset T_K(x) \quad .$

These definitions are motivated by the following theorems.

Theorem 2.2 (G. Haddad, 1981)

If F has convex values and if $K \subset \text{Dom}(F)$ is a closed viability domain of F, then for any $x_0 \in K$, there exists a viable solution to the differential inclusion (2.5) (viability property)

Theorem 2.3 (F.H. Clarke, 1975)

If F is Lipschitz and if $K \subset \text{Dom } F$ is a closed invariant subset by F, then for any $x_0 \in K$, all the solutions to the differential inclusion (2.5) are viable (invariance property).

The concept of invariance in the above sense requires the knowledge of F outside K. Let us mention a more intrinsic result.

Theorem 2.4 (J.P. Aubin and F.H. Clarke, 1977)

If F is continuous and if $K \subset \text{Dom } F$ is a closed invariant subset by F, then the viability property holds true.

We now provide an example of viability domains.
Let us consider "limit sets"

(2.8) $L(x(\cdot)) := \bigcap_{T>0} \overline{x([T,\infty[)}$

of solutions $x(\cdot)$ to the differential inclusion $x'(t) \in F(x(t))$.

Theorem 2.5

If F has convex values, the limit sets of the solutions $x(\cdot)$ to the differential inclusion $x'(t) \in F(x(t))$ are closed viability domains.

This theorem provides many examples of viability domains. Equilibria (solutions to $0 \in F(\bar{x})$), trajectories of periodic solutions, etc., are closed viability domains. The question arises whether there exists a largest closed viability domain. Such a largest closed viability domain would then contain all the interesting features of the differential inclusion.

Theorem 2.6

If F has convex values, there exists a largest closed viability domain of F.

Let us mention a consequence of the "coincidence theorem" due to Ky Fan.

Theorem 2.7 (Ky Fan)

If F has convex values, any compact convex viability domain of F contains an equilibrium.

We also observe that for the set-valued analogues of linear operators the concepts of closed viability convex cones and invariant closed convex cones are "dual".

We recall that closed convex processes A are the set-valued maps whose graphs are closed convex cones.

When P is a cone, we denote by P^+ its (positive) polar cone, defined by

$$P^+ := \{p \in X^* \mid \forall x \in P , \langle p,x \rangle \geq 0\} \quad . \tag{2.9}$$

We can "transpose" closed convex processes in the following way: A^*, the transpose of A, is defined by

$$\begin{aligned} p \in A^*(q) &\iff \forall x \in \text{Dom } A,\ \forall y \in A(x),\ \langle p,x \rangle \leq \langle q,y \rangle \\ &\iff (-p,q) \in (\text{Graph } A)^+ \end{aligned} \tag{2.10}$$

Theorem 2.8 (J.P. Aubin, H. Frankowska, C. Olech, 1985)

Let A be a closed convex process defined on the whole space X. The two following properties are equivalent

$$(2.11) \quad \begin{cases} \text{i)} & \text{a closed convex cone } P \text{ is invariant by } A \\ \text{ii)} & P^+ \text{ is a viability domain of } A^* . \end{cases}$$

This result plays a crucial role in the study of controllability of the differential inclusion

$$(2.12) \qquad x'(t) \in A(x(t)) \quad ,$$

and the observability of the adjoint differential inclusion

$$(2.13) \qquad -q'(t) \in A^*(q(t)) \quad .$$

It plays also a role in existence theorems of eigenvalues and eigenvector, as a consequence of Ky Fan's theorem.

<u>Theorem 2.9</u> (J.P. Aubin, H. Frankowska, C. Olech, 1985).

Let A be a closed convex process defined on the whole space X and P be a closed convex cone with nonempty interior. If P is invariant by A, the two following equivalent conditions

$$(2.14) \quad \begin{cases} \text{i)} & \exists \lambda \in \mathbb{R} \text{ such that } \operatorname{Im}(A-\lambda I) \neq X \\ \text{ii)} & \exists q \neq 0,\ q \in P^+ \text{ such that } \lambda q \in A^*(q) \end{cases}$$

holds true.

We can say that a solution λ to (2.14)i) is an <u>eigenvalue</u> of A and that a solution q of (2.14)ii) is an <u>eigenvector</u> of A^*.

3. SMOOTH SOLUTIONS TO CONTROL PROBLEMS

Let us return now to our control problem (1.1), which reduces to the differential inclusion $x' \in F(x)$ where F is the set-valued map defined by

$$F(x) := f(x, U(x)) \quad .$$

Let us introduce the feedback map R associated to a subset $K \subset \operatorname{Dom} U$ in the following way:

(3.1) $R(x) := \{u \in U(x) \mid f(x,u) \in T_K(x)\}$.

Then Viability Theorem 2.1 implies the following theorem.

Theorem 3.1

Let us assume that U has a closed graph and compact values, that $f : \text{Graph } U \to X$ is continuous, that

(3.2) $\forall x \in \text{Dom } U \ , \quad \sup_{u \in U(x)} \| f(x,u) \| \leq a\|x\| + b$

and that the subsets $f(x,U(x))$ of velocities are convex. Let K be a closed subset of Dom U. Then the viability property holds true if and only if

(3.3) $\forall x \in K \ , \quad R(x) \neq \phi$.

When this tangential condition is satisfied, viable controls evolve according to the law

(3.4) for almost all t, $u(t) \in R(x(t))$.

The measurable selection theorem allows to state that we can find such viable controls which are measurable.

Since the definition of heavy viable solutions involves the derivatives of viable controls, we have to find sufficient conditions for having absolutely continuous viable controls. For that purpose, we can think to impose an a priori bound on the velocity of the viable controls, requiring for instance that

(3.5) for almost all t, $\|u'(t)\| \leq c(\|x(t)\| + \|u(t)\| + 1)$.

Theorem 3.5

Let us assume that the graph of U is closed and that $f : \text{Graph } U \to X$ is continuous and satisfies for some $c_0 \in \mathbb{R}_+$:

(3.6) $\forall (x,u) \in \text{Graph}(F) \ , \quad \|f(x,u)\| \leq c_0(\|x\| + \|u\| + 1)$.

Then we can associate with any $c \geq c_0$ a set-valued map $R_c \subset U$ having the following property:

$\forall x_0 \in K$, $\forall u_0 \in R_c(x)$, there exists a smooth solution to the control problem

$$(3.7) \quad \begin{cases} \text{i)} & x'(t) = f(x(t),u(t)) \ , \\ \text{ii)} & u'(t) \in c(\|x(t)\|+\|u(t)\|+1)B \ , \ B \text{ is the unit ball} \end{cases}$$

which are viable in the sense that

$$(3.8) \quad \forall t \geq 0 \ , \quad x(t) \in K \quad \text{and} \quad u(t) \in R_c(x(t)) \ .$$

Furthermore, R_c is the largest of the set-valued maps satisfying the above property.

If we introduce the set-valued map G_c defined by

$$(3.9) \quad G_c(x,u) := \begin{cases} \{f(x,u)\} \times c(\|x\|+\|u\|+1)B & \text{if } x \in K \text{ and } u \in U(x) \\ \phi & \text{if not} \end{cases}$$

Then the graph of set-valued map R_c satisfying properties (3.7) and (3.8) is the largest closed viability domain of this set-valued map G_c.

We observe that if $c_1 \leq c_2$, then

$$(3.10) \quad R_{c_1}(x) \subset R_{c_2}(x) \subset R(x)$$

and that the set-valued map $c \to \text{Graph } R_c$ is upper semicontinuous.

Hence smooth viable solutions of the control problem are governed by controls u(t) evolving according to the feedback law

$$(3.11) \quad \text{for all } t \geq 0 \ , \quad u(t) \in R_c(x(t)) \ .$$

Since the definition of heavy viable solutions to control problems involves the knowledge of the derivative u'(t) of the controls u(t) governing (smooth) viable solutions, we are led to "differentiate" the feedback law (3.11) and, for that purpose, to "differentiate" the set-valued map R_c.

4. CONTINGENT DERIVATIVES OF SET-VALUED MAPS

We choose the concept of contingent derivatives (see Aubin (1981), Aubin and Ekeland (1984)). When F is a set-valued map from a Banach space X to a Banach space Y and when (x,y) belongs to the graph of F, then we define the contingent derivative DF(x,y) as the closed process from X to Y whose graph is equal to the contingent cone to Graph(F) at (x,y):

$$\text{(4.1)} \qquad \text{Graph } DF(x,y) := T_{\text{Graph}(F)}(x,y) \quad .$$

In other words,

$$\text{(4.2)} \qquad v \in DF(x,y)(u) \iff (u,v) \in T_{\text{Graph}(F)}(x,y) \quad .$$

We can check that

$$\text{(4.3)} \qquad v \in DF(x,y)(u) \iff \liminf_{\substack{h \to 0+ \\ u' \to u}} d\left(v\ ,\ \frac{F(x+hu')-y}{h}\right) = 0 \quad .$$

This concept of contingent derivative captures many of the properties of the Gâteaux derivative of single-valued differential maps. We just mention here the "chain rule" property which is relevant to our problem.

Let $x(\cdot)$ and $y(\cdot)$ be two absolutely continuous functions of t satisfying the relation

$$\text{(4.4)} \qquad \text{for all } t, \quad y(t) \in F(x(t)) \quad .$$

Then

$$\text{(4.5)} \qquad \text{for almost all } t, \quad y'(t) \in DF(x(t),y(t))(x'(t)) \quad .$$

5. HEAVY VIABLE SOLUTIONS TO A CONTROL PROBLEM

Since smooth viable solutions $x(t)$ to the control problem (1.1) and (3.5) are governed by absolutely continuous controls $u(t)$ obeying the feedback law (3.11) we know that the velocity $u(t)$ obeys the law

$$\text{(5.1)} \qquad \text{for almost all } t \geq 0,\ u'(t) \in DR_C(x(t),u(t))(f(x(t),u(t))).$$

Therefore, heavy viable solutions $x(t)$ are governed by controls $u(t)$ which are solutions to the differential inclusion

$$\text{(5.2)} \quad \text{for almost all } t \geq 0,\ u'(t) \in m(DR_C(x(t),u(t))(f(x(t),u(t)))),$$

where, when A is a subset of a vector space,

$$\text{(5.2)} \qquad m(a) := \{x \in A \mid \|x\| = \inf_{y \in A} \|y\|\}$$

Theorem 5.1 (Aubin-Frankowska)

Heavy viable solutions to the control problem (1.1) and (3.5) are solutions to the differential inclusions

$$\text{(5.3)} \qquad \begin{cases} \text{i)} & x'(t) = f(x(t),u(t)) \\ \text{ii)} & u'(t) \in d(0, DR_C(x(t),u(t))(f(x(t),u(t)))) \end{cases}$$

which are viable in the sense that

$$\text{(5.4)} \qquad \forall t \geq 0, \quad u(t) \in R_C(x(t)) \quad .$$

If we assume that

$$\text{(5.5)} \qquad (x,u,v) \to DR_C(x,u)(v) \text{ is lower semicontinuous,}$$

then for any $x_0 \in \text{Dom } R_C$ and any $u_0 \in R_C(x_0)$, there exists a heavy viable solution to the control problem (1.1) and (5.2).

REFERENCES

Aubin, J.P. (1981a) Contingent derivatives of set-valued maps and existence of solutions to nonlinear inclusions and differential inclusions. Advances in Mathematics. Supplementary Studies. Ed. L. Nachbin. Academic Press. 160-232.

Aubin, J.P. (1981b) A dynamical, pure exchange economy with feedback pricing. J. Economic Behavior and Organizations 2, 95-127.

Aubin, J.P. and A. Cellina (1984) Differential inclusions. Springer Verlag.

Aubin, J.P. and F.H. Clarke (1977) Monotone invariant solutions to differential inclusions. J. London Math. Soc. 16, 357-366.

Aubin, J.P. and H. Frankowska (1985) Heavy viable trajectories of controlled systems. Ann. Int. Henri Poincaré. Analyse Nonlinéaire.

Aubin, J.P., H. Frankowska and C. Olech (1985) Contrôlabilité des processus convexes. C.R.A.S. (Controllability of convex processes. To appear.)

Aubin, J.P. and I. Ekeland (1984) Applied Nonlinear Analysis. Wiley Interscience.

Clarke, F.H. (1975) Generalized gradients and applications. Trans. A.M.S. 205, 247-262.

Fan, Ky (1972) A minimax inequality and applications. In Inequalities III. O. Sisha Ed. Academic Press. 103-113.

Haddad, G. (1981) Monotone trajectories of differential inclusions and functional differential inclusions with memory. Israel J. Math. 39, 83-100.

Williamson, P.G. (1985) Palaeontological documentation of speciation in Cenezoic Molluscs from Turkana Basin. Nature 293, p. 437.

Footnotes

(1)

Optimal control theory does assume implicitly

(1) the existence of a decision-maker operating the controls of the system (there may be more than one decision-maker in a game-theoretical setting)

(2) the availability of information (deterministic or stochastic) on the future of the system; this is necessary to define the costs associated with the trajectories

(3) that decisions (even if they are conditional) are taken once and for all the initial time.

(2)

Palaeontological concepts such as punctuated equilibria proposed by Elredge and Gould are consistent with the concept of heavy viable trajectories.

Indeed, for the first time, excavations at Kenya's Lake Turkana have provided clear fossil evidence of evolution from one species to another. The rock strata there contain a series of fossils that show every small step of an evolutionary journey that seems to have proceeded in fits and starts. Williamson (1981) examined 3.300 fossils showing how thirteen species of molluscs changed over several million years. What the record indicated was that the animals stayed much the same for immensely long stretches of time. But twice, about 2 million years ago and then again 700.000 years ago, the pool of life seemed to explode - set off, apparently, by a drop in the lake's water level. In an instant of geologic time, as the changing lake environment allowed new types of molluscs to win the race for survival, all of the species evolved into varieties sharply different from their ancestors. That immediate forms appeared so quickly, with new species suddenly evolving in 5.000 to 50.000 years after millions of years of constancy, challenges the traditional theories of Darwin's disciples since the fossils of Lake Turkana don't record any gradual change; rather, they seem to reflect eons of stasis interrupted by brief evolutionary "revolutions".

On Simplified Proofs of Theorems of von Neumann, Heinz, and Ky Fan, and Their Extended Versions

GONG-NING CHEN Department of Mathematics, Beijing Normal University, Beijing, People's Republic of China

Dedicated to Professor Ky Fan on the occasion of his retirement in June, 1985.

ABSTRACT

This note introduces the use of a weak maximum principle in the simplified discussion of Theorems of Von Neumann, Heinz, and Ky Fan, and their extended versions.

1. INTRODUCTION AND NOTATION

Let H be a Hilbert space over the complex field C, $H \neq \{0\}$, and denote by $B(H)$ the Banach algebra of all bounded linear operators T on H with respect to the operator norm $\|\cdot\|$, by $\widetilde{U}$ the set of all unitary operators of $B(H)$. If $T \in B(H)$, we denote by $\sigma(T)$ the spectrum of T. For a complex holomorphic function $u(\lambda)$ on a domain Ω containing $\sigma(T)$ in C, $u(T)$ will denote an operator of $B(H)$ defined by the Riesz-Dunford integral [1, p. 568]

$$u(T) = (2\pi i)^{-1}\int_{\Gamma} u(\lambda)(\lambda I - T)^{-1}d\lambda, \tag{1}$$

where I stands for the identity of $B(H)$ and Γ is any contour that surrounds $\sigma(T)$ in Ω. In [7] (see also [9, p. 437]), Von Neumann proved the following

<u>THEOREM I</u>. If T is a contraction of $B(H)$ (i.e., $\|T\| \le 1$) and if u is a complex holomorphic function on a domain Ω containing the closed unit disk $\bar{\Delta} = \{\lambda : |\lambda| \le 1\}$ such that $|u(\lambda)| \le 1$ for all λ in $\bar{\Delta}$, then $\|u(T)\| \le 1$.

In his original proof, he showed, by a result of Schur, that the general case of functions $u(\lambda)$ in question can be reduced to the particular case of Möbius transformations of the form

$\lambda \longrightarrow (\lambda+a)/(1+\bar{a}\lambda), \quad |a| < 1,$

for which the theorem is demonstrated by an immediate calculation. It is well known that there exist other elegant proofs of Theorem I, most of which can be found in [1, p. 933], [2] for references. Among these are the proofs due to Heinz [4] and Ky Fan [2] by deriving Theorem I from their theorems which are stated here as Theorems II and III, respectively.

THEOREM II. If $\|T\| \leq 1$ and if v is a complex holomorphic function on a domain Ω containing $\bar{\Delta}$ such that $\mathrm{Re}\, v(\lambda) \geq 0$ for all λ in $\bar{\Delta}$, then $\mathrm{Re}\, v(T) \geq 0$.

Here $\mathrm{Re}\, T$ is the real part of T in $B(H)$, i.e., $\mathrm{Re}\, T = \frac{1}{2}(T+T^*)$. For two self-adjoint operators T_1 and T_2 in $B(H)$ we write $T_1 \geq T_2$ if $((T_1-T_2)w,w) \geq 0$ for all w in H; write $T_1 > T_2$ if $((T_1-T_2)w,w) > 0$ for all $w \neq 0$ in H.

THEOREM III. If $\|T\| < 1$ and if h and k are complex holomorphic on the open unit disk $\Delta = \{\lambda : |\lambda| < 1\}$ such that $|h(\lambda)| < 1$ and $\mathrm{Re}\, k(\lambda) > 0$ for all λ in Δ, then $\|h(T)\| < 1$ and $\mathrm{Re}\, k(T) > 0$.

Besides the preceding methods, there exists a simple proof proposed by Harris [3] by considering Theorem I as an immediate consequence of a maximum principle and the spectral theorem for unitary operators on a Hilbert space. The novelty of this approach, in contrast with the original Von Neumann's proof, consists in that it refers the problem to make an estimate of the norm of $u(T)$ for a general $\|T\| \leq 1$ to the one of $u(V)$, where V is in $\tilde{U}$, for which Theorem I is valid.

In this note, we investigate a weak maximum principle (Theorem 1 in Sect. 2) and its corollary and use them to give the simplified proofs of Theorems I, II, III and their extended versions of Tao [11]. All the proofs here are surprisingly simple and are almost uniform.

2. A WEAKER FORM OF THE MAXIMUM PRINCIPLE

Let A denote a Banach algebra over C having an identity e and an involution $*$. An element x in A is said to be <u>self-adjoint</u> if $x = x^*$. An element x in A is said to be <u>unitary</u> if $xx^* = x^*x = e$. Denote by U the set of all unitary elements of A, by U_e the identity component of U. If $x \in A$, we denote by $\sigma(x)$ the spectrum of x and by $|x|_\sigma$ the spectral radius of x. An involution for A is said to be <u>hermitian</u> if self-adjoint elements have real spectra, and is said to be <u>continuous</u> if $\|x^*\| \leq M x$ for all x in A and for some constant M. Define $p(x) = |x^*x|_\sigma^{1/2}$ for x in A. Let D be an open subset of A. An A-valued function $g : D \longrightarrow A$ is said to be holomorphic on D if the Fréchet derivative of g at each x in D exists as a bounded complex linear self-mapping of A [10, p. 248]. It is known [3] that holomorphic functions are continuous and that the chain rule holds. For a subset S of A, let $\overline{\text{co}}\ S$ denote the closed convex hull of S, and Co(S) the (closed) holomorphic convex hull of S (see [3, p. 11]).

The following basic theorem contains a weakened form of the maximum principle of Harris [3, Theorem 8], however, it is quite enough for our purposes.

THEOREM 1. Put $A_1 = \{x \in A : |x|_\sigma \le 1,\ p(x) \le 1\}$ and let D be an open subset of A containing $r\ \overline{\text{co}}\ U_e$ for some r, $0 < r \le 1$. If $h : D \longrightarrow A$ is a holomorphic function, then

$$A_1 \subset \overline{\text{co}}\ U_e, \tag{2}$$

$$h(rA_1) \subset \overline{\text{co}}\ h(rU_e), \tag{3}$$

$$\|h(rx)\| \le \sup\{\|h(ry)\| : y \in U_e\} \le \infty \quad \text{for all } x \text{ in } A_1, \tag{4}$$

and

$$p(h(rx)) \le \sup\{p(h(ry)) : y \in U_e\} \le \infty \quad \text{for all } x \text{ in } A_1 \text{ whenever the involution for } A \text{ is hermitian and continuous.} \tag{5}$$

Moreover, $h|_{rA_1}$ is completely determined by its values on rU_e.

Proof. Observe that both (2) and (3) of Theorem 1 with $r = 1$ can be deduced from Theorem 8 of [3] together with the fact that $\text{Co}(S) \subset \overline{\text{co}}\ S$ for each subset S of A. We, however, prefer to sketch a direct proof as follows. Clearly we may assume that $|x|_\sigma < 1$ and $p(x) < 1$. To prove (3), consider first the case of $r = 1$. Then, $D \supset \overline{\text{co}}\ U_e$. Let $\ell \in A^*$, the conjugate space of A, and define $F(\lambda) = \exp(\ell(h(f(\lambda))))$ for $|\lambda| \le 1$. (First take as h the identity function on A to obtain (2).) Here f is the characteristic function defined by

$$f(\lambda) = (e-xx^*)^{-1/2}(\lambda e+x)(e+\lambda x^*)^{-1}(e-x^*x)^{1/2}, \tag{6}$$

which is holomorphic in a neighborhood of $\overline{\Delta}$, $f(0) = x$, and maps the unit circle in C into U_e [3, p. 2]. By the chain rule and the maximum modulus principle,

$$|F(0)| \le \sup_{|\lambda|=1} |F(\lambda)|,$$

and therefore

$$\operatorname{Re} \ell(h(x)) \leq \sup_{|\lambda|=1} \operatorname{Re} \ell(h(f(\lambda))) \leq \sup \operatorname{Re} \ell(h(U_e)).$$

It then follows that $h(x) \in \overline{co}\, h(U_e)$, for, if not, by a separation theorem [1, p. 417] there exist constants c and t, $t > 0$, and $\ell_1 \in A^*$ such that

$$\sup \operatorname{Re} \ell_1(\overline{co}\, h(U_e)) \leq c - t < c \leq \operatorname{Re} \ell_1(h(x)),$$

a contradiction, so that (3) holds when $r = 1$. In the general case of $0 < r \leq 1$, the function k defined by $k(y) = h(ry)$ is holomorphic on $r^{-1}D$ into A. (3) now follows directly from the result just proved and from the convexity of balls. Obviously, (4) is a simple consequence of (3) whether $\overline{co}\, U_e$ is bounded in A or not. If the involution for A is hermitian and continuous, we have ([3], [8]) that p is an algebra pseudonorm on A, and is continuous on A, whence (5) is valid by (3). For the last part of the theorem, as before, it will be enough to verify the result in the case $r = 1$. But this follows directly from the mean value property for vector-valued holomorphic functions [5, p. 99]:

$$h(x) = h(f(0)) = (2\pi)^{-1}\int_0^{2\pi} h(f(e^{i\theta}))\,d\theta \tag{7}$$

for each x in A with $|x|_\sigma < 1$ and $p(x) < 1$, where f is as in (6). That completes the proof of the theorem.

REMARK. We note that there exists an alternative proof of Theorem 1 by deriving it from the formula (7).

COROLLARY 1. Suppose that A has hermitian and continuous involution. Let Ω be a domain containing $r\bar{\Delta}$ for some r, $0 < r \leq 1$, and define A_1 as in Theorem 1 and $A_\Omega = \{x \in A : \sigma(x) \subset \Omega\}$. If k is a complex holomorphic function on Ω, then (3)-(5) of Theorem 1 are valid with D replaced by A_Ω and $h(x)$ by the function $k(x)$ defined by

$$k(x) = (2\pi i)^{-1}\int_\Gamma k(\lambda)(\lambda e - x)^{-1}\,d\lambda, \tag{8}$$

where Γ is any contour that surrounds $\sigma(x)$ in Ω.

Proof. It is well known [10] that A_Ω is open in A and that the function k defined by (8) is a continuously differentiable mapping of A_Ω into A, so that $k : A_\Omega \rightarrow A$ is holomorphic by definition. In virtue of the properties of p given above, we have $A_\Omega \supset r\, \overline{co}\, U_e$ whenever $\Omega \supset r\bar{\Delta}$. Corollary 1 now follows from Theorem 1.

3. SIMPLIFIED DISCUSSION OF THEOREMS I, II, III, AND THEIR EXTENDED VERSIONS

In this section we shall introduce the use of the results of the preceding section in the simplified proofs of Theorems I, II, III, and their extended versions of Tao [11]. A result (see [6, p. 109] and [10, p. 309]) of fundamental importance for us is the following

LEMMA I. Let E be the spectral resolution of the identity for a normal operator V in $B(H)$, and suppose that g is a complex holomorphic function on a domain Ω containing $\sigma(rV)$ in C for some r, $0 < r \leq 1$. Then

$$g(rV) = \int_{\sigma(V)} g(rt)\,dE(t), \tag{9}$$

where the integral exists with respect to the norm topology in $B(H)$, and

$$\|g(rV)\| = \sup\{|g(r\lambda)| : \lambda \in \sigma(V)\}. \tag{10}$$

We note that the involution on $B(H)$, the Hilbert space adjoint, is clearly hermitian and continuous, and that $|T|_\sigma \leq p(T) = \|T\|$, so that $B(H)_1 = \{T \in B(H) : \|T\| \leq 1\}$.

Proof of Theorem I. Observe that $\Omega \supset \bar{\Delta} \supset \sigma(V)$ for each V in $\tilde{U}$ since $\sigma(V)$ lies on the unit circle. By Corollary 1 with $r = 1$ and by (10),

$$\|u(T)\| \leq \sup\{\|u(V)\| : V \in \tilde{U}\} \leq \sup\{|u(\lambda)| : |\lambda| = 1\} \leq 1,$$

as desired.

Proof of Theorem II. By (9), with $r = 1$ and since $v(V)^* = \bar{v}(V)$ for each V in $\tilde{U}$, we have

$$\operatorname{Re} v(V) = \int_0^{2\pi} \operatorname{Re} v(e^{i\theta})\,dE(\theta),$$

and therefore

$$(\operatorname{Re} v(V)w,w) = \int_0^{2\pi} \operatorname{Re} v(e^{i\theta})\,dE_{w,w}(\theta)$$

for all w in H, where $E_{w,w}(\theta) = (E(\theta)w,w)$. It is well known [10] that each $E_{w,w}$ is a positive measure on $\sigma(V)$ whose total variation is $\|w\|^2$. Since $\operatorname{Re} v(\lambda) \geq 0$ is continuous on the compact set $\sigma(V)$ in C, $(\operatorname{Re} v(V)w,w) \geq 0$ for each w in H, i.e., $\operatorname{Re} v(V) \geq 0$ by definition. The result now follows directly from (3) of Corollary 1.

Proof of Theorem III. If T is a proper contraction of $B(H)$, there exist real r, $0 < r < 1$, and a contraction $\hat{T}$ of $B(H)$ such that $T = r\hat{T}$. Thus, as before, $\|h(T)\| \leq \sup\{|h(r\lambda)| : |\lambda| = 1\} < 1$. By applying a similar argument to that used in the preceding theorem, we can obtain $\operatorname{Re} k(T) > 0$ for each $\|T\| < 1$.

It is natural to expect that the proofs of the extended versions of Theorems I, II, and III, given by Tao [11], should be also greatly simplified by the same approach mentioned above. These are indeed the case. For notations, let Ω be a domain in C and denote by $N_H(\Omega)$ the set of all B(H)-valued holomorphic functions $\hat{g}$ on Ω such that $\hat{g}(\sigma)\hat{g}(\lambda) = \hat{g}(\lambda)\hat{g}(\sigma)$ and $\hat{g}(\lambda)$ is normal in B(H) for all λ,σ in Ω. Recall [11] that $\hat{g}(T)$ denotes a B(H)-valued function, with domain $B(H)_\Omega = \{T \in B(H) : \sigma(T) \subset \Omega\}$, that arise from a B(H)-valued function $\hat{g}$ on Ω by the formula

$$\hat{g}(T) = (2\pi i)^{-1}\int_\Gamma \hat{g}(\lambda)(\lambda I-T)^{-1}d\lambda, \tag{11}$$

where Γ is as in (1).

<u>THEOREM IV</u>. Suppose that $\hat{u},\hat{v}$ are in $N_H(\Omega)$ and $\hat{h}$ is in $N_H(\Delta)$, each of which commutes with T in B(H). Then Theorems I, II, and III hold if u, v, and h therein are replaced by $\hat{u}$, $\hat{v}$, and $\hat{h}$, respectively.

<u>Proof</u>. Consider only the extended result with respect to Theorem I. In a similar way one can verify the others. By a slight extension of Theorem 10.38 in [10], we can assert that the function $\hat{g}(T)$ defined by (11) is a holomorphic mapping of $B(H)_\Omega$ into B(H). Instead of Lemma I we use an analogous lemma, that is Lemma 6.2 of [11]. Thus, as before, we are able to conclude that the extended result is valid.

We conclude with the remark that each of Theorems 3.3, 3.4 (2)-(4), and 3.5 of [11] may be shown to be valid by means of the same methods, and is merely an immediate consequence of Theorem 1 and Corollary 1 together with either Lemma I or Lemma 6.2 of [11].

REFERENCES

1. N. Dunford and J.T. Schwartz, "Linear Operators," Interscience, New York, pt. I, 1958; pt. II, 1963.
2. Ky Fan, Analysis functions of a proper contraction, Math. Z. 160 (1978), 275-290.
3. L. Harris, Banach algebras with involution and Möbius transformations, J. Functional Analysis 11(1972), 1-16.
4. E. Heinz, Ein v. Neumannscher Satz über beschränkte Operatoren im Hilbertschen Raum, Nachr. Akad. Wiss. Göttingen Math.-Phys. KI.II (1952), 5-6.
5. E. Hille and R.S. Phillips, "Functional Analysis and Semigroups," Rv. Ed., Amer. Math. Soc., Providence, R.I. 1957.
6. B. Sz.-Nagy and C. Foias, "Harmonic Analysis of Operators on a Hilbert Space," North-Holland, Amsterdam, 1970.
7. J. Von Neumann, Eine Spektraltheorie für allgemeine Operatoren eines unitären Raumes, Math. Nachr. 4(1950/1951), 258-281.
8. V. Pták, On the spectral radius in Banach algebras with involution, Bull. London Math. Soc. 2(1970), 327-334.

9. F. Riesz and B. Sz.-Nagy, "Functional Analysis," Fredrick Ungar, New York, 1955.

10. W. Rudin, "Functional Analysis," McGraw-Hill, New York, 1973.

11. Tao Zhiguang, Analytic operator functions, J. Math. Anal. Appl. 103(1984), 293-320.

Vandermonde Determinant and Lagrange Interpolation in R^s

CHARLES K. CHUI Department of Mathematics, Texas A&M University, College Station, Texas

HANG-CHIN LAI Institute of Mathematics, National Tsing Hua University, Hsinchu, Taiwan, Republic of China

The problem of multivariate polynomial interpolation is very old. Among the papers published during the last decade, we only include [1-16] in the References. Let $N_n^s = \binom{n+s}{s}$. The problem can be stated as follows: study the location of nodes (or sample points) $\{x_i : i = 1, \ldots, N_n^s\}$ in R^s such that for every data $\{f_i : i = 1, \ldots, N_n^s\}$, there is a unique polynomial

$$p(w) = \sum_{|j| \leq n} a_j w^j$$

with total degree n which interpolates the given data at the nodes, namely: $p(x_i) = f_i$, $i = 1, \ldots, N_n^s$. Here and throughout, we use the usual multivariate notation $w^j = w_1^{j_1} \ldots w_s^{j_s}$, $|j| = j_1 + \ldots + j_s$ $(j_1, \ldots, j_s \in Z_+)$, etc. Of course, the problem is equivalent to the study of the nonsingularity of the square matrix

$$\left[\phi_1 \vdots \ldots \vdots \phi_{N_n^s}\right]$$

where $\phi_i = [x_i^j]^T$, $|j| \leq n$, is the ith column of the matrix. While the determinant of this matrix in the case $s = 1$ is the wellknown Vandermonde determinant, which is always nonzero for arbitrary distinct nodes, not much seems to be known in the multidimensional setting. The purpose of this paper is to give an expression of this determinant, which we will also call a Vandermonde determinant, so that we could conclude from it the possibility of multivariate Hermite and Birkhoff interpolations by considering coalescence of nodes, lines (of nodes), planes (of nodes), etc. This approach is elementary but is different from those considered in the literature on multivariate polynomial interpolation.

1. Preliminaries

For convenience, we first consider the bivariate setting and set $N_n = N_n^2$. Also, let

$$\hat{X}_n = \{x_i : i = 1, \ldots, N_n\}$$

be a set of distinct nodes. The following distribution of nodes guarantees the existence and uniqueness of Lagrange interpolants as studied in several recent papers (cf. [4, 10, 12]):

Supported by the U. S. Army Research Office under Contract No. DAAG 29-84-K-0154

Node Configuration A. *There exist distinct lines* $\gamma_0, \ldots, \gamma_n$ *such that* $n+1$ *nodes of* $\hat{X}_n$ *lie on* γ_n, n *nodes of* $\hat{X}_n$ *lie on* $\gamma_{n-1}\backslash\gamma_n, \ldots,$ 1 *node of* $\hat{X}_n$ *lies on* $\gamma_0\backslash(\gamma_1 \cup \ldots \cup \gamma_n)$.

If NCA (Node Configuration A) is satisfied, we relabel the nodes, if necessary, so that

$$x_{N_{j-1}+1}, \ldots, x_{N_j} \in \gamma_j \backslash (\gamma_{j+1} \cup \ldots \cup \gamma_n),$$

for $j = 0, \ldots, n$, where $N_{-1} := 0$ and $\gamma_n \backslash (\gamma_{n+1} \cup \ldots \cup \gamma_n) := \gamma_n$.

We will use $w = (u, v)$ as the variables in $\mathbf{R}^2$ and let φ_m denote the monomials:

$$\varphi_m(w) = \varphi_m(u, v) = u^{k-m+N_{k-1}+1} v^{m-N_{k-1}-1}$$

for $N_{k-1} < m \le N_k$, $k = 0, 1, \ldots$. Note that with the notation $N_{-1} := 0$, we have $\varphi_1 = 1$. Hence, the multivariate Vandermonde determinant we will study can be formulated as follows:

$$VD_n\begin{pmatrix} \varphi_1, \ldots, \varphi_{N_n} \\ x_1, \ldots, x_{N_n} \end{pmatrix} = \det[\phi_1 \quad \ldots \quad \phi_{N_n}]$$

where

$$\phi_i = [\varphi_1(x_i) \ldots \varphi_{N_n}(x_i)]^T.$$

If a node distribution guarantees the existence and uniqueness of a Lagrange interpolant to any given data, we say that the set of nodes admits unique Lagrange interpolation. Hence, $\hat{X}_n$ admits unique Lagrange interpolation if and only if

$$VD_n\begin{pmatrix} \varphi_1, \ldots, \varphi_{N_n} \\ x_1, \ldots, x_{N_n} \end{pmatrix} \neq 0.$$

To allow coalescence of nodes along the lines $\gamma_0, \ldots, \gamma_n$, we consider the following node distribution.

Node Configuration B. *There exist distinct lines* $\gamma_0, \ldots, \gamma_n$ *such that*

$$x_{N_{j-1}+1}, \ldots, x_{N_j} \in \gamma_j \backslash (\gamma_{j+1} \cup \ldots \cup \gamma_n)$$

$j = 0, \ldots, n$ *as in NCA,where*

$$x_{N_{j-1}+1}, \ldots, x_{N_j} = \underbrace{y_{j1}, \ldots, y_{j1}}_{\ell_{j1}}, \ldots, \underbrace{y_{jk_j}, \ldots, y_{jk_j}}_{\ell_{jk_j}}$$

with $\ell_{j1} + \ldots + \ell_{jk_j} = N_j - N_{j-1} = j + 1$, $j = 0, \ldots, n$.

Node coalescence along γ_j corresponds to Hermite interpolation with directional derivatives $D^k_{\gamma_j} f$ ($D^0_{\gamma_j} := I$, the identity operator) along γ_j. We denote the column vectors by

$$D^k_{\gamma_j}\phi_i = [D^k_{\gamma_j}\varphi_1(x_i)\dots D^k_{\gamma_j}\varphi_{N_n}(x_i)]^T.$$

Hence, the "Vandermonde" determinant corresponding to this Hermite interpolation problem on the nodes $\hat{X}_n$ satisfying NCB becomes:

$$HD_n\begin{pmatrix}\varphi_1,\dots,\varphi_{N_n}\\ x_1,\dots,x_{N_n}\end{pmatrix}$$
$$= \det\left[\phi_1\vdots\dots\underbrace{\vdots\phi_{j_1}\vdots D_{\gamma_j}\phi_{j_1}\vdots\dots\vdots D^{\ell_{j1}-1}_{\gamma_j}\phi_{j_1}\vdots\dots\vdots\phi_{jk_j}\vdots D_{\gamma_j}\phi_{jk_j}\vdots\dots\vdots D^{\ell_{jk_j}-1}_{\gamma_j}\phi_{jk_j}\vdots}_{\text{(for points on } \gamma_j)}\dots\right].$$

To allow coalescence of the lines $\gamma_0,\dots\gamma_n$, we consider the following node distribution.

Node Configuration C. *The set $\hat{X}_n$ consists of distinct nodes $x_1,\dots,x_{N_n}$, and there exist lines $\gamma_0,\dots,\gamma_n$ where*

$$\gamma_0,\dots,\gamma_n = \underbrace{\beta_1,\dots,\beta_1}_{m_1},\dots,\underbrace{\beta_d,\dots\beta_d}_{m_d},$$

$m_1+\dots+m_d = n+1$ and $\beta_1,\dots\beta_d$ distinct, such that

$$x_{N_{j-1}+1},\dots,x_{N_j}$$

lie on γ_j but not on those of $\gamma_{j+1},\dots,\gamma_n$ different from γ_j.

The corresponding interpolation problem involves interpolating values of normal derivatives. Let $N^k_{\gamma_j}$ denote the kth order differential operator along the direction of the line obtained by rotating γ_j by 90° in the counterclockwise direction, and consider the column vectors

$$N^k_{\gamma_j}\phi_i = [N^k_{\gamma_j}\varphi_1(x_i)\dots N^k_{\gamma_j}\varphi_{N_n}(x_i)]^T.$$

The interpolation problem is:

$$N^{m_j-k-1}_{\beta_j}(p-f)(x_i) = 0, \quad 0 \le k \le m_j - 1,$$

$i = N_{m_1+\dots+m_{j-1}+k-1}+1,\dots,N_{m_1+\dots+m_{j-1}+k}$, and $j = 1,\dots,d$, with the usual notation $m_1+\dots+m_{j-1} := 0$ for $j=1$. Since this is a Birkhoff interpolation problem (where normal derivatives instead of function values are interpolated at the nodes on the line which contains less nodes and is brought to coincide with a line with more nodes), we use the notation BD_n to denote the determinant of the coefficient matrix. That is, the "Vandermonde" determinant becomes:

$$BD_n\begin{pmatrix}\varphi_1,\dots,\varphi_N\\ x_1,\dots x_{N_n}\end{pmatrix}$$

$$= \det\Bigg[\underbrace{\dots\vdots N_{\beta_j}^{m_j-1}\phi_{N_{m_1+\dots+m_{j-1}-1}+1}\vdots\dots\vdots N_{\beta_j}^{m_j-1}\phi_{N_{m_1+\dots+m_{j-1}}}\vdots\dots}_{\text{for } k=0}$$

$$\underbrace{\vdots\phi_{N_{m_1+\dots+m_{j-2}}+1}\vdots\dots\vdots\phi_{N_{m_1+\dots+m_{j-1}}}\vdots\dots}_{\text{for } k=m_j-1}\Bigg]$$

where we have listed the columns for the points on β_j.

Of course, we may also consider the general case where coalescent nodes lie on coalescent lines, but the notation becomes more complicated and we will not discuss the detail in this paper.

For $s > 2$, there is probably no natural way to order the monomials; so we just pick an arbitrary, but fixed, order: $\varphi_1^s,\dots,\varphi_{N_n^s}^s$. The Vandermonde determinant is now

$$VD_n^s\begin{pmatrix}\varphi_1^s,\dots,\varphi_{N_n^s}^s\\ x_1,\dots,x_{N_n^s}\end{pmatrix}$$

where $\hat{X}_n^s = \{x_i : i = 1,\dots,N_n^s\}$ is a set of distinct points satisfying some configuration. One configuration could be obtained inductively as follows.

Node Configuration A **in** $\mathbf{R}^s$. Let $\hat{X}_n^s$ be a set of distinct points $x_1,\dots,x_{N_n^s}$ in $\mathbf{R}^s$ such that there exist $n+1$ hyperplanes K_i^s, $i = 0,\dots,n$, with

$$x_{N_{n-1}^s+1},\dots,x_{N_n^s} \in K_n^s$$

and

$$x_{N_{j-1}^s+1},\dots,x_{N_j^s} \in K_j^s\backslash(K_{j+1}^s \cup \dots \cup K_n^s),$$

for $j = 0,\dots,n-1$, and that each set of points

$$\{x_{N_{j-1}^s+1},\dots,x_{N_j^s}\},$$

$0 \le j \le n$, considered as points in $\mathbf{R}^{s-1}$ satisfies NCA in $\mathbf{R}^{s-1}$.

To discuss the Vandermonde determinant VD_n^s, we first note that $(N_j^s - N_{j-1}^s) = N_j^{s-1}$. Hence, for each set of nodes $\hat{X}_j^{s-1} = \{x_{N_{j-1}^s+1},\dots,x_{N_{N_j}^s}\}$ on K_j^s, we can find an orthogonal transformation A_j that rotates K_j^s to the hyperplane which is normal to the first coordinate axis, so that for $i = N_{j-1}^s + 1,\dots,N_{N_j}^s$,

$$A_j x_s = (y_0, \hat{y}_i)$$

where y_0 is a constant and $\hat{y}_i \in \mathbf{R}^{s-1}$, and the determinant of the appropriate submatrix of the coefficient matix is

$$VD_j^{s-1}\begin{pmatrix}\varphi_1^{s-1}\dots\varphi_{N_j^{s-1}}^{s-1}\\ \hat{y}_{N_{j-1}^s+1}\dots\hat{y}_{N_j^s}\end{pmatrix}.$$

2. Vandermonde determinant with nodes in $\mathbf{R}^2$

In the bivariate setting, we have the following result.

Theorem 1. *Let $\hat{X}_n = \{x_i : i = 1, \ldots, N_n\}$ be a set of distinct nodes in $\mathbf{R}^2$ satisfying NCA. Then*

$$VD_n \begin{pmatrix} \varphi_1, \ldots, \varphi_{N_n} \\ x_1, \ldots x_{N_n} \end{pmatrix} = c \prod_{k=1}^{n} \prod_{N_{k-1} < j < i \le N_k} d(x_i, x_j) \prod_{p=1}^{n} \prod_{q=1}^{N_{p-1}} d(x_q, \gamma_p)$$

where $d(x_i, x_j)$ and $d(x_q, \gamma_p)$ denote euclidean distances between points x_i, x_j and point x_q to line γ_p respectively, and $c = 1$ or -1.

We need some preliminary lemmas. Let $\alpha, \beta, \gamma, \theta$ be real numbers with $\alpha\theta - \gamma\beta = 1$ and consider the $(n+1) \times (n+1)$ matrix $L_n = [a_{ij}^n]$, where a_{ij}^n are the coefficients in the expansion

$$(\alpha u + \beta v)^{n+1-i} (\gamma u + \theta v)^{i-1} = \sum_{j=1}^{n+1} a_{ij}^n u^{n+1-j} v^{j-1}.$$

We have the following result.

Lemma 1. det $L_n = 1$ *for all* $n = 1, 2, \ldots$.

To prove this lemma, we set $a_{ij}^n = 0$ if one of the two indices i and j is less than 1 or greater than $n+1$. Then we have

$$\begin{aligned} \sum_{j=1}^{n+1} a_{ij}^n u^{n+1-j} v^{j-1} &= (\alpha u + \beta v) \Big[\sum_{j=1}^{n} a_{ij}^{n-1} u^{n-j} v^{j-1} \Big] \\ &= \sum_{j=1}^{n+1} (\alpha a_{ij}^{n-1} + \beta a_{i,j-1}^{n-1}) u^{n+1-j} v^{j-1}, \end{aligned}$$

so that

$$a_{ij}^n = \alpha a_{ij}^{n-1} + \beta a_{i,j-1}^{n-1}.$$

Similarly, we can also show

$$a_{ij}^n = \gamma a_{i-1,j}^{n-1} + \theta a_{i-1,j-1}^{n-1}.$$

Without loss of generality, we assume that $\alpha \ne 0$. Hence, we have:

$$\det\ L_n = \det \begin{bmatrix} 1 & 0 & \ldots & \ldots & 0 \\ -\gamma/\alpha & 1 & & & \vdots \\ 0 & & & & 0 \\ \vdots & & & & \\ 0 & \ldots & 0 & -\gamma/\alpha & 1 \end{bmatrix} \det\ L_n$$

$$= \det \begin{bmatrix} 1 & 0 & \cdots & \cdots & 0 \\ -\gamma/\alpha & 1 & & & \vdots \\ 0 & & & & 0 \\ \vdots & & & & \\ 0 & \cdots & 0 & -\gamma/\alpha & 1 \end{bmatrix} \begin{bmatrix} a^n_{11} & \cdots & a^n_{1n+1} \\ \vdots & & \vdots \\ a^n_{n+1,1} & \cdots & a^n_{n+1,n+1} \end{bmatrix}$$

$$= \det \begin{bmatrix} a^n_{11} & a^n_{12} & \cdots & a^n_{1,n+1} \\ a^n_{21} - \frac{\gamma}{\alpha} a^n_{11} & a^n_{22} - \frac{\gamma}{\alpha} a^n_{12} & \cdots & a^n_{2,n+1} - \frac{\gamma}{\alpha} a^n_{1,n+1} \\ \cdots & \cdots & \cdots & \cdots \\ a^n_{n+1,1} - \frac{\gamma}{\alpha} a^n_{n1} & a^n_{n+1,2} - \frac{\gamma}{\alpha} a^n_{n2} & \cdots & a^n_{n+1,n+1} - \frac{\gamma}{\alpha} a^n_{n,n+1} \end{bmatrix}$$

where for $k = 1, \ldots, n$ and $j = 1, \ldots, n+1$,

$$a^n_{k+1,j} - \frac{\gamma}{\alpha} a^n_{kj} = (\gamma a^{n-1}_{kj} + \theta a^{n-1}_{k,j-1}) - \frac{\gamma}{\alpha}(\alpha a^{n-1}_{kj} + \beta a^{n-1}_{k,j-1})$$
$$= \theta a^{n-1}_{k,j-1} - \frac{\gamma\beta}{\alpha} a^{n-1}_{k,j-1} = \frac{1}{\alpha} a^{n-1}_{k,j-1}.$$

That is, we can write

$$\det\ L_n = \det \begin{bmatrix} a^n_{11} & a^n_{12} & \cdots & a^n_{1,n+1} \\ 0 & \frac{1}{\alpha} a^{n-1}_{11} & \cdots & \frac{1}{\alpha} a^{n-1}_{1n} \\ \vdots & & \cdots & \\ 0 & \frac{1}{\alpha} a^{n-1}_{n1} & \cdots & \frac{1}{\alpha} a^{n-1}_{nn} \end{bmatrix}$$

$$= \left(\frac{1}{\alpha}\right)^n a^n_{11} \det\ L_{n-1}.$$

Since $a^n_{11} = \alpha^n$ and $L_1 = \left[\begin{smallmatrix} \alpha & \beta \\ \gamma & \theta \end{smallmatrix}\right]$, we can conclude that $\det\ L_n = \det\ L_{n-1} = \ldots = \det\ L_1 = \alpha\theta - \beta\gamma = 1$ for all n, completing the proof of the lemma.

Lemma 2. *Let $x_i = L_1 y_i$, $i = 1, \ldots, N_n$. Then*

$$\begin{bmatrix} \varphi_{N_{k-1}+1}(x_1) & \cdots & \varphi_{N_{k-1}+1}(x_{N_n}) \\ \cdots & \cdots & \cdots \\ \varphi_{N_k}(x_1) & \cdots & \varphi_{N_k}(x_{N_n}) \end{bmatrix} = L_k \begin{bmatrix} \varphi_{N_{k-1}+1}(y_1) & \cdots & \varphi_{N_{k-1}+1}(y_{N_n}) \\ \cdots & \cdots & \cdots \\ \varphi_{N_k}(y_1) & \cdots & \varphi_{N_k}(y_{N_n}) \end{bmatrix}$$

for $k = 1, 2, \ldots$.

This lemma is a consequence of the definition of L_k. We also need the following lemma.

Lemma 3. *Let $z_1, \ldots, z_{n+2}$ be distinct nodes in $\mathbf{R}^2$ that lie on the line parallel to the u-axis. Then*

$$VD_{n+1}\begin{pmatrix} \varphi_1, \ldots, \varphi_{N_n}, & \varphi_{N_n+1}, \ldots, \varphi_{N_{n+1}} \\ x_1, \ldots, x_{N_n}, & z_1, \ldots, z_{n+2} \end{pmatrix}$$

$$= \pm \prod_{k=1}^{N_n} (v_k - v_0) \prod_{1 \le i < j \le n+2} (u_{N_n+j} - u_{N_n+i}) VD_n \begin{pmatrix} \varphi_1, \ldots, \varphi_{N_n} \\ x_1, \ldots, x_{N_n} \end{pmatrix},$$

where $x_i = (u_i, v_i)$ *and* $z_j = (u_{N_n+j}, v_0)$, $i = 1, \ldots, N_n$ *and* $j = 1, \ldots, n+2$. *(Note:* $N_n + n + 2 = N_{n+1}$.*)*

Instead of giving a formal proof, we indicate the method of proof by considering the special case $n = 2$, namely:

$$VD_3\begin{pmatrix}\varphi_1, \ldots, \varphi_6, \varphi_7, \ldots, \varphi_{10} \\ x_1, \ldots, x_6, z_1, \ldots, z_4\end{pmatrix}$$

$$= \begin{vmatrix}
1 & \ldots & 1 & 1 & 1 & 1 & 1 \\
u_1 & \ldots & u_6 & u_7 & u_8 & u_9 & u_{10} \\
v_1 & \ldots & v_6 & v_0 & v_0 & v_0 & v_0 \\
u_1^2 & \ldots & u_6^2 & u_7^2 & u_8^2 & u_9^2 & u_{10}^2 \\
u_1v_1 & \ldots & u_6v_6 & u_7v_0 & u_8v_0 & u_9v_0 & u_{10}v_0 \\
v_1^2 & \ldots & v_6^2 & v_0^2 & v_0^2 & v_0^2 & v_0^2 \\
u_1^3 & \ldots & u_6^3 & u_7^3 & u_8^3 & u_9^3 & u_{10}^3 \\
u_1^2v_1 & \ldots & u_6^2v_6 & u_7^2v_0 & u_8^2v_0 & u_9^2v_0 & u_{10}^2v_0 \\
u_1v_1^2 & \ldots & u_6v_6^2 & u_7v_0^2 & v_8v_0^2 & u_9v_0^2 & u_{10}v_0^2 \\
v_1^3 & \ldots & v_6^3 & v_0^3 & v_0^3 & v_0^3 & v_0^3
\end{vmatrix}$$

$$= \begin{vmatrix}
1 & \ldots & 1 & 1 & 1 & 1 & 1 \\
u_1 & \ldots & u_6 & u_7 & u_8 & u_9 & u_{10} \\
v_1 - v_0 & \ldots & v_6 - v_0 & 0 & 0 & 0 & 0 \\
u_1^2 & \ldots & u_6^2 & u_7^2 & u_8^2 & u_9^2 & u_{10}^2 \\
u_1(v_1 - v_0) & \ldots & u_6(v_6 - v_0) & 0 & 0 & 0 & 0 \\
v_1(v_1 - v_0) & \ldots & v_6(v_6 - v_0) & 0 & 0 & 0 & 0 \\
u_1^3 & \ldots & u_6^3 & u_7^3 & u_8^3 & u_9^3 & u_{10}^3 \\
u_1^2(v_1 - v_0) & \ldots & u_6^2(v_6 - v_0) & 0 & 0 & 0 & 0 \\
u_1v_1(v_1 - v_0) & \ldots & u_6v_6(v_6 - v_0) & 0 & 0 & 0 & 0 \\
v_1^2(v_1 - v_0) & \ldots & v_6^2(v_6 - v_0) & 0 & 0 & 0 & 0
\end{vmatrix}$$

$$= -\begin{vmatrix}
1 & 1 & 1 & 1 \\
u_7 & u_8 & u_9 & u_{10} \\
u_7^2 & u_8^2 & u_9^2 & u_{10}^2 \\
u_7^3 & u_8^3 & u_9^3 & u_{10}^3
\end{vmatrix}
\begin{vmatrix}
v_1 - v_0 & \ldots & v_6 - v_0 \\
u_1(v_1 - v_0) & \ldots & u_6(v_6 - v_0) \\
v_1(v_1 - v_0) & \ldots & v_6(v_6 - v_0) \\
u_1^2(v_1 - v_0) & \ldots & u_6^2(v_6 - v_0) \\
u_1v_1(v_1 - v_0) & \ldots & u_6v_6(v_6 - v_0) \\
v_1^2(v_1 - v_0) & \ldots & v_6^2(v_6 - v_0)
\end{vmatrix}$$

$$= -\prod_{k=1}^{6}(v_k - v_0) \prod_{1 \le i < j \le 4} (u_{6+j} - u_{6+i}) VD_2\begin{pmatrix}\varphi_1, \ldots, \varphi_6 \\ x_1, \ldots, x_6\end{pmatrix}.$$

We are now ready to prove Theorem 1. Let α, β, γ, θ be chosen such that $L_1^{-1}\gamma_n$ is parallel to the u-axis. Write $L_1^{-1}x_i = y_i = (u_i, v_i)$ for $i = 1, \ldots, N_n$, so that $v_j = v_0$, say, for $j = N_{n-1} + 1, \ldots, N_n$. Then applying Lemmas 2, 1, and 3 consecutively, we have:

$$
\begin{aligned}
VD_n\begin{pmatrix}\varphi_1,\ldots,\varphi_{N_n}\\ x_1,\ldots,x_{N_n}\end{pmatrix}
&= \det\begin{bmatrix} L_1 & & & \\ & L_2 & & \\ & & \ddots & \\ & & & L_n\end{bmatrix} VD_n\begin{pmatrix}\varphi_1,\ldots,\varphi_{N_n}\\ y_1,\ldots,y_{N_n}\end{pmatrix}\\
&= \pm\prod_{k=1}^{N_{n-1}}(v_k - v_0)\prod_{1\le i<j\le n+1}(u_{N_{n-1}+j}-u_{N_{n-1}+i})VD_{n-1}\begin{pmatrix}\varphi_1,\ldots,\varphi_{N_{n-1}}\\ y_1,\ldots,y_{N_{n-1}}\end{pmatrix}.
\end{aligned}
$$

Since distances are preserved under rotation and

$$
\begin{aligned}
VD_{n-1}\begin{pmatrix}\varphi_1,\ldots\varphi_{N_{n-1}}\\ y_1,\ldots,y_{N_{n-1}}\end{pmatrix}
&= \det\begin{bmatrix} L_1^{-1} & & & \\ & L_2^{-1} & & \\ & & \ddots & \\ & & & L_{n-1}^{-1}\end{bmatrix} VD_{n-1}\begin{pmatrix}\varphi_1,\ldots,\varphi_{N_{n-1}}\\ x_1,\ldots,x_{N_{n-1}}\end{pmatrix}\\
&= VD_{n-1}\begin{pmatrix}\varphi_1,\ldots,\varphi_{N_{n-1}}\\ x_1,\ldots,x_{N_{n-1}}\end{pmatrix},
\end{aligned}
$$

we have

$$
VD_n\begin{pmatrix}\varphi_1,\ldots,\varphi_{N_n}\\ x_1,\ldots,x_{N_n}\end{pmatrix} = \pm\prod_{k=1}^{N_{n-1}} d(x_k,\gamma_n)\prod_{N_{n-1}<i<j\le N_n} d(x_i,x_j)VD_{n-1}\begin{pmatrix}\varphi_1,\ldots,\varphi_{N_{n-1}}\\ x_1,\ldots,x_{N_{n-1}}\end{pmatrix}.
$$

Hence, the proof of Theorem 1 is complete by using mathematical induction and the fact that

$$
VD_0\begin{pmatrix}\varphi_1\\ x_1\end{pmatrix} = 1.
$$

Remark 1. *Let $\hat{X}_n = \{x_i : i = 1,\ldots,N_n\}$ be a set of distinct nodes in $\mathbf{R}^2$ such that distinct lines $\gamma_{n-k-1}, \gamma_{n-k},\ldots,\gamma_n$ $(0 < k < n-1)$ exist with $n-j+1$ points of $\hat{X}_n$ lying on*

$$
\gamma_{n-j}\backslash(\gamma_{n-j+1}\cup\ldots\cup\gamma_n)
$$

for each $j = 0,\ldots,k$. Suppose that more than $n-k$ points of $\hat{X}_n$ lie on

$$
\gamma_{n-k+1}\backslash(\gamma_{n-k}\cup\ldots\cup\gamma_n)
$$

Then the set $\hat{X}_n$ does not admit unique Lagrange interpolation.

To see this, we again consider the determinant (of the transpose) of the coefficient matrix. From the recurrence relation we obtained in the proof of Theorem 1, we see that VD_{n-k+1} is a factor of VD_n. Since the "zero block" in VD_{n-k+1} (obtained after appropriate row reductions in the induction of the proof of Lemma 3 and row rearrangement) has more columns than rows due to the extra points lying on γ_{n-k-1}, this factor is zero. Hence, $VD_n = 0$.

Let us consider a simple example where

$$\begin{aligned}
&x_1 = (u_0, v_0)\\
&x_2 = (u_0, v_1), \quad x_3 = (u_1, v_1)\\
&x_4 = (u_0, v_2), \quad x_5 = (u_1, v_2), \quad x_6 = (u_2, v_2)\\
&\cdots\cdots\cdots\cdots\cdots\cdots\\
&x_{N_{n-1}+1} = (u_0, v_n), \quad x_{N_{n-1}+2} = (u_1, v_n), \ldots, x_{N_n} = (u_n, v_n).
\end{aligned}$$

Then clearly $\hat{X}_n$ satisfies NCA with $\gamma_j : v = v_j$, and the result in Theorem 1 says that the determinant of the coefficient matrix in Lagrange interpolation at $\hat{X}_n$ is

$$VD_n\begin{pmatrix}\varphi_1, \ldots, \varphi_{N_n}\\ x_1, \ldots, x_{N_n}\end{pmatrix} = c \prod_{0 \le i < j \le n} (u_j - u_i)^{n-j+1} \prod_{0 \le i < j \le n} (v_j - v_i)^{n-j+1},$$

where c = 1 or -1.

We also remark that the error in Lagrange interpolation at nodes satisfying NCA is given by

$$(f - P_n)(x) = \frac{1}{VD_n\binom{\varphi_1, \ldots, \varphi_{N_n}}{x_1, \ldots, x_{N_n}}} \begin{vmatrix} \varphi_1(x_1) & \cdots & \varphi_1(x_{N_n}) & \varphi_1(x)\\ \varphi_2(x_1) & \cdots & \varphi_2(x_{N_n}) & \varphi_2(x)\\ \cdots & \cdots & \cdots & \cdots\\ \varphi_{N_n}(x_1) & \cdots & \varphi_{N_n}(x_{N_n}) & \varphi_{N_n}(x)\\ f(x_1) & \cdots & f(x_{N_n}) & f(x) \end{vmatrix}$$

3. Hermite and Birkhoff interpolation in $\mathbf{R}^2$

It is well known that coalescence of nodes in one-dimension corresponds to Hermite interpolation. For instance, if we have $x_1 < \ldots < x_k$, and

$$\hat{X}_n = \{\underbrace{x_1, \ldots, x_1}_{n_1}, \underbrace{x_2, \ldots, x_2}_{n_2}, \ldots, \underbrace{x_k, \ldots, x_k}_{n_k}\},$$

$n_1 + \ldots + n_k = n + 1$, is a set of nodes in $\mathbf{R}^1$, then the Lagrange interpolation problem becomes the Hermite interpolation problem:

$$(f - p_n)^{(\ell_i)}(x_i) = 0,$$

$\ell_i = 0, \ldots, n_i - 1$, $i = 1, \ldots, k$, where p_n is a polynomial of degree n. This idea naturally carries over to the two dimensional setting if the Node Configuration B is regarded as a consequence of coalescence of nodes that originally satisfied NCA. By using the recurrence relation derived in the proof of Theorem 1, the determinant of the coefficient matrix of the Hermite interpolation problem is obtained by taking directional derivatives with respect to the coalescent nodes along the lines they lie on. That is, we have the following result.

Theorem 2. *Let $\hat{X}_n = \{x_i : i = 1, \ldots, N_n\}$ satisfy NCB. Then the determinant of the coefficient matrix of the corresponding Hermite interpolation problem is given by*

$$HD_n \begin{pmatrix} \varphi_1, \ldots, \varphi_{N_n} \\ x_1, \ldots x_{N_n} \end{pmatrix}$$

$$= c \left[\prod_{j=1}^{n} \prod_{1 \le s < t \le k_j} [d(y_{js}, y_{jt})]^{\ell_{js}\ell_{jt}} \prod_{u=1}^{k_j} \prod_{\rho=1}^{\ell_{ju}-1} \rho! \right] \left[\prod_{j=1}^{n} \prod_{v=1}^{k_j} [d(y_{jv}, \gamma_j)]^{\ell_{jv}} \right]$$

where $c = 1$ or -1. Consequently, $\hat{X}_n$ admits unique Hermite interpolation from polynomials of total degree n.

We next consider the Node Configuration C. The simplest example is the following case where $n = 1$ and $N_n = 3$:

Let $u_1 < u_2 < u_3$ and $\hat{X}_1 = \{x_1, x_2, x_3\}$ with $x_i = (u_i, 0)$. Obviously $\hat{X}_1$ does not admit unique Lagrange interpolation from linear polynomials $p_1(u,v) = a + bu + cv$ $(p_1(x_i) = a + bu_i$, $i = 1,2,3$, and c is not affected). To determine c, one way is to take the derivative with respect to v. This gives rise to the Birkhoff interpolation problem:

$$(f - p)(x_i) = 0, \quad i = 1,2$$

and

$$\frac{\partial}{\partial v}(f - p)(x_3) = 0.$$

This example suggests that the interpolation at x_3 results from a limit as $\epsilon \to 0$ of Lagrange interpolation at x_1, x_2, and $x_{3,\epsilon} = (u_3, \epsilon)$ where $\epsilon \neq 0$, or equivalently coalescence of two lines, one containing x_1 and x_2, and the other containing x_3.

In general we consider d distinct lines $\beta_1, \ldots, \beta_d$ and a set $\hat{X}_n = \{x_i : i = 1, \ldots, N_n\}$ of distinct nodes in $\mathbf{R}^2$ lying on these lines. We could imagine that Node Configuration C is satisfied. That is, we classify the lines $\beta_1, \ldots \beta_d$ as

$$\gamma_0, \ldots, \gamma_n = \underbrace{\beta_1, \ldots \beta_1}_{m_1}, \ldots, \underbrace{\beta_d, \ldots, \beta_d}_{m_d}$$

where $m_1 + \ldots + m_d = n + 1$, such that

$$x_{N_{j-1}+1}, \ldots, x_{N_j}$$

lie on γ_j but not on those $\gamma_{j+1}, \ldots, \gamma_n$ different from γ_j. For the nodes $x_{N_{n-1}+1}, \ldots, x_{N_n}$ on β_d (we call it the "last" β_d), we assign Lagrange interpolation: $(f - p)(x_i) = 0$, $i = N_{n-1}+1, \ldots, N_n$. For the nodes on the next to the last β_d, we assign interpolation of the first directional derivative along the direction which is orthogonal to β_d and so on, and finally to the last set of nodes on the first β_d, we assign interpolation of the $(m_d - 1)^{st}$ directional derivative in the direction orthogonal to β_d. The same Birkhoff interpolation scheme is performed at the nodes on the other lines $\beta_{d-1}, \ldots, \beta_1$. The determinant of the coefficient matrix of this interpolation problem is

$$BD_n \begin{pmatrix} \varphi_1, \ldots, \varphi_n \\ x_1, \ldots, x_n \end{pmatrix}$$

and is obtained by taking the appropriate "normal" derivatives as discussed above. That is, using Theorem 1, we have obtained the following result.

Theorem 3. *Let $\hat{X}_n = \{x_1, \ldots, x_{N_n}\}$ be a set of distinct nodes satisfying NCC. Then the determinant of the coefficient matrix of the Birkhoff interpolation problem described above is given by*

$$BD_n \begin{pmatrix} \varphi_1, \ldots, \varphi_{N_n} \\ x_1, \ldots, x_{N_n} \end{pmatrix}$$
$$= c \prod_{k=1}^{n} \prod_{N_{k-1}<j<i\leq N_k} d(x_i, x_j) \prod_{p=2}^{d} \prod_{x_q \in B_p} d(x_q, \beta_p) \prod_{p=1}^{d} \prod_{s=1}^{m_p - 1} (s!)^{m_1+\ldots+m_p - s}$$

where $B_p = \beta_1 \cup \ldots \cup \beta_{p-1}$ and $c = 1$ or -1. Consequently, $\hat{X}_n$ admits unique Birkhoff interpolation from polynomials of total degree n.

Remark 2. Suppose that $\hat{X}_n$ satisfies NCC. We can now consider coalescence of nodes along each β_p. This means that at some nodes on β_p both directional derivatives along and orthogonal to β_p are interpolated. Since the determinant of the coefficient matrix of this interpolation problem can be obtained by taking derivatives at the nodes along β_p of the formula in Theorem 3 and is certainly nonzero, the new set of nodes admits unique interpolation from polynomials of total degree n.

Remark 3. Coalescence of nodes can also be regarded as a result of line coalescence. In other words, normal instead of tangential derivative data can be interpolated. For instance, let $\{y_0, \ldots, y_n\}$ be a set of distinct points on the u-axis. We could consider the u-axis to be γ_n and $y_0 = x_{N_{n-1}+1}, \ldots, y_n = x_{N_n}$. Now, consider $\gamma_{n-1}, \gamma_{n-2}, \ldots, \gamma_0$, again to be the u-axis, but were brought in to coincide with γ_n consecutively from parallel lines, so that the nodes on γ_{n-j} now become $y_{n-j}, \ldots, y_n$, $j = 1, \ldots, n$. That is, we consider the interpolation problem

$$D_v^{k_j}(f - P_n)(y_j) = 0, \quad k_j = 0, \ldots, j \quad \text{and} \quad j = 0, \ldots, n,$$

where D_v is the partial derivative with respect to v. The determinant of the coefficient matrix of this interpolation problem can be obtained by taking the same derivatives of the expression of the determinant in Theorem 1, yielding:

$$\prod_{k=0}^{n-1} [(n-k)!]^{k+1} \prod_{\ell=0}^{n-1} \prod_{\ell \leq j < i \leq n} (y_i - y_j)$$

which is non-zero. Hence, the set of nodes admits unique interpolation as discussed above.

4. Extensions to $\mathbf{R}^s$, $s > 2$

All the results we discussed in Sections 2 and 3 can be extended to arbitrarily higher dimension. Since the notation would become too complicated, we only consider the NCA in $\mathbf{R}^s$. The extension of Lemma 3 is the following

Lemma 4. *Let $\hat{X} = \{x_i : i = 1, \ldots, N^s_{n-1}\}$ be a set of nodes in $\mathbf{R}^s$ and $\{z_1, \ldots, z_{N^{s-1}_n}\}$ lie on a hyperplane K the normal of which is parallel to the first coordinate axis. Write $x_i = (u_{i1}, \ldots, u_{is}) = (u_{i1}, \overline{v}_i)$ and $z_j = (u_0, w_{j2}, \ldots, w_{js}) = (u_0, \overline{w}_j)$. Then*

$$VD^s_n \begin{pmatrix} \varphi^s_1, \ldots, \varphi^s_{N^s_{n-1}}, & \varphi^s_{N^s_{n-1}+1}, \ldots, \varphi^s_{N^s_n} \\ x_1, \ldots x_{N^s_{n-1}}, & z_1, \ldots, z_{N^{s-1}_n} \end{pmatrix}$$

$$= \pm \prod_{i=1}^{N^s_{n-1}} (u_{i1} - u_0) VD^{s-1}_{n-1} \begin{pmatrix} \varphi^{s-1}_1, \ldots, \varphi^{s-1}_{N^{s-1}_n} \\ \overline{w}_1, \ldots, \overline{w}_{N^{s-1}_n} \end{pmatrix} VD^s_{n-1} \begin{pmatrix} \varphi^s_1, \ldots, \varphi^s_{N^s_{n-1}} \\ x_1, \ldots, x_{N^s_{n-1}} \end{pmatrix}$$

$$= \pm \prod_{i=1}^{N^s_{n-1}} d(x_i, K) VD^{s-1}_{n-1} \begin{pmatrix} \varphi^{s-1}_1, \ldots, \varphi^{s-1}_{N^{s-1}_n} \\ z_1, \ldots, z_{N^{s-1}_n} \end{pmatrix}\Bigg|_K VD^s_{n-1} \begin{pmatrix} \varphi^s_1, \ldots, \varphi^s_{N^s_{n-1}} \\ x_1, \ldots, x_{N^s_{n-1}} \end{pmatrix}.$$

As in the two-dimensional setting, an application of this lemma yields the following result.

Theorem 4. *Let $\hat{X} = \{x_i : i = 1, \ldots, N^s_{n-1}\}$ be a set of nodes satisfying NCA in $\mathbf{R}^s$. Then*

$$VD^s_n = \begin{pmatrix} \varphi^s_1, \ldots, \varphi^s_{N^s_n} \\ x_1, \ldots, x_{N^s_n} \end{pmatrix}$$

$$= c \prod_{\ell=1}^{n} \prod_{i=1}^{N^s_{\ell-1}} d(x_i, K^s_\ell) \prod_{p=1}^{n} VD^{s-1}_p \begin{pmatrix} \varphi^{s-1}_1, \ldots, \varphi^{s-1}_{N^{s-1}_p} \\ x_{N^s_{p-1}+1}, \ldots, x_{N^s_p} \end{pmatrix}\Bigg|_{K^s_p},$$

where $c = 1$ or -1. Consequently, $\hat{X}$ admits unique Lagrange interpolation from polynomials of total degree n in s variables.

References

1. K. C. Chung and T. H. Yao, On lattices admitting unique Largange interpolation, SIAM J. Numer. Anal. **14**, 735-743, 1977.

2. P. G. Ciarlet and P. A. Raviart, General Lagrange and Hermite interpolation in $\mathbf{R}^n$ with applications to finite elements, Arch. Rational Mech. Anal. **46**, 177-199, 1972.

3. F. J. Delvos, *d*-variate Boolean interpolation, J. Approx. Theory **34**, 99-114, 1982.

4. M. Gasca and J. I. Maezta, On Lagrange and Hermite interpolation in $\mathbf{R}^k$, Numer. Math. **39**, 1-14, 1982.

5. M. Gasca and V. Ramirez, Interpolation systems in $\mathbf{R}^k$, J. Appprox. Theory **42**, 36-51, 1984.

6. M. Gasca and A. Lopez-Carmona, A general recurrence interpolation formula and its applications to multivariate interpolation, J. Approx. Theory **34**, 361-374, 1982.

7. F. Hack, Eindeutigkeit saussagen und Konstructive Methoden bei mehrdimensionalen interpolations aufgaben, Doctoral Thesis, Univ. of Duisburg, Duisburg, 1983.

8. H. Hakopian, Multivariate divided differences and multivariate interpolation of Lagrange and Hermite type, J. Approx. Theory **34**, 286-305, 1982.

9. W. Haussmann and H. B. Knoop, On two dimensional interpolation, in: *Coll. Math. Soc. Bolyai* **19** *Fourier Analysis and Approximation Theory*, Budapest, 1976.

10. K. Jetter, Some contributions to bivariate interpolation and cubature, in *Approximation Theory IV*, C. K. Chui, L. L. Schumaker and J. D. Ward, eds., Academic Press, New York, 533-538, 1984.

11. P. Kergin, A natural interpolation of C^k function, J. Approx. Theory **29**, 278-293, 1980.

12. Liang, Xuezhang, Properly posed nodes for bivariate interpolation and the superposed interpolation, Jilin Daxue Xuebao (Chinese) **1**, 27-32, 1979.

13. G. G. Lorentz and R. A. Lorentz, Multivariate interpolation, in *Rational Approximation and Interpolation*, P. Graves-Morris et al. eds., Lecture Notes in Math. No. 1105, Springer-Verlag, Berlin, 136-144, 1984.

14. R. A. Lorentz, Some regular problems of bivariate interpolation, in *Constructive Theory of Functions*, Publishing house of the Bulgarian Academy of Sciences, Sofia, 549-562, 1984.

15. C. Micchelli and P. Milman, A formula for Kergin interpolation in R^k, J. Approx. Theory, **29**, 294-296, 1980.

16. H. Werner, Remarks on Newton type multivariate interpolation for subset of grids, Computing, **25**, 181-191, 1980.

Applications of Nonstandard Theory of Locally Convex Spaces

CARL L. DeVITO* Department of Mathematics, University of Arizona, Tucson, Arizona

Our purpose here is to show how one can use some results from the non-standard theory of locally convex spaces to investigate the continuous, linear operators on these spaces. This approach is successful in that it enables us to answer one question and shed light on several other questions in the theory of linear operators. Unfortunately, we cannot be more specific until after we have developed some terminology. Let us stress, however, that while the results from the non-standard theory of locally convex spaces play a crucial role in all that follows, no knowledge of non-standard analysis is needed either to understand our questions or our discussion of them.

1. We shall work with Hausdorff, locally convex, topological vector spaces over $\mathbb{C}$, the field of complex numbers. The letter E will always denote such a space. A subset S of E is said to be balanced if $\lambda S \subseteq S$ whenever λ is a scalar with $|\lambda| \leq 1$. The family of all balanced, convex neighborhoods of zero in E will be denoted by $\hat{U}$. Recall that a subset B of E is said to be a bounded set if, for every U in $\hat{U}$, there is a scalar λ such that $B \subseteq \lambda U$. We shall work with both nets in E and filters on E. A net of points of E whose domain is the directed set A will be written $\{x_\alpha | \alpha \in A\}$. We refer the reader to [7,p.65] for our terminology regarding nets and to [8,pp.11-12] for our terminology regarding filters.

Definition 1. A filter $\hat{F}$ on E is said to be an ultimately bounded filter if for every $U \in \hat{U}$ there is a scalar λ such that $\lambda U \in \hat{F}$. We shall say that a filter base $\hat{B}$ on E is ultimately bounded if it generates a filter which is ultimately bounded.

Ultimately bounded filters arise in the non-standard theory of locally convex spaces [4, theorem 1.4, p.140] as we shall see below. Ultimately bounded nets were introduced in [2] and used there and in [3] to study certain classes of locally convex spaces. There is a canonical way of associating a net to a filter and vice versa and one can show that a filter is ultimately bounded if and only if the corresponding net is ultimately bounded.

In the non-standard theory of locally convex spaces one associates with each such space, call it E, a new space, $\hat{E}$ say, which is called

*Current affiliation: Naval Postgraduate School, Monterey, California

the non-standard hull of E. In general, $\hat{E}$ depends both on E and on the way we construct the non-standard hull. There is, however, a class of spaces whose non-standard hulls are independent of the particular way in which they are constructed. Such spaces are said to have an invariant non-standard hull; see [4,p.416 and 5]. These spaces can be characterized as follows: E has an invariant non-standard hull if and only if every ultrafilter on E which is ultimately bounded is a Cauchy filter [4, Theorem 4.1, p.416].

In particular, every Schwartz space [6, Definition 2, p.275] has an invariant non-standard hull [5, Theorem 4, p.201] and a Frechet space (i.e., a complete, metrizable locally convex space) has an invariant non-standard hull if and and only if it is a Frechet-Montel space [6, Definition 1, p.231 and 5, Theorem 1, p.196].

Suppose now that E is complete and also has an invariant non-standard hull. Then every continuous, linear operator on E is of the following type:

Definition 2. A continuous, linear operator T on E is said to be a convergence compact operator (abbreviated c.c. operator) if for every filter base $\hat{B}$ on E which is ultimately bounded, the filter base $T(\hat{B})$ has an adherent point in E.

We recall that a linear operator T on a normed space E is said to be a compact operator if the set T(B), where B is the unit ball of E, is relatively compact (that is, its closure is compact) in E. This concept has been generalized to locally convex spaces in at least two ways. One way is to require that the operator map some neighborhood of zero onto a relatively compact set. We shall call operators with this property compact operators. It is clear that every compact operator is a c.c. operator. Another generalization is obtained by requiring that the operators be continuous, and that they map any bounded set onto a relatively compact set. We shall call such operators mildly compact operators. Clearly every c.c. operator is a mildly compact operator.

There are two important aspects to the discussion of compact operators on a normed space. First we have the set of theorems collectively known as the Riesz theory, and, secondly, the theorem of Schauder which states that the transpose (defined below) of any compact operator is also a compact operator. Now it is well-known that the Riesz theory is valid for compact operators on locally convex spaces [10, Chap.VIII, p.142]. Our first question is this: Are the results of Riesz theory valid for mildly compact operators? We shall answer this question, in the negative, by showing that no analogue of the Riesz theory is valid even for c.c. operators. However, before we can do that, we need some more terminology.

Let S be a continuous, linear operator on E, let I be the identity operator on E, let $\lambda \in \mathbb{C}$ and let n be a positive integer. Then the null space of $(\lambda I-S)^n$, which we shall denote by $\mathrm{Ker}(\lambda I-S)^n$, is a closed, linear subspace of E, and clearly $\mathrm{Ker}(\lambda I-S)^n \subseteq \mathrm{Ker}(\lambda I-S)^{n+1}$ for $n \geq 1$. We shall say that λ is an eigenvalue of S if $\mathrm{Ker}(\lambda I-S) \neq \{0\}$. Given an eigenvalue λ of S there are two possibilities. Either there is an integer m such that $\mathrm{Ker}(\lambda I-S)^m = \mathrm{Ker}(\lambda I-S)^{m+1}$, in which case we shall say that the pair $\{S,\lambda\}$ has finite ascent, or no such integer exists. In the latter case we shall say that $\{S,\lambda\}$ has infinite ascent. For any compact operator T on E and any non-zero eigenvalue λ of T we have [10, Prop. 1, p.144 and Prop. 2, p.146]: (a) for each n the space $\mathrm{Ker}(\lambda I-T)^n$ is finite dimensional; (b) $\{T,\lambda\}$ has finite ascent.

We shall now construct a c.c. operator T on a locally convex space E such that 1 is an eigenvalue of T and the pair $\{T,1\}$ has infinite ascent. We shall also show that the space Ker(I-T) is not finite dimensional. For notational convenience the space $\mathbb{C}^k$, k a positive integer, will be denoted by $\mathbb{C}(k)$. The matrix theory used can be found in [1,pp.189-196].

Choose an integer K_1 and a $k_1 x k_1$ matrix S_1 satisfying $\{0\} \neq \mathrm{Ker}(I-S_1) \neq \mathrm{Ker}(I-S_1)^2 = \mathrm{Ker}(I-S)^{2+j} \neq \mathbb{C}(k_1)$ for all $j \geq 1$. Next choose k_2 and a $k_2 x k_2$ matrix S_2 which satisfies $\{0\} \neq \mathrm{Ker}(I-S_2) \neq \mathrm{Ker}(I-S_2)^2 \neq \mathrm{Ker}(I-S_2)^3 = \mathrm{Ker}(I-S_2)^{3+j} \neq \mathbb{C}(k_3)$ for all $j \geq 1$. Continue in this way. After $k_1, k_2, \ldots, k_{n-1}$ and $S_1, S_2, \ldots, S_{n-1}$ have been chosen, we choose k_n and a $k_n x k_n$ matrix S_n which satisfies $\{0\} \neq \mathrm{Ker}(I-S_n) \neq \ldots \neq \mathrm{Ker}(I-S_n)^{n+1} = \mathrm{Ker}(I-S_n)^{N+1+j^n} \neq \mathbb{C}(k_n)$ for all $j \geq 1$. In this way we generate a sequence of integers and a sequence of matrices such that for each n we can choose vectors ν_n, y_n in $\mathbb{C}(k_n)$ for which $\nu_n \in \mathrm{Ker}(I-S_n)$, $y_n \in \mathrm{Ker}(I-S_n)^{n+1}$ and $y_n \in \mathrm{Ker}(I-S_n)^n$. Let $E = \prod_{n=1}^{\infty} \mathbb{C}(k_n)$ with the product topology. Then E is a complete Schwartz space [6, Prop. 6b, p.278). Thus E is complete and has an invariant non-standard hull [5, Theorem 4, p.201].

Note that each z in E can be written uniquely as a sequence $(z_1, 3_2, \ldots)$ where $z_n \in \mathbb{C}(k_n)$ for every n. Define a linear operator T on E by $T(z) = (S_1 z_1, S_2 z_2, \ldots)$. It is clear that T is continuous on E and that T is a c.c. operator on E. However, Ker(I-T) contains the infinite, linearly independent set $(\nu_1,0,0,\ldots)$, $(0,\nu_2,0,0,\ldots)\ldots$, and for each n the vector $(0,0,\ldots,y_n,0,0,\ldots)$, where the y_n is in the n-th place, is in $\mathrm{Ker}(I-T)^{n+1}$ but is not in $\mathrm{Ker}(I-T)^n$. Hence $\{T,1\}$ has infinite ascent.

The results from the non-standard theory of locally convex spaces which enabled us to answer the question posed above can also be used to construct a class of operators which properly contains the class of compact operators and is properly contained in the class of c.c. operators. This class was investigated by the author and Ana M. Suchanek. We were able to show that both the Riesz theory and a version of Schauders' theorem are valid for these operators. Let us begin by constructing an example.

If a locally convex space E is both a Frechet space and a Schwartz space then (a) E is a Montel space [6, corollary to Prop. 4, p.277], (b) E/M, where M is any closed, linear subspace of E, is also both a Frechet space and a Schwartz space [6, corollary to Prop.7, p.279]. There is an example [8, 31(5),pp.433-434] of a Frechet-Montel space F which has a closed, linear subspace M such that E/M is not a Montel space. Clearly then, by (a), (b) and the results mentioned in our first section, F is complete, has an invariant non-standard hull, and is not a Schwartz space. Let us work now with this specific space F. For any $U \epsilon \hat{U}$ let q_u be the gauge function of U[6, pp.94,208] and let F_u be the quotient space of F by $\{x \epsilon F | q_u(x)=0\}$. Then (F_u,q_u) is a normed space and the canonical map from F onto this space is continuous. Since F is not a Schwartz space we may choose $U \epsilon \hat{U}$ such that there is no $V \epsilon \hat{U}$ whose canonical image in (F_u,q_u) is precompact [6, Prop.3, p.275]. Hence the canonical map, call it P_u, is continuous, it takes the neighborhood U onto a bounded set, and yet it is not a compact map. However, the image under P_u of any ultimately bounded filter base in F has adherence in F_u because F is complete and has an invariant non-standard hull. Finally, we construct an operator by first letting $G = F \times F_u$ with the product topology and then letting $T(x,y) = (0, P_u(x))$ for all $(x,y) \epsilon G$. Clearly T is a non-compact, c.c. operator which maps the neighborhood $U \times F_u$ onto a bounded set.

<u>Definition 3</u>. A c.c. operator on E is said to be a bounded, convergence compact operator (abbreviated b.c.c. operator) if it maps some $U \epsilon \hat{U}$ onto a bounded set.

It is clear that any compact operator is a b.c.c. operator and we have just seen that a b.c.c. operator need not be a compact operator. Note however, that if T is a b.c.c. operator then T^2 is a compact operator. It follows immediately from this [10, p.143] that the Riesz theory is valid for b.c.c. operators. Schauders' theorem has been generalized to operators on locally convex spaces in several ways [9, 542, pp.200-204] and [10, Prop.5, p.153]. We shall show that this theorem can be generalized to b.c.c. operators. However, we must first recall some important results about locally convex spaces and some more terminology.

Whenever we want to call attention to a specific locally convex topology t on E we will write E[t]. A bounded subset of E[t] will be called a t-bounded set. Similarly, we will speak of t-ultimately bounded nets, t-c.c. and t-b.c.c. operators, etc. The vector space of all t-continuous, linear functionals on E will be denoted by E', and will be called the dual of E[t]. The weakest topology on E for which each element of E' is continuous is called the weak topology, and will be denoted by $\sigma(E,E')$. It turns out that the dual of $E[\sigma(E.E')]$ is also E'. There is a strongest locally convex topology on E for which this is true. It is called the Mackey topology [6, Paragraph 5, p.203].

We shall need two topologies on E'. The first of these is the weak* topology which we shall denote by $\sigma(E',E)$. This is the weakest topology on E' for which each element of E, regarded as a linear functional on E', is continuous. We will also need the strong topology, $\beta(E',E)$, on E'. This is the topology of uniform convergence on the $\sigma(E,E')$-bounded subsets of E[6, Definition 2, p.201]. Any linear operator T on E has a transpose, T*, defined as follows: For each $f\in E'$, $(T^*f)(x) = f(Tx)$ for all $x\in E$. When T is continuous on E, T* is continuous on both $E'[\sigma(E',E)]$ and $E'[\beta(E',E)]$ [6, Corollary to Proposition 3,p.256].

Schauders' theorem states that the transpose of a compact operator defined on a normed space is a compact operator for the norm (strong) topology of the dual space. This is not true in general, see [9, § 42, p. 200-204] or [10, Chapter VIII, § 2, pp.151-154]. For certain spaces E it is true that the transpose of a mildly compact operator is mildly compact on $E'[\beta(E',E)]$ [10, Prop. 5, p.153]. We shall show that for certain spaces E, to be defined presently, the transpose of any b.c.c. operator (and hence of any compact operator) is a b.c.c. operator on $E'[\beta(E',E)]$.

The dual of the space $E'[\beta(E',E)]$ will be denoted by E" and called the bidual of E. There is a natural way of embedding E into E"[6, Paragraph 8, p.226] and it is customary to simply regard E as a subspace of its bidual. On E" we have the topologies $\sigma(E'',E')$ and $\beta(E'',E')$. The restriction of the latter topology to E is denoted by $\beta^*(E,E')$. In general, this topology is stronger than the Mackey topology on E. However, there are many spaces for which these two topologies coincide. Such spaces are said to be quasi-barrelled [6, Definition 2, p.217]. We shall need one fact about $E''[\sigma(E'',E')]$. If B is any bounded subset of E, then the closure of B in $E''[\sigma(E'',E')]$ is a compact set [6, Theorem 1, p.201].

<u>Definition 4</u>. A locally convex space E is said to be an ab-space if it has its Mackey topology and every net in E' which is both $\sigma(E',E)$-Cauchy and $\beta(E',E)$-ultimately bounded is $\sigma(E',E)$-covergent to a point of E'.

These spaces were introduced in [2], where it is shown that every bornological space [6, Definition 1, p.220] is an ab-space and that every ab-space is quasi-barrelled. The fact that any normed space is an ab-space also follows from the well-known theorem of Alaoglu [8, § 20, 9(5), p.249].

Theorem 1. Let E be an ab-space and let T be a b.c.c. operator on E. Then T* is a b.c.c. operator on $E'[\beta(E',E)]$.

Proof. Let U be a convex neighborhood of zero in E for which T(U) is a bounded set. The $U_1 \equiv T(U)^0 = \{f \in E' \mid |f(x)| \leq 1 \text{ for all } x \in T(U)\}$ is a $\beta(E',E)$-neighborhood of zero in E'. Furthermore, one can easily show that $T^*(U_1) \subseteq U^0$. Thus, since E is a quasi-barrelled space, $T^*(U_1)$ is a $\beta(E',E)$-bounded set. So in order to prove that T* is a b.c.c. operator on $E'[\beta(E',E)]$ it suffices to prove that it is a c.c. operator on this space.

We begin by showing that any $\beta(E',E)$-ultimately bounded net in E' has a subnet which is $\sigma(E',E)$-convergent to a point of E'. This is most conveniently done for filters. Recall that a filter $\hat{F}$ on E' is $\beta(E',E)$-ultimately bounded if for every strong neighborhood U of zero there is a scalar λ such that $\lambda U \in \hat{F}$. Clearly, any ultrafilter $\hat{G}$ which contains $\hat{F}$ is also $\beta(E',E)$-ultimately bounded. Now $\sigma(E'E)$ is weaker than $\beta(E',E)$ and so $\hat{G}$ is also $\sigma(E',E)$-ultimately bounded. It follows that $\hat{G}$ is $\sigma(E',E)$-Cauchy ([5. Theorem 4, p.201], [6, Example 1, p.280] and [5, Lemma 1, (vi), p. 195]). Translating this into the language of nets we have: a $\beta(E',E)$-ultimately bounded net in E' has a subnet which is $\sigma(E',E)$-Cauchy. However, we are working here with an ab-space, and so our claim is proved.

Now let $\{f_\alpha \mid \alpha \varepsilon A\}$ be a net in E' which is $\beta(E',E)$-ultimately bounded and assume, as we have just shown we may, that this net is $\sigma(E',E)$-convergent to a point $g \varepsilon E'$. Then there is a scalar λ and $\alpha_0 \varepsilon A$ such that $f_\alpha \varepsilon \lambda U_1$ for all $\alpha \geq \alpha_0$. Hence $T^* f_\alpha \in \lambda T^*(U_1) \subseteq \lambda U^0$ for all $\alpha \geq \alpha_0$. It follows that the set $\{T^* f_\alpha \mid \alpha \geq \alpha_0\}$ is both $\beta(E',E)$-bounded in E' and equicontinuous on E[6, pp. 198-199 and Proposition 6, p.200]. Let B be any balanced, convex, bounded subset of E and let K denote the closure of the set T(B). Then K is a compact subset of E because T is, in particular, a c.c. operator. Now the restriction of the members of the family $\{T^* f_\alpha \mid \alpha \geq \alpha_0\}$ to K gives us a family of functions which is both bounded and equicontinuous on the compact set K. By the Ascoli-Arzela Theorem [7, Theorem 17, p.233], $\{T^*f_\alpha | \alpha \geq \alpha_0\}$ has a subnet which converges uniformly over K. Clearly this subnet converges to T*g uniformly over K.

At this point we must digress briefly to prove two technical facts: (i) $T^{**}(\tilde{B}) = K$; here $T^{**} = (T^*)^*$ and $\tilde{B}$ is the closure of B in $E''[\sigma(E'',E')]$.

Let $\phi \in \tilde{B} \setminus B$ and let $\{b_\gamma | \gamma \in \Gamma\}$ be a net of points of B which converges to ϕ for $\sigma(E'',E')$. Then $\{T^{**}b_\gamma\}$ will converge to $T^{**}\phi$ for this topology. However, $T^{**}b_\gamma = Tb_\gamma \in T(B)$ for every $\gamma \in \Gamma$, and T(B) is relatively compact in E(recall that the closure of T(B) is K). It follows that some subnet of $\{Tb_\gamma\}$ converges to a point $y \in K$, and hence that $T^{**}\phi = y$. We conclude that $T^{**}(B) \subseteq K$.

Now $T(B) \subseteq T^{**}(\tilde{B})$ and this last set is, as we have just seen, in E. Since T** is $\sigma(E'',E')$-continuous and $\tilde{B}$ is $\sigma(E'',E')$-compact, the set $T^{**}(\tilde{B})$ is $\sigma(E'',E')$-compact. But this set is in E so it must be compact for the restriction of $\sigma(E'',E'')$ to E; i.e., for $\sigma(E,E')$. In particular $T^{**}(\tilde{B})$ is $\sigma(E,E')$-closed in E and so closed for the original topology of E because it is a convex set [8, Paragraph 20(6), p.245]. Thus $K \subseteq T^{**}(\tilde{B})$.

(ii) The fact that $\{T^*f_\alpha\}$ converges to T*g uniformly over K implies that this net converges to T*g uniformly over B.

Note that $T^*f_\alpha(K) = T^*f_\alpha[T^{**}(\tilde{B})] = T^{***}f_\alpha(\tilde{B})$ by (i) and the definition of T***. Hence $\{T^{***}f_\alpha\}$ converges to T***g uniformly over $\tilde{B}$. But $B \subseteq \tilde{B}$ and, on B, $T^{***}f_\alpha = T^*f_\alpha$.

We now return to the proof of the theorem. Recall that T(U) is a bounded set in E, let C be the closure of T(U) and let C_1 be the closure of T(C). Choose a subnet of $\{T^*f_\alpha | \alpha \geq \alpha_0\}$ which converges to T*g uniformly over the compact set C_1, hence (by (ii)) uniformly over C. We shall show now that this subnet converges to T*g uniformly over any bounded subset of E; i.e., it converges to T*g for $\beta(E',E)$. For any bounded set B, $B \subseteq nU$ for some integer n. Thus $T(B) \subseteq n\ T(U)$ and so the closure of T(B) is contained in nC. Our subnet of $T^*f_\alpha | \alpha \geq \alpha_0\}$ converges to T*g uniformly over C hence, by linearity, uniformly over nC. But then it converges to T*g uniformly over the closure of T(B) hence, by (ii), uniformly over B. So starting with an arbitrary $\beta(E',E)$-ultimately bounded net $\{f_\alpha | \alpha \in A\}$ we have shown that $\{T^*f_\alpha\}$ has a subnet which converges to a point of E' for $\beta(E',E)$. In other words, we have shown that T* is a c.c. operator on $E'[\beta(E',E)]$.

Corollary 1. Let E be an ab-space which is complete, and suppose that $E'[\beta(E',E)]$ is also an ab-space. Then a continuous, linear operator T on E is a b.c.c. operator if, and only if, T^* is a b.c.c. operator on $E'[\beta(E',E)]$.

Proof. Suppose that T^* is a b.c.c. operator on $E'[\beta(E',E)]$. Then, by the theorem, T^{**} is a b.c.c. operator on $E''[\beta(E'',E')]$. Now the restriction of T^{**} to E is just T, and the restriction of $\beta(E'',E')$ to E is just $\beta^*(E,E')$. The latter topology coincides with the original topology of E because every ab-space is quasi-barrelled. Since E is complete, and hence closed as a subspace of $E''[\beta(E'',E')]$, it follows that T is a c.c. operator on E.

Now let U be a convex, $\beta(E'',E')$-neighborhood of zero in E" which is mapped by T^{**} onto a bounded set. We may assume that $U = \{y \in E'' \mid |y(f)| \leq 1 \text{ for all } f \in C\}$ where C is a convex, bounded subset of $E'[\beta(E',E)]$. Let $W = \{x \in E \mid |f(x)| \leq 1 \text{ for all } f \in C\}$. Then W is a $\beta^*(E,E')$-neighborhood of zero in E, and $T^{**}(U \cap E) = T(W)$ is a bounded set. Thus T is a b.c.c. operator on E.

Corollary 2. Let T be a compact operator on an ab-space E. Then $(T^*)^2$ is a compact operator on $E'[\beta(E',E)]$.

References

1. R. Bronson, Matrix Methods, Academic Press, New York, 1970.
2. C. L. DeVito, 'On Alaoglu's Theorem, Bornological Spaces and the Mackey-Ulam Theorem'; Math. Ann., 192(1971), 83-89.
3. __________, 'On Grothendieck's Characterization of the Completion of a Locally Convex Space'; J. London Math. Soc. (2), 5(1972), 293-297.
4. C. W. Henson and L.C. Moore, Jr., 'The non-standard theory of topological vector spaces'; Trans. Amer. Math. Soc., 172 (1972), 405-435.
5. __________, 'Invariance of the non-standard hulls of locally convex spaces'; Duke Math. J., 40 (1973), 193-205.
6. J. Horvath, 'Topological vector spaces and distributions', Vol. I Addison-Wesley, Reading, Massachusetts, 1966.
7. J. Kelley, 'General topology', D. Van Nostrand Co. Inc., Princeton, New Jersey, 1955.
8. G. Kothe, 'Topological vector spaces' I, Springer, Berlin, 1969.

9. __________, 'Topological vector spaces' II, Springer-Verlag, New York, 1979.
10. A. P. Robertson and W. Robertson, 'Topological vector spaces', Second edition, Cambridge University Press, Cambridge, 1973.

Local Invertibility of Set-Valued Maps

HALINA FRANKOWSKA Mathematical Research Center, University of Paris-Dauphine, Paris, France

I dedicate this paper to Professor Ky-Fan, who has greatly influenced me, in particular, when I met him in CEREMADE during the fall of 1982.

ABSTRACT

We prove several equivalent versions of the inverse function theorem: an inverse function theorem for smooth maps on closed subsets, one for set-valued maps, a generalized implicit function theorem for set-valued maps. We provide applications to the problem of local controllability of differential inclusions.

1. The Inverse Function Theorem

Let X be a Banach space $K \subset X$ be a subset of X. We recall the definition of the *tangent cone* to a subset K at x_o introduced in Clarke [1975].

$$C_K(x_o) := \{v \in X \mid \lim_{\substack{h \to o+ \\ x \to x_o \\ x \in K}} \frac{d(x+hv,K)}{h} = 0\}$$

We state now our basic result.

Theorem 1.1.

Let X be a Banach space, Y be a finite dimensional space, $K \subset X$ be a closed subset of X and x_o belong to K. Let A be a differentiable map from a neighborhood of K to Y. We assume that A' is continuous at x_0 and that the following *surjectivity assumption* holds true

$$(1) \qquad A'(x_0)C_K(x_0) = Y \quad .$$

Then $A(x_0)$ belongs to the interior of $A(K)$ and there exist constants ρ and ℓ such that, for all

$$(2) \qquad \begin{cases} y_1, y_2 \in A(x_0) + \rho B \text{ and any solution } x_1 \in K \text{ to the} \\ \text{equation } A(x_1) = y_1 \text{ satisfying } \|x_0 - x_1\| \leq \ell\rho, \text{ there} \\ \text{exists a solution } x_2 \in K \text{ to the equation } A(x_2) = y_2 \\ \text{satisfying } \|x_1 - x_2\| \leq \ell \|y_1 - y_2\|. \quad \blacktriangle \end{cases}$$

We recall

Definition

A set-valued map G from Y to X is *pseudo-Lipschitz* around $(y_0, x_0) \in \text{Graph}(G)$ if there exist neighborhoods V of y_0 and W of x_0 and a constant ℓ such that

$$\begin{cases} \text{i)} \quad \forall y \in V, \ G(y) \neq \emptyset \\ \text{ii)} \quad \forall y_1, y_2 \in V, \ G(y_1) \cap W \subset G(y_2) + \ell \|y_1 - y_2\| B \quad . \quad \blacktriangle \end{cases}$$

The above definition was introduced in Aubin [1982], [1984]. (See also Rockafellar [to appear] d) for a thorough study of pseudo-Lipschitz maps.)

Hence, the second statement of Theorem 1.1. reads:

$$(2)' \qquad \begin{cases} \text{the map } y \to A^{-1}(y) \cap K \text{ is pseudo-Lipschitz around} \\ (Ax_0, x_0). \end{cases}$$

Remark

If x_0 belongs to the interior of K, then $C_K(x_0) = X$. Then assumption (1) states that $A'(x_0)$ is surjective, and we obtain the usual "inverse function theorem", also called "Liusternik theorem".

We deduce a characterization of the interior of a closed subset of a finite-dimensional space given by Clarke [1983]:

$$x_o \in \mathrm{Int}(K) \Leftrightarrow C_K(x_o) = X \quad .$$

(We take X = Y and A to be the identity). ▲

The proof of Theorem 2.1 is based on the Ekeland variational principle [1974] and is given in Aubin-Frankowska [1985].

Corollary 1.2.

We posit the assumptions of Theorem 1.1. Let $M := A^{-1}(A(x_o)) \cap K$ be the set of solutions $x \in K$ to the equation $A(x) = A(x_o)$. Then there exist a neighborhood U of x_o and a constant ℓ such that

$$\forall x \in K \cap U,\ d(x,M) \leq \ell \,\|A(x) - A(x_o)\| \quad .$$

Furthermore

$$C_K(x_o) \cap \mathrm{Ker}\, A'(x_o) \subset C_M(x_o) \quad . \quad ▲$$

We shall derive the extension to set-valued maps of the inverse function theorem. Let X,Y be Banach spaces and F be a map from X into the subsets of Y.

The *derivative* $CF(x_o,y_o)$ of F at $(x_o,y_o) \in \mathrm{Graph}\ (F)$ is the set-valued map from X to Y associating to any $u \in X$ elements $v \in Y$ such that (u,v) is tangent to Graph (F) at (x_o,y_o):

$$v \in CF(x_o,y_o)(u) \Leftrightarrow (u,v) \in C_{\mathrm{Graph}\ (F)}(x_o,y_o)$$

Theorem 1.3.

Let F be a set-valued map from a Banach space X to a finite dimensional space Y and (x_o,y_o) belongs to the graph of F. If

graph F is closed and CF (x_o,y_o) is surjective,

then F^{-1} is pseudo-Lipschitz around $(y_o,x_o) \in \mathrm{Graph}\ (F^{-1})$.

▲

Proof

We apply Theorem 1.1 when X is replaced by $X \times Y$, K is the graph of F and A is the projection from $X \times Y$ to Y. ∎

Remark: A dual formulation.

Since the dimension of Y is finite, assumption (1) is equivalent to

$$A'(x_0)\, C_K(x_0) \text{ is dense in } Y$$

which can be translated as

$$\text{if } A'(x_0)^* q \text{ belongs to } C_K(x_0)^-, \text{ then } q = 0 \quad .$$

If F is a set-valued map from X to Y, we define the coderivative $CF(x_0,y_0)^*$ of F at $(x_0,y_0) \in \text{Graph}(F)$ as the "transpose" of $CF(x_0,y_0)$, from Y^* to X^* defined by

$$p \in CF(x_0,y_0)^*(q) \Leftrightarrow \sup_{(u,v) \in \text{Graph } CF(x_0,y_0)} (\langle p,u\rangle - \langle q,v\rangle) = 0 \quad .$$

Therefore, in Theorem 1.3, we can replace the surjectivity assumption by the "dual assumption"

$$CF(x_0,y_0)^{*-1}(0) = \{0\} \quad .$$

2. Applications to Local Controllability

Let us consider a set-valued map F from $\mathbb{R}^n$ into compact subsets of $\mathbb{R}^n$. We associate with F the *differential inclusion*

(3) $$x' \in F(x) \quad .$$

A particular case of (3) is the parametrized system (also called "control system")

(4) $$x' = f(x,u(t)) \quad , \quad u(t) \in U$$

where U is a given set of controls; then F is defined by

$$F(x) = \{f(x,u) : u \in U\}$$

Let $T > 0$. A function $x \in W^{1,1}(0,T)$ (Sobolev space) is called a *solution of differential inclusion* (3) if

$$x'(t) \in F(x(t)) \text{ a.e. in } [0,T] \quad .$$

For a point $\xi \in \mathbb{R}^n$ denote by $S_T(\xi)$ the set of solutions to (3) issued from ξ and defined on the time interval $[0,T]$. The *reachable set* to (3) at time T from ξ is denoted by $R(T,\xi)$, i.e.

$$R(T,\xi) = \{x(T) : x \in S_T(\xi)\} \quad .$$

The system (3) is called *locally controllable* around ξ if for some time $T > 0$

(5) $$\xi \in \text{Int } R(T,\xi) \quad .$$

The purpose of this section is to provide a sufficient condition for (5) when ξ is an equilibrium of F, i.e. $0 \in F(\xi)$.

We shall apply results of Section 1. The set of solutions $S_T(\xi)$ is closed in $W^{1,1}(0,T)$ whenever Graph (F) is closed in $\mathbb{R}^n \times \mathbb{R}^n$. Consider the continuous linear operator A from the Banach space $W^{1,1}(0,T)$ into the finite dimensional space $\mathbb{R}^n$ defined by

$$A(x) = x(T) \text{ for all } x \in W^{1,1}(0,T) \quad .$$

Theorem 1.1. states then that if x_o denotes the constant trajectory $x_o(\cdot) \equiv \xi$ and $\{w(T) : w \in C_{S_T(\xi)}(x_o)\} = \mathbb{R}^n$ then the relation (5) holds true.

Let B denote the closed unit ball in $\mathbb{R}^n$. We say that a set-valued map F is Lipschitzian (in the Hausdorff metric) on an open neighborhood V of ξ if for a constant $L \geq 0$ and all $x,y \in V$

$$F(x) \subset F(y) + L\|x-y\|B \quad .$$

Thanks to this property we can compute a subset of $C_{S_T(\xi)}(x_o)$:

Theorem 2.1. Assume that F has closed graph and is Lipschitzian around the equilibrium ξ. Then every solution of the differential inclusion

$$(6) \qquad \begin{cases} w'(t) \in CF(\xi,0)\ w(t) & \text{a.e. in } [0,T] \\ w\ (0) = 0 \end{cases}$$

belongs to $C_{S_T(\xi)}(x_o)$. ▲

The proof of the last result is based on a Filippov Theorem [1967].

We say that the inclusion (6) is *controllable* if its reachable set at some time $T > 0$ is equal to the whole space.

Theorems 1.1. and 2.1. together imply

Theorem 2.2. Assume that F has closed graph and is Lipschitzian around the equilibrium ξ. The inclusion (3) is locally controllable around ξ if the inclusion (6) is controllable. ▲

Remark. Actually the idea of the proof of Theorem 1.1. allows to prove a stronger result: We denote by co $F(\xi)$ the closed convex hull of the set $F(\xi)$.

Theorem 2.3. Assume that F has closed graph and is Lipschitzian around the equilibrium ξ. The inclusion (3) is locally controllable around ξ if the inclusion

$$(7) \qquad \begin{cases} w' \in \text{cl}\ [CF(\xi,0)w + C_{\text{co}F(\xi)}(0)] \\ w(0) = 0 \end{cases}$$

is controllable. ▲

The proof requires a very careful calculation of variations of solutions (see Frankowska [1984]).

A necessary condition for the controllability of the inclusions (6), (7) is

$$\text{Dom } CF\ (\xi,0) := \{w \in \mathbb{R}^n : CF\ (\xi,0)w \neq \emptyset\} = \mathbb{R}^n \quad .$$

Whenever it holds true the right-hand sides of (6), (7) are set-valued maps whose graphs are *closed convex cones*. Such maps, called "closed convex processes", are set-valued analogues of linear operators. The controllability of such differential inclusions is the issues of the next section.

Before, we provide the following

Example. Using Theorem 2.3 one can obtain a classical result on local controllability of control system (4) without assuming too much regularity. Let U be a compact in $\mathbb{R}^m$ and let $f : \mathbb{R}^n \times U \to \mathbb{R}^n$ be a continuous function. Assume that for some $(\xi,\bar{u}) \in \mathbb{R}^n \times U$, $f(\xi,\bar{u}) = 0$ and for some $\beta > 0$, $L > 0$ and all $u \in U$; $x,y \in \xi + \beta B$

$$\begin{cases} \|f(x,u) - f(y,u)\| \leq L\|x-y\| \\ \frac{\partial f}{\partial x}(\cdot,\bar{u}) \text{ is continuous on } \xi + \beta B \end{cases} .$$

Theorem 2.4. If the sublinearized differential inclusion

$$\begin{cases} w' \in \frac{\partial f}{\partial x}(\xi,\bar{u})w + C_{co\, f(\xi,U)}(0) \\ w(0) = 0 \end{cases}$$

is controllable, then the system (4) is locally controllable around ξ. ▲

3. Controllability of Convex Processes

A *convex process* A from $\mathbb{R}^n$ to itself is a set-valued map satisfying

$$\forall x, y \in \text{Dom } A \text{ , } \lambda,\mu \geq 0 \text{ , } \lambda A(x) + \mu A(y) \subset A(\lambda x + \mu y)$$

or, equivalently, a set-valued map whose graph is a convex cone. Convex processes are the set-valued analogues of linear operators. We shall say that a convex process is *closed* if its graph is closed and that it is *strict* if its domain is the whole space. Convex processes were introduced and studied in

Rockafellar [1967], [1970], [1974] (see also Aubin-Ekeland [1984]). We associate with a strict closed convex process A the Cauchy problem for the differential inclusion

$$(8) \qquad \begin{cases} x'(t) \in A(x(t)) & \text{a.e.} \\ x(0) = 0 & . \end{cases}$$

We say that the differential inclusion (8) is controllable if the *reachable set*

$$R := \{x(t) : x \in W^{1,1}(0,t) \text{ is a solution of (8)}, \ t \geq 0\}$$

is equal to the whole space $\mathbb{R}^n$.

A particular case of (8) is a linear control system

$$(9) \qquad \begin{cases} x' = Fx + GU & u \in U \\ x(0) = 0 \end{cases}$$

where U is an m-dimensional space and $F \in L(\mathbb{R}^n, \mathbb{R}^n)$, $G \in L(\mathbb{R}^m, \mathbb{R}^n)$ are linear operators.

We observe that the reachable set $R(T,0)$ of (8) at time T is convex. Since $0 \in A(0)$ the family $\{R(T,0)\}_{T>0}$ is increasing. Moreover, $R = \bigcup_{T>0} R(T,0)$. Hence (8) is controllable if and only if it is controllable at some time $T > 0$, i.e. $\exists\, T > 0$ such that

$$R(T,0) = \mathbb{R}^n \quad .$$

a) <u>The rank condition</u>

Let A be a strict closed convex process. Set $A^1(0) = A(0)$ and for all integer $i \geq 2$ set

$$A^i(0) = A(A^{i-1}(0)) \quad .$$

<u>Theorem 3.1</u>. The differential inclusion (8) is controllable if and only if

$$\text{for some } m \geq 1 \quad A^m(0) = (-A)^m(0) = \mathbb{R}^n \quad .$$

▲

In the case of system (9) for all $x \in \mathbb{R}^n$ $Ax = Fx + \operatorname{Im} G$. Thus

$$A^m(0) = (-A)^m(0) = \operatorname{Im} G + F(\operatorname{Im} G) + \ldots + F^{m-1}(\operatorname{Im} G) \quad .$$

The Cayley-Hamilton theorem implies then the *Kalman rank condition* for the controllability of the linear system (9):

$$\operatorname{rk}[G, FG, \ldots F^{n-1}G] = n \quad .$$

Theorem 3.1. is a consequence of the following

b) "Eigenvalue" criterion for controllability

We say that a subspace P of $\mathbb{R}^n$ is *invariant* by a strict closed convex process A if $A(P) \subset P$.

A real number λ is called an eigenvalue of A if $\operatorname{Im}(A-\lambda I) \neq \mathbb{R}^n$, where I denotes the identity operator.

Theorem 3.2. The differential inclusion (8) is controllable if and only if A has neither proper invariant subspace nor eigenvalues. ▲

It is more convenient to write the above criterion in a *"dual form"*:

c) "Eigenvector" criterion for controllability

The convex processes can be transposed as linear operators. Let A be a convex process; we define its *transpose* A^* by

$$p \in A^*(q) \Leftrightarrow \forall (x,y) \in \text{Graph } A, \quad \langle p,x\rangle \leq \langle q,y\rangle \quad .$$

It can easily be shown that λ is an eigenvalue of A if and only if for some $q \in \operatorname{Im}(A-\lambda I)^{\perp}$, $q \neq 0$

$$\lambda q \in A^* q \quad .$$

We call such a vector $q \neq 0$ an eigenvector of A^*. Theorem 3.2 is equivalent then to

Theorem 3.3. The differential inclusion (8) is controllable if and only if A^* has neither proper invariant subspace nor eigenvectors. ▲

The proof of Theorem 3.3 is based on a separation theorem and the KY-FAN coincidence theorem [1972]. (See Aubin-Frankowska-Olech [1985]).

Examples: a) Let F be a linear operator from $\mathbb{R}^n$ to itself, L be a closed convex cone of controls and A be the strict closed convex process defined by

$$A(x) := Fx + L \quad .$$

Then its transpose is equal to

$$A^*(q) = \begin{cases} F^*q & \text{if } q \in L^+ \\ \emptyset & \text{if } q \notin L^+ \end{cases}$$

When $L = \{0\}$, i.e., when $A = F$, we deduce that $A^* = F^*$, so that transposition of convex processes is a legitimate extension of transposition of linear operators.

Consider the control system

$$(10) \qquad \begin{cases} x' = Ax + u, & u \in L \\ x(0) = 0 \end{cases}$$

Corollary 3.4.

The following conditions are equivalent.

a) the system (10) is controllable

b) For some $m \geq 1$ $L + F(L) + \ldots + F^m(L) = L - F(L) + \ldots + (-1)^m F^m(L) = \mathbb{R}^n$ (see Korobov [1980]).

c) F has neither proper invariant subspace containing L nor eigenvalue λ satisfying $Im(F-\lambda I) + L \neq \mathbb{R}^n$.

d) F^* has neither proper invariant suspace contained in L^+ nor eigenvector in L^+.

e) the subspace spanned by L, $F(L), \ldots, F^{n-1}(L)$ is equal to $\mathbb{R}^n$ and F^* has no eigenvector in L^+ (see Brammer [1972]) ▲

b) Consider the control system with feedback in $\mathbb{R}^2$:

$$(11)\quad \begin{cases} x' = xv + y + u + xu & u,w \in U = [0,1] \\ y' = -x + w & v \in V(x) = \begin{cases} +1 & x \geq 0 \\ -1 & x < 0 \end{cases} \\ x(0) = y(0) = 0 \end{cases}$$

Set $F(x,y) = \{(xv + y + u + xu, -x + w) : (u,w,v) \in U \times U \times V(x)\}$.

Then $0 \in F(0)$, i.e. zero is a point of equilibrium. The direct computation gives

$$CF(0,0)(x,y) = (|x| + y + \mathbb{R}_+ , -x + \mathbb{R}_+)$$

Set $A(x,y) = CF(0,0)(x,y)$. Then

$$A(0) = \mathbb{R}_+ \times \mathbb{R}_+; \ -A(0) = \mathbb{R}_- \times \mathbb{R}_-$$

$$A^2(0) = \mathbb{R}_+ \times \mathbb{R}; \ (-A)^2(0) = \mathbb{R} \times \mathbb{R}_-$$

$$A^3(0) = \mathbb{R}^2 \quad ; \ (-A)^3(0) = \mathbb{R}^2 \quad .$$

Thus by Theorem 2.2 and 3.1 the control system (11) is locally controllable around zero.

For further references and bibliographical comments, we refer to the book, "Optimization and Nonsmooth Analysis," by F.H. Clarke [1983].

REFERENCES

Aubin J.P.

[1982] Comportement Lipschitzien des solutions de problèmes de minimisation convexes. CRAS 295, 235-238.

[1984] Lipschitz behavior of solutions to convex minimization problems. Math. Op. Res. 9, 87-111.

Aubin, J.P. and A. Cellina

[1984] Differential Inclusions, Springer Verlag.

Aubin, J.P. and I. Ekeland

[1984] Applied Nonlinear Analysis, Wiley Interscience, New York.

Aubin, J.P. and H. Frankowska

[1985] On inverse function theorems for set-valued maps, J. Math. Pure. Appl. (to appear).

Aubin, J.P., H. Frankowska and C. Olech

[1985] Controllability of convex processes. SIAM J. of Control, (to appear).

Brammer, R.F.

[1972] Controllability in linear autonomous systems with positive controllers. SIAM J. Control, 10, 339-353.

Clarke, F.H.

[1975] Generalized Gradient and Applications, Trans. Amer. Math. Soc. 205:247-262.

[1983] Optimization and Nonsmooth Analysis. Wiley Interscience.

Ekeland, I.

[1974] On the Variational Princiople. J. Math. Anal. Appl. 47, 324-353.

Fan, Ky

[1972] A minimax inequality and applications. In Inequalities III, O. Sisha Ed. Academic Press 103-113.

Filippov, A.F.

[1967] Classical solutions of differential equations with multivalued right-hand side. English translation: SIAM J. Control, 5, 609-621.

Frankowska, H.

[1984] Local controllability and infinitesimal generators of semi-groups of set-valued maps (to appear).

Ioffe, A.E.

[to appear] On the local surjection property.

Korobov,V.I.

[1980] A geometric criterion of local controllability of dynamical systems in the presence of constraints on the control, Diff. Eqs., 15, No. 9, 1136-1142.

Rockafellar, R.T.

[1967] Monotone processes of convex and concave type. Mem. of Ann. Math. Soc. 77.

[1970] Convex Analysis. Princeton University Press.

[1974] Convex algebra and duality in dynamic models of production. In Mathematical models in Economics, Łoś (Ed.), North-Holland.

[to appear] a) Lagrange multipliers and subderivatives of optimal value functions in nonlinear programming. Math. Prog. Study No. 5, R. Wets (ed).

b) Extensions of subgradient calculus with applications to optimization.

c) Maximal monotone relations and the generalized second derivatives of non-smooth functions.

d) Lipschitzian properties of multifunctions.

Saperstone S.M. and J.A. Yorke

[1971] Controllability of linear oscillatory systems using positive controls. SIAM J. Control 9, 253-262.

Some Minimax Theorems Without Convexity

ANDRZEJ GRANAS Department of Mathematics, University of Montreal, Montreal, Quebec, Canada

FON-CHE LIU Institute of Mathematics, Academia Sinica, Taipei, Taiwan, Republic of China

Dedicated to Ky Fan.

In numerous extensions of the von Neumann's Minimax Theorem, one of the main purposes was to eliminate as much as possible the underlying convexity structure from the original hypothesis. In this note we discuss some new general minimax results without convexity. In the first part, we extend the well-known version of the von Neumann Principle due to Ky Fan [1]. In the second part (as an application) we give a further generalization of the Ky Fan-Nikaido-Kneser Theorem concerning systems of inequalities. Some of the main results presented here were announced without proof in [4]. We remark that the proof of the main results is based on the von Neumann Minimax Theorem in R^n.

1. Preliminaries. We begin with some preliminary results concerning inf sup.

(1.1) Lemma. Let X and Y be two arbitrary sets and $f,g : X \times Y \to R$ be real-valued functions defined on $X \times Y$. Then the following statements are equivalent:

(i) $\operatorname{Inf}_{x\in X} \operatorname{Sup}_{y\in Y} f(x,y) \le \operatorname{Sup}_{y\in Y} \operatorname{Inf}_{x\in X} g(x,y)$

(ii) For each $\lambda \in R$ and $\varepsilon > 0$ either

(a) there is $x_0 \in X$ such that $f(x_0,y) \le \lambda$ for all $y \in Y$,

or

(b) there is $y_0 \in Y$ such that $g(x,y_0) \ge \lambda - \varepsilon$ for all $x \in X$.

Proof. (i) => (ii) : If (ii) (a) does not hold, then for each $x \in X$, $\operatorname{Sup}_{y\in Y} f(x,y) > \lambda$, hence $\operatorname{Inf}_{x\in X} \operatorname{Sup}_{y\in Y} f(x,y) \ge \lambda$, which implies by (i) that

---

This research was supported in part by a grant from the National Research Council of Canada.

$\operatorname{Sup}_{y\in Y} \operatorname{Inf}_{x\in X} g(x,y) \geq \lambda$. Thus there is $y_0 \in Y$ such that $\operatorname{Inf}_{x\in X} g(x,y_0) \geq \lambda - \varepsilon$, i.e., $g(x,y_0) \geq \lambda - \varepsilon$ for all $x \in X$. So (ii) (b) holds.

(ii) => (i) : Let $\gamma = \operatorname{Inf}_{x\in X} \operatorname{Sup}_{y\in Y} f(x,y)$. We may assume that $\gamma > -\infty$. If $\gamma < +\infty$, we apply (ii) with $\lambda = \gamma - \varepsilon$. Since (ii)(a) does not hold for this λ, (ii)(b) holds, i.e., there is $y_0 \in Y$ such that $g(x,y_0) \geq \lambda - \varepsilon = \gamma - 2\varepsilon$ for all $x \in X$, or $\operatorname{Sup}_{y\in Y} \operatorname{Inf}_{x\in X} g(x,y) \geq \gamma - 2\varepsilon$. But since $\varepsilon > 0$ is arbitrary, we have $\operatorname{Sup}_{y\in Y} \operatorname{Inf}_{x\in X} g(x,y) \geq \gamma$. Thus (i) holds. If $\gamma = +\infty$, then for any $\lambda \in R$, (ii)(a) does not hold, hence there is $y_0 \in Y$ such that $g(x,y_0) \geq \lambda - \varepsilon$ for all $x \in X$. Thus

$$\operatorname{Sup}_{y\in Y} \operatorname{Inf}_{x\in X} g(x,y) \geq \lambda - \varepsilon \ .$$

If we let $\lambda \to +\infty$, then $\operatorname{Sup}_{y\in Y} \operatorname{Inf}_{x\in X} g(x,y) = +\infty$. Hence (i) holds. The proof is complete.

(1.2) <u>Lemma</u>. Let $f,g : X \times Y \to R$ be two numerical functions and suppose that one of the following conditions holds :

(*) X is a compact topological space and $x \to f(x,y)$ is lower semi-continuous on X for each $y \in Y$.

(**) Y is a compact topological space and $y \to g(x,y)$ is upper semi-continuous on Y for each $x \in X$.

Then the following two statements are equivalent :

(i) $\operatorname{Inf}_{x\in X} \operatorname{Sup}_{y\in Y} f(x,y) \leq \operatorname{Sup}_{y\in Y} \operatorname{Inf}_{x\in X} g(x,y)$

(ii) For each $\lambda \in R$, either

(a) there is $x_0 \in X$ such that $f(x_0,y) \leq \lambda$ for all $y \in Y$,

or

(b) there is $y_0 \in Y$ such that $g(x,y_0) \geq \lambda$ for all $x \in X$.

Proof. Suppose that (*) holds. The implication (II) => (i) follows in a straightforward manner from (1.1). To show that (i) => (ii), let $\lambda \in R$ be given and assume that (ii)(b) does not hold. By Lemma (1.1), for each $n = 1,2,\ldots$, we have $\inf_{x\in X} \sup_{y\in Y} f(x,y) \le \lambda + \frac{1}{n}$. But, since $x \to \sup_{y\in Y} f(x,y)$ is lower semicontinuous on X, there is $x_n \in X$ such that $\sup_{y\in Y} f(x_n,y) = \inf_{x\in X} \sup_{y\in Y} f(x,y) \le \lambda + \frac{1}{n}$. From this letting $n \to \infty$, and using the compactness of X and the lower-semicontinuity on X of $x \to \sup_{y\in Y} f(x,y)$, we get a point $x_0 \in X$ such that $\sup_{y\in Y} f(x_0,y) \le \lambda$. Thus (ii)(a) holds and we have shown that (i) => (ii). The proof of our assertion in the case (**) being similar is omitted.

Now we are able to formulate the main result of this section representing a generalization of a known theorem of Ky Fan [1].

(1.3) Theorem. Let X,Y be two compact topological spaces and $f,g : X \times Y \to R$ two real-valued functions such that :

(i) $x \to f(x,y)$ is lower-semicontinuous on X for each $y \in Y$;

(ii) $y \to g(x,y)$ is upper-semicontinuous on Y for each $x \in X$.

Then the following conditions are equivalent:

A. For any two finite sets $\{x_0,\ldots,x_n\} \subset X$, $\{y_1,\ldots,y_m\} \subset Y$, there exists $x_0 \in X$ and $y_0 \in Y$ such that

$$f(x_0,y_k) \le g(x_i,y_0) \quad \text{for all} \quad 1 \le i \le n \quad \text{and} \quad 1 \le k \le m .$$

B. For each $\lambda \in R$, either there is $x_0 \in X$ such that

$$f(x_0,y) \le \lambda \quad \text{for all} \quad y \in Y$$

or there is $y_0 \in Y$ such that

$$g(x,y_0) \ge \lambda \quad \text{for all} \quad x \in X .$$

C. The following minimax inequality holds

$$\operatorname*{Min}_{x \in X} \operatorname*{Sup}_{y \in Y} f(x,y) \leq \operatorname*{Max}_{y \in Y} \operatorname*{Inf}_{x \in X} g(x,y) .$$

Proof. We already know from Lemma (1.2) that B and C are equivalent. The implication C => A being obvious, it remains only to prove that A implies B .

Assume that A is verified ; we are now going to prove that B holds. Let $\lambda \in R$ be given. Define

$$L(y) = \{x \in X \mid f(x,y) \leq \lambda\} \text{ for each } y \in Y ,$$

$$R(x) = \{y \in Y \mid g(x,y) \geq \lambda\} \text{ for each } x \in X .$$

To show that B holds is equivalent to show that either $\bigcap_{y \in Y} L(y) \neq \phi$ or $\bigcap_{x \in X} R(x) \neq \phi$. Since the sets $L(y)$ and $R(x)$ are closed subsets of X and Y respectively, it is sufficient to show that either $\{L(y)\}_{y \in Y}$ or $\{R(x)\}_{x \in X}$ has the finite intersection property. Suppose that $\{L(y)\}_{y \in Y}$ does not have the finite intersection property : there are $y_1, \ldots, y_m \in Y$ such that $\bigcap_{k=1}^{m} L(y_k) = \phi$, i.e.,

$$(*) \qquad \operatorname*{Max}_{1 \leq k \leq m} f(x,y_i) > \lambda \text{ for each } x \in X .$$

Let now $\{x_1, \ldots, x_n\}$ be any finite subset of X . By (i) there exist $x_0 \in X$ and $y_0 \in Y$ such that

$$f(x_0,y_k) \leq g(x_i,y_0) \text{ for all } 1 \leq i \leq n , \ 1 \leq k \leq m ,$$

or

$$\operatorname*{Max}_{1 \leq k \leq m} f(x_0,y_k) \leq g(x_i,y_0) \text{ for all } 1 \leq i \leq n .$$

From (*) we have then

$$g(x_i,y_0) \geq \operatorname*{Max}_{1 \leq k \leq m} f(x_0,y_k) > \quad , \ 1 \leq i \leq n .$$

Thus $y_0 \in \bigcap_{i=1}^{n} R(x_i)$. Hence $\{R(x)\}_{x \in X}$ has the finite intersection property and the proof is complete.

2. A Minimax Inequality of the von Neumann type.

In this section, we prove a minimax theorem which represents a further generalization of a minimax theorem of Ky Fan [1] under "convexity conditions" milder than in our previous note [4].

In what follows given positive integer ℓ, we let $[\ell] = \{i \in N \mid i \le i \le \ell\}$ and denote by $\Delta^{\ell-1} = \{(x_1,\ldots,x_\ell) \in R^\ell \mid x_1,\ldots,x_\ell \ge 0 , \sum_{i=1}^{\ell} x_i = 1\}$ the standard simplex in R^ℓ. We introduce first some terminology.

(2.1) Definition. Let X be a set and $F = \{f\}$ a family of real-valued function on X. We shall say that X is finitely-convex with respect to F provided for any $f_1,f_2,\ldots,f_n \in F$ and any $x_1,x_2 \in X$ there is a point $x \in X$ such that

$$f_i(x) \le \frac{1}{2}[f_i(x_1) + f_i(x_2)]$$

for all $i \in [n]$; X is called finitely-concave with respect to F, if in the above definition the symbol "$\le$" is replaced by "$\ge$".

(2.2) Definition. Let X,Y be sets and $f : X \times Y \to R$ a real-valued function; consider the family $\{f_y\}_{y\in Y}$ of functions $f_y : X \to R$ where $f_y(x) = f(x,y)$ for $x \in X$. We shall say that X is finitely f-convex (resp. f-concave) if X is finitely convex (resp. concave) with respect to the family $\{f_y\}_{y\in Y}$. Similar definitions can also be given for Y, by considering the family $\{f_x\}_{x\in X}$ of functions $f_x : Y \to R$ where $f_x(y) = f(x,y)$ for $y \in Y$.

We can now state and prove the main result of this section :

(2.3) Theorem. Suppose that X and Y are two non-empty compact spaces and that $f,s,t,g : X \times Y \to R$ are four real-valued functions such that :

(i) $f(x,y) \le s(x,y) \le t(x,y) \le g(x,y)$ for all $(x,y) \in X \times Y$;

(ii) $x \to f(x,y)$ is lower semicontinuous on X for each $y \in Y$;

(iii) X is finitely s-convex ; i.e., for any $x_1,x_2 \in X$ and $y_1,\ldots,y_n \in Y$ there is $\bar{x} \in X$ such that $s(\bar{x},y_i) \le \frac{1}{2}[s(x_1,y_i) + s(x_2,y_i)]$ for all $i = 1,2,\ldots,n$;

(iv) Y is finitely t-concave ; i.e., for any $y_1,y_2 \in Y$ and $x_1,\ldots,x_m \in X$ there exist $\bar{y} \in Y$ such that $t(x_j,\bar{y}) \ge \frac{1}{2}[t(x_j,y_1) + t(x_j,y_2)]$ for all $j = 1,2,\ldots,m$;

(v) $y \to g(x,y)$ is upper semicontinuous on Y for each $x \in X$.

Then

$$\operatorname*{Min}_X \operatorname*{Sup}_Y f(x,y) \le \operatorname*{Max}_Y \operatorname*{Inf}_X g(x,y) .$$

Proof. Let $\varepsilon > 0$ be given. To establish our assertion, it is clearly sufficient to show that

$$\operatorname*{Min}_X \operatorname*{Sup}_Y f(x,y) \le \operatorname*{Max}_Y \operatorname*{Inf}_X \{g(x,y)\} + \varepsilon = \operatorname*{Max}_Y \{\operatorname*{Inf}_X g(x,y) + \varepsilon\} .$$

Hence, in view of Theorem (1.3), we need to show only that for any two finite sets $x_1,\dots,x_n \in X$ and $y_1,\dots,y_m \in Y$ there exist $x_0 \in X$ and $y_0 \in Y$ such that

(1) $$f(x_0,y_k) \le g(x_i,y_0) + \varepsilon$$

for all $i \in [n]$, $k \in [m]$. For this purpose define a function $G : \Delta^{n-1} \times \Delta^{m-1} \to R$ by

$$G(\alpha,\beta) = \sum_{i=1}^{n} \sum_{k=1}^{m} \alpha_i\beta_k \; s(x_i,y_k)$$

for $\alpha = (\alpha_1,\dots,\alpha_n) \in \Delta^{n-1}$, $\beta = (\beta_1,\dots,\beta_m) \in \Delta^{m-1}$. By the von Neumann Minimax Theorem in R^n , there exist

$$\hat{\alpha} = (\hat{\alpha}_1,\dots,\hat{\alpha}_n) \in \Delta^{n-1} \quad \text{and} \quad \hat{\beta} = (\hat{\beta}_1,\dots,\hat{\beta}_m) \in \Delta^{m-1}$$

such that

(2) $$G(\hat{\alpha},\beta) \le G(\alpha,\hat{\beta}) \quad \text{for all } (\alpha,\beta) \in \Delta^{n-1} \times \Delta^{m-1} .$$

From the definition of G and (2) we get

(3) $$\operatorname*{Max}_{k\in[m]} \sum_{i=1}^{n} \hat{\alpha}_i \; s(x_i,y_k) \le \operatorname*{Min}_{i\in[n]} \sum_{k=1}^{m} \hat{\beta}_k \; s(x_i,y_k) .$$

Next we choose $\tilde{\alpha} = (\tilde{\alpha}_1,\dots,\tilde{\alpha}_n) \in \Delta^{n-1}$ and $\tilde{\beta} = (\tilde{\beta}_1,\dots,\tilde{\beta}_m)$ with dyadic rationals as their components such that

(4) $$\sum_{i=1}^{n} \tilde{\alpha}_i \; s(x_i,y_k) \le \sum_{i=1}^{n} \hat{\alpha}_i \; s(x_i,y_k) + \frac{\varepsilon}{2} \quad \text{for all } k \in [[m] \;;$$

(5) $$\sum_{k=1}^{m} \hat{\beta}_k \; t(x_i,y_i) \le \sum_{k=1}^{m} \tilde{\beta}_k \; t(x_i,y_k) + \frac{\varepsilon}{2} \quad \text{for all } i \in [n] .$$

Now, because X is finitely s-convex and Y is finitely t-concave, there exist $x_0 \in X$ and $y_0 \in Y$ such that

$$(6) \qquad s(x_0,y_k) \le \sum_{i=1}^{n} \tilde{\alpha}_i \, s(x_i,y_k) \quad \text{for all} \quad k \in [m] \ ;$$

$$(7) \qquad t(x_i,y_0) \ge \sum_{k=1}^{m} \tilde{\beta}_i \, t(x_i,y_k) \quad \text{for all} \quad i \in [n] \ .$$

Let $i \in [n]$ and $k \in [m]$ be given. Then using the assumption (i) of the Theorem and (4), (5,), (6), and (7) we get

$$(8) \qquad f(x_0,y_k) \le s(x_0,y_k) \le \sum_{i=1}^{n} \tilde{\alpha}_i \, s(x_i,y_k) \le \sum_{i=1}^{n} \hat{\alpha}_i \, s(x_i,y_k) + \frac{\varepsilon}{2}$$

$$\le \operatorname*{Max}_{k\in[m]} \sum_{i=1}^{n} \hat{\alpha}_i \, s(x_i,y_k) + \frac{\varepsilon}{2} \ ;$$

$$(9) \qquad \operatorname*{Min}_{i\in[n]} \sum_{k=1}^{m} \hat{\beta}_k \, s(x_i,y_k) \le \operatorname*{Min}_{i\in[n]} \sum_{k=1}^{m} \hat{\beta}_k \, t(x_i,y_k) \le \sum_{k=1}^{m} \hat{\beta}_k \, t(x_i,y_k)$$

$$\le \sum_{k=1}^{m} \tilde{\beta}_k \, t(x_i,y_k) + \frac{\varepsilon}{2} \le t(x_i,y_0) + \frac{\varepsilon}{2} \ .$$

Finally, from (3), (8), and (9) we conclude that

$$f(x_0,y_k) \le t(x_i,y_0) + \varepsilon \ .$$

Since $i \in [n]$ and $k \in [m]$ were fixed arbitrarily, this implies (1), and the proof of the theorem is complete.

As an immediate consequence we obtain the following result established in our Note [4] :

(2.4) <u>Theorem</u>. Let X,Y be two compact spaces and let $f,s,t,g : X \times Y \to R$ be four real-valued functions satisfying conditions (i), (ii) and (v) of Theorem (2.3). Assume furthermore that

(iii)$_\star$ For any $x_1,x_2,\ldots,x_n \in X$ and $(\alpha_1,\alpha_2,\ldots,\alpha_n) \in \Delta^{n-1}$ there is a point $\hat{x} \in X$ such that for all $y \in Y$

$$s(\hat{x},y) \le \sum_{i=1}^{n} \alpha_i \, s(x_i,y) \ ;$$

(iv)$_*$ For any $y_1,y_2,\dots,y_m \in Y$ and $(\beta_1,\beta_2,\dots,\beta_m) \in \Delta^{m-1}$ there is a point $\hat{y} \in Y$ such that all $x \in X$

$$t(x,\hat{y}) \geq \sum_{j=1}^{m} \beta_j \; t(x,y_j) .$$

Then

$$\underset{X}{\text{Min}} \; \underset{Y}{\text{Sup}} \; f(x,y) \leq \underset{Y}{\text{Max}} \; \underset{X}{\text{Inf}} \; g(x,y) .$$

Proof. It is enough to observe that (iii)$_*$ (resp. (iv)$_*$) implies the condition (iii) (resp. (iv)) of Theorem (2.3) .

By taking in (2.3) $f = s = t = g$ we obtain among special cases of Theorem (2.3) the following result :

(2.5) Theorem. Let X,Y be two non-empty compact spaces and $f : X \times Y \to R$ be a real-valued function such that :

(i) $x \to f(x,y)$ is lower semi-continuous on X for each $y \in Y$;

(ii) $y \to f(x,y)$ is upper semi-continuous on Y for each $x \in X$;

(iii) For any $x_1,x_2 \in X$ and $y_1,y_2,\dots,y_n \in Y$ there is $\hat{x}$ such that

$$f(\hat{x},y_i) \leq \frac{1}{2} [f(x_1,y_i) + f(x_2,y_i)] \quad \text{for all} \quad i \in [n] ;$$

(iv) For any $y_1,y_2 \in X$ and any $x_1,x_2,\dots,x_m \in X$ there is $\hat{y} \in Y$ such that

$$f(x_j,\hat{y}) \geq \frac{1}{2} [f(x_j,y_1) + f(x_j,y_2)] \quad \text{for all} \quad j \in [m].$$

The the following minimax equality holds

$$\underset{X}{\text{Min}} \; \underset{Y}{\text{Max}} \; f(x,y) = \underset{Y}{\text{Max}} \; \underset{X}{\text{Min}} \; f(x,y) .$$

As a special case we have the following :

(2.6) Corollary (Nikaido- von Neumann).: Let X and Y be two compact convex subsets in linear topological spaces and $f : X \times Y \to R$ a real-valued function such that

(a) $x \to f(x,y)$ is lower semi-continuous and convex on X for each $y \in Y$;

(b) $y \to f(x,y)$ is upper semi-continuous and concave on Y for each $x \in X$.

Thenthe following minimax equality holds :

$$\operatorname*{Min}_X \operatorname*{Max}_Y f(x,y) = \operatorname*{Max}_Y \operatorname*{Min}_X f(x,y) .$$

3. A theorem concerning systems of inequalities.

In this section we establish a further generalization of a theorem of Ky Fan [2] concerning systems of inequalities ; our result extends three previous generalizations given in [3],[4] and [5].

Let X be a non-empty set and $F = \{f\}$, $G = \{g\}$ two families of real-valued functions defined on X ; we write $F \leq G$ provided for each $f \in F$ there is $g \in G$ such that $f(x) \leq g(x)$ for all $x \in X$.

We need the following :

(3.1) Definition. A family $F = \{f\}$ is weakly F-concave on X provided for any $f_1, f_2 \in F$ there is an $f \in F$ such that

$$f(x) \geq \frac{1}{2} [f_1(x) + f_2(x)]$$

for all $x \in X$. *)

We may now formulate our second main result :

(3.2) Theorem. Let X be a compact space and F,G,H be three non-empty families of real-valued functions defined on X satisfying :

1) $F \leq G \leq H$;

*) We recall that a family $F = \{f\}$ is called F-concave on X provided for any $f_1, f_2, \ldots, f_n \in F$ and $\alpha = (\alpha_1, \alpha_2, \ldots, \alpha_n) \in \Delta^{n-1}$ there is $f \in F$ such that $f(x) \geq \sum_{i=1}^{n} \alpha_i f_i(x)$ for all $x \in X$. This notion was first formulated by Ky Fan [1].

2) Each function in F is lower-semicontinuous on X ;

3) X is finitely convex with respect to G ;

4) The family H is weakly F-concave.

Then :

(I) For each $\lambda \in R$, the following alternative holds :

(a) There is $h \in H$ such that $h(x) > \lambda$ for all $x \in X$;

(b) There is $x_0 \in X$ such that $f(x_0) \leq \lambda$ for all $f \in F$.

(II) $\operatorname{Min}_{x\in X} \operatorname{Sup}_{f\in F} f(x) \leq \operatorname{Sup}_{h\in H} \operatorname{Inf}_{x\in X} h(x)$.

Proof. Since (II) follows easily from (I), we only need to prove (I).

To show (I), let $\lambda \in R$ be given ; we assume that the condition (I)(a) is not verified and proceed to show that (I)(b) holds. For each positive integer n and each $f \in F$, we let

$$S_n(f) = \{x \in X \mid f(x) \leq \lambda + \tfrac{1}{n}\} .$$

Clearly each $S_n(f)$ is closed in X and therefore compact. We are going to show first that for each fixed n , the family of compact sets $\{S_n(f)\}_{f\in F}$ has the finite intersection property. Let $f_1,\ldots,f_m \in F$ be given. Since $F \leq G \leq H$, there are $g_1,\ldots,g_m \in G$ and $h_1,\ldots,h_m \in H$ such that

$$f_i(x) \leq g(x) \leq h_i(x) \quad \text{for all } x \in X \text{ and } i \in [m] .$$

Consider now the functions F , G, $H : X \times \Delta^{m-1} \to R$ defined by

$$F(x,\xi) = \sum_{i=1}^{m} \xi_i f_i(x) ,$$

$$G(x,\xi) = \sum_{i=1}^{m} \xi_i g_i(x) ,$$

$$H(x,\xi) = \sum_{i=1}^{m} \xi_i h_i(x) ,$$

for $x \in X$ and $\xi = (\xi_1,\ldots,\xi_m) \in \Delta^{m-1}$.

We are going to apply Theorem (2.3) with $Y = \Delta^{m-1}$, $f = F$, $s = t = G$ and $g = H$. We observe first that

$$(*)\quad \begin{cases} x \to F(x,\xi) & \text{is lower semi-continuous on } X \text{ for each } \xi \in \Delta^{m-1}\ ; \\ \xi \to G(x,\xi) & \text{is concave on } \Delta^{m-1} \text{ for each } x \in X\ ; \\ \xi \to H(s,\xi) & \text{is upper semi-continuous on } \Delta^{m-1} \text{ for each } x \in X\ . \end{cases}$$

Furthermore since (by the assumption) X is finitely convex with respect to G, given $x_1, x_2 \in X$ there is $x \in X$ such that

$$g_i(x) \le \frac{1}{2}[g_i(x_1) + g_i(x_2)] \quad \text{for} \quad i \in [m]\ ,$$

and this implies

$$G(x,\xi) \le \frac{1}{2}[G(x_1,\xi) + G(x_2,\xi)]$$

for all $\xi \in \Delta^{m-1}$. Thus X is finitely G-convex. Because of this fact and (*), the hypotheses of Theorem (2.3) are verified with $Y = \Delta^{m-1}$, $f = F$, $s = t = G$ and $g = H$. Consequently by Theorem (2.3) we get

$$(**)\qquad \operatorname*{Min}_{X} \operatorname*{Max}_{\Delta^{m-1}} F(x,\xi) \le \operatorname*{Max}_{\Delta^{m-1}} \operatorname*{Inf}_{X} H(x,\xi)\ .$$

Let $n \in N$ be a given integer. From (**), by applying Theorem (1.3), we have the following alternative, either :

(a) there is $x_0 \in X$ such that $F(x_0,\xi) \le \lambda + \frac{1}{n}$ for all $\xi \in \Delta^{m-1}$,

or

(b) there is $\hat{\xi} = (\hat{\xi}_1,\ldots,\hat{\xi}_m) \in \Delta^{m-1}$ such that $H(x,\hat{\xi}) \ge \lambda + \frac{1}{n}$ for all $x \in X$.

By adapting arguments of S. Simons [5] and those of our earlier Note [3] we are going to show that (b) is not true. Suppose to the contrary that the condition (b) holds. By restricting the following set-up to some face of Δ^{m-1} if necessary, we may assume that $\hat{\xi}$ is in the interior of Δ^{m-1}. Define $h : \Delta^{m-1} \to R \cup \{-\infty\}$ by

$$k(\xi) = \operatorname*{Inf}_{x \in X} H(x,\xi)\ , \quad \text{for} \quad \xi \in \Delta^{m-1}\ .$$

Since $H(x,\xi) \ge F(x,\xi)$ and $x \to F(x,\xi)$ is lower semi-continuous on X for each ξ, k is real-valued and is bounded from below on Δ^{m-1}. Furthermore, since k

is concave on the convex Δ^{m-1}, k is continuous at $\hat{\xi}$. Because $k(\hat{\xi}) \geq \lambda + \frac{1}{n}$ and k is continuous at $\hat{\xi}$, there is a point $\eta = (\eta_1,\ldots,\eta_m)$ with dyadic rational coordinates $\eta_1,\eta_2,\ldots,\eta_m$ such that $\eta \in \operatorname{Int} \Delta^{m-1}$ and $k(\eta) > \lambda$. Finally, since the family $H = \{h\}$ is weakly F-concave, there is $h \in H$ such that

$$h(x) \geq \sum_{i=1}^{m} \eta_i h_i(x) = H(x,\eta) \geq k(\eta) > \lambda$$

for all $x \in X$. This contradicts our assumption that (I)(a) is not verified. Thus the condition (b) does not hold and hence in view of (a) there is $x_0 \in X$ such that $F(x_0,\xi) \leq \lambda + \frac{1}{n}$ for all $\xi \in \Delta^{m-1}$. This means that $x_0 \in \bigcap_{i=1}^{m} S_n(f_i)$, i.e., the family $\{S_n(f)\}_{f\in F}$ has the finite intersection property.

From this for each $n = 1,2,\ldots,$ $A_n = \bigcap_{f\in F} S_n(f) \neq \phi$. Next it is clear (since each A_n is compact and $A_n \supset A_{n+1}$) that $\bigcap_{n=1} A_n \neq \phi$. Obviously, if $x_0 \in \bigcap_{n=1} A_n$, then $f(x_0) \leq \lambda$ for all $f \in F$. Thus (I)(b) holds and the proof of the theorem is complete.

As an immediate consequence of (3.2) we obtain the following result :

(3.3) Theorem. Let X be a compact space, Y an arbitrary set and $f,g,h : X \times Y \to R$ three real-valued functions verifying :

1. $f(x,y) \leq g(x,y) \leq h(x,y)$ for all $(x,y) \in X \times Y$;
2. $x \to f(x,y)$ is lower semi-continuous on X for each $y \in Y$.

Assume furthermore that one of the following conditions (A), (B), C) is satisfied :

(A)

A3. For any $x_1,x_2 \in X$ and $y_1,y_2,\ldots,y_n \in Y$ there is $\hat{x} \in X$ such that

$$g(\hat{x},y_i) \leq \frac{1}{2}[g(x_1,y_i) + g(x_2,y_i)] \text{ for all } i \in [n].$$

A4. For any $y_1,y_2 \in Y$ there is $\hat{y} \in Y$ such that

$$h(x,\hat{y}) \geq \frac{1}{2}[h(x,y_1) + h(x,y_2)] \text{ for all } x \in X.$$

(B)

- B3. For any $x_1, x_2, \ldots, x_m \in X$ and $(\alpha_1, \alpha_2, \ldots, \alpha_m) \in \Delta^{m-1}$ there is $\hat{x} \in X$ such that

$$g(\hat{x}, y) \leq \sum_{i=1}^{m} \alpha_i g(x_i, y) \quad \text{for all } y \in Y$$

- B4. For any $y_1, y_2, \ldots, y_n \in Y$ and $(\beta_1, \beta_2, \ldots, \beta_n) \in \Delta^{n-1}$ there is $\hat{y} \in Y$ such that

$$h(x, \hat{y}) \geq \sum_{j=1}^{n} \beta_j h(x, y_j) \quad \text{for all } x \in X .$$

(C)

- C3. X is convex
- C4. $x \to g(x,y)$ is convex on X for each $y \in Y$.
- C5. The family $\{h_y\}_{y \in Y}$ is F-concave.

Then the following two assertions hold :

(I) For any $\lambda \in R$ either

(α) there is $x_0 \in X$ such that

$$f(x_0, y) \leq \lambda \quad \text{for all } y \in Y ;$$

or

(β) there is $y_0 \in Y$ such that

$$h(x, y_0) > \lambda \quad \text{for all } x \in X .$$

(II) $\operatorname{Inf}_x \operatorname{Sup}_y f(x,y) \leq \operatorname{Sup}_y \operatorname{Inf}_x h(x,y)$.

By taking in Theorem (3.3), $f = g = h$ we obtain as a special case the following result, which should be compared with Theorem (2.5) :

(3.4) Theorem. Let X be a compact space, Y an arbitrary set and $f : X \times Y \to R$ a function such that $x \to f(x,y)$ is lower semi-continuous on X for each $y \in Y$. Assume that the following two additional conditions are verified :

1. For any $x_1, x_2 \in X$ and $y_1, y_2, \ldots, y_n \in Y$ there is $\hat{x} \in X$ such that

$$f(\hat{x}, y_i) \leq \frac{1}{2} [f(x_1, y_i) + f(x_2, y_i)] \quad \text{for all } i \in [n] .$$

2. For any $y_1, y_2 \in Y$ there is $\hat{y} \in Y$ such that

$$f(x,\hat{y}) \geq [f(x,y_1) + f(x,y_2)] \quad \text{for all} \quad x \in X .$$

Under the above hypotheses we have :

(I) For any $\lambda \in R$ either

(α) there is $x_0 \in X$ such that

$$f(x_0,y) \leq \lambda \quad \text{for all} \quad y \in Y \;;$$

or

(β) there is $y_0 \in Y$ such that

$$f(x,y_0) > \lambda \quad \text{for all} \quad x \in X .$$

(II) $\operatorname*{Inf}_x \operatorname*{Sup}_y f(x,y) = \operatorname*{Sup}_y \operatorname*{Inf}_x f(x,y)$.

As a special case we get the following :

(3.5) Corollary. (Ky Fan-Nikaido-Kneser). Let X be a compact convex space, Y an arbitrary set and $f : X \times Y \to R$ a real-valued function such that $x \to f(x,y)$ is lower-semicontinuous on X for each $y \in Y$. Assume that one of the following additional hypotheses is verified :

1. (Ky Fan). The family $\{f_y\}_{y \in Y}$ is F-concave.

2. (Nikaido) Y is a convex subset of a vector space and $y \to f(x,y)$ is concave on Y for each $x \in X$.

3. (Kneser) Y is a vector space and $y \to f(x,y)$ is affine on Y for each $x \in X$.

Then

$$\operatorname*{Inf}_x \operatorname*{Sup}_y f(x,y) = \operatorname*{Sup}_y \operatorname*{Inf}_x f(x,y) .$$

References

1. Ky Fan, Minimax Theorems, Proc. Nat. Acad. Sci., U.S.A., 39 (1953), 42-47.

2. ______, Existence theorems and extreme solutions for inequalities concerning convex functions or linear transformations, Math. Z. 68 (1957), 205-217.

3. A. Granas & F.C. Liu, Remark on a theorem of Ky Fan concerning systems of inequalities, Bull. Inst. Math. Acad. Sinica 11 (1983), 639-643.

4. ______________, Théorèmes du minimax sans convexité, C.R. Acad. Sci. Paris, t. 300, no 11 (1985), 347-350.

5. S. Simons, Remark on a remark of Granas and Liu concerning systems of inequalities, Bull. Inst. Math. Acad. Sinica (to appear).

Weak Compactness and the Minimax Equality

CHUNG-WEI HA Department of Mathematics, National Tsing Hua University, Hsinchu, Taiwan, Republic of China

Let E be a real locally convex Hausdorff topological vector space, E' be the topological dual of E and let X be a bounded subset of E. (We suppose throughout that X and Y are nonempty.) We denote by $\sigma(E,E')$ and $\tau(E,E')$, respectively, the weak topology and the Mackey topology on E induced by E'. It is easy to show (see, e.g., [5], Remark 6) that if X is $\sigma(E,E')$-compact, then, for any subset Y of E',

$$\inf_{\eta \in \mathscr{F}(Y)} \sup_{x \in X} \inf_{y \in \eta} \langle y,x \rangle \leqq \sup_{\xi \in \mathscr{F}(X)} \inf_{y \in Y} \sup_{x \in \xi} \langle y,x \rangle \tag{1}$$

where $\mathscr{F}(X)$ and $\mathscr{F}(Y)$ denote the families of nonempty finite subsets of X and Y, respectively. Moreover, if X is <u>convex</u> and $\sigma(E,E')$-compact, then for any convex subset Y of E',

$$\inf_{y \in Y} \sup_{x \in X} \langle y,x \rangle \leq \sup_{x \in X} \inf_{y \in Y} \langle y,x \rangle \tag{2}$$

(2) follows immediately from (1) and the fact that for any convex sets $X \subset E$ and $Y \subset E'$, we always have

$$\inf_{\eta \in \mathscr{F}(Y)} \sup_{x \in X} \inf_{y \in \eta} \langle y,x \rangle = \inf_{y \in Y} \sup_{x \in X} \langle y,x \rangle \tag{3}$$

and

$$\sup_{\xi \in \mathscr{F}(X)} \inf_{y \in Y} \sup_{x \in \xi} \langle y,x \rangle = \sup_{x \in X} \inf_{y \in Y} \langle y,x \rangle \tag{4}$$

(see, e.g., [5], Lemma 11). The relation (2) is the <u>minimax equality</u> referred to in the title; obviously (2) with the inequality sign reversed is always true. In this paper, we shall show that each of the properties (1) and (2) in a way characterizes the weak compactness of

X. Our main results are Theorems 1 and 2 below. As applications, we obtain directly from Theorem 2 DeBranges's characterization [1] of the $\tau(E',E)$-open convex sets in E' and a theorem of Fan [2] generalizing the Alaoglu-Bourbaki theorem. If E is assumed to be $\tau(E,E')$-complete, then Theorem 2 becomes a result of Simons [5]. In the following we shall assume the notations given above. The reader is referred to [4] for the terminology not herein defined.

Theorem 1. *Let X be a bounded set in E. If (1) holds for any convex set Y in E' for which the left-hand side of (1) is positive, then X is relatively $\sigma(E,E')$-compact in E.*

Proof: Suppose that X is not relatively $\sigma(E,E')$-compact in E, then the $\sigma(E'^*,E')$-closure of X in the algebraic dual E'^* of E' contains an element ϕ which is not in E. For any elements $x_1,\dots,x_m$ in X, there exists $y \in E'$ satisfying $\phi(y) > 1$ but $|\langle y,x_i\rangle| < 1/3$ for all $1 \leq i \leq m$. Let

$$Y = \{y \in E' : \phi(y) \geq 1\} \tag{5}$$

Then Y is convex in E', and

$$\sup_{\xi \in \mathscr{F}(X)} \inf_{y \in Y} \sup_{x \in \xi} |\langle y,x\rangle| \leq 1/3.$$

On the other hand, since ϕ is in the $\sigma(E'^*,E')$-closure of X, for any elements $y_1,\dots,y_n$ in Y, there exists $x \in X$ satisfying $|\phi(y_j) - \langle y_j,x\rangle| < 1/3$ and so $\langle y_j,x\rangle > 2/3$ for all $1 \leq j \leq n$. Thus,

$$\inf_{\eta \in \mathscr{F}(Y)} \sup_{x \in X} \inf_{y \in \eta} \langle y,x\rangle \geq 2/3. \tag{6}$$

Hence (1) does not hold for the convex set Y defined in (5), which by (6) makes the left-hand side of (1) positive. This completes the proof.

Theorem 2. *Let X be a bounded convex set in E. If (2) holds for any convex set Y in E' for which the left-hand side of (2) is positive, then X is relatively $\sigma(E,E')$-compact in E.*

Clearly Theorem 2 follows from Theorem 1, (3) and (4). Now we shall use Theorem 2 to prove the following result of DeBranges ([1], Theorem 4):

Theorem 3. *Let U be a convex set in E'. U is $\tau(E',E)$-open if and only if $\bar{H} \cap U = \emptyset$ for any convex set H in E' such that $H \cap U = \emptyset$, where $\bar{H}$ denotes the $\sigma(E',E)$-closure of H in E'.*

Proof: Since the $\sigma(E'E)$-closure and the $\tau(E',E)$-closure of a convex set H in E' are the same, the condition is clearly necessary. Conversely, by a translation we may assume that the origin 0 of E' belongs to U. It suffices to prove that 0 is in the $\tau(E',E)$-interior of U. Let X be the polar U^{o} of U in E, that is,

$$X = \{x \in E : \langle u,x \rangle \leqq 1 \quad \text{for} \quad u \in U\}$$

then X is bounded, closed, convex and contains 0. (X is bounded because U is radial round all of its points.) We shall show that X is $\sigma(E,E')$-compact. To this end, let Y be a convex set in E' and $\alpha > 0$ be a real number such that

$$\alpha < \inf_{y \in Y} \sup_{x \in X} \langle y,x \rangle \ . \tag{7}$$

By dividing both sides of (7) by α, we may assume that $\alpha = 1$. Then (7) implies that $Y \cap U = \emptyset$. By Zorn's lemma, there exists a maximal convex set H in E' containing Y but disjoint from U. Clearly H is $\sigma(E',E)$-closed. It follows from the separation theorem that there exists an element $x \in E$ and a real number β such that

$$\langle y,x \rangle \geqq \beta > \langle u,x \rangle \tag{8}$$

for all $y \in Y$ and $u \in U$. Since $0 \in U$, $\beta > 0$ and so we can assume $\beta = 1$. Thus, (8) implies that $x \in X$ and so

$$1 \leqq \sup_{x \in X} \inf_{y \in Y} \langle y,x \rangle \ . \tag{9}$$

By Theorem 2, X is $\sigma(E,E')$-compact. Now if $y \notin U$, then by applying the preceding argument to $\{y\}$ in place of Y, we obtain an element $x \in X$ such that $\langle y,x \rangle \geqq 1$. Hence

$$\{u \in E' : \langle u,x \rangle < 1 \quad \text{for} \quad x \in X\} \subset U$$

which shows that 0 is in the $\tau(E',E)$-interior of U. The proof is completed.

Next we shall show that the following result of Fan ([2], Theorem 1) is a consequence of Theorem 2.

Theorem 4. *Let X be a bounded set in E. If the polar*

$$U = \{u \in E' : \langle u,x \rangle \leqq 1 \quad \text{for} \quad x \in X\} \tag{10}$$

of X in E' has a nonempty $\tau(E',E)$-*interior, then X is relatively* $\sigma(E,E')$-*compact in E.*

Proof: Since X and the closed convex hull of $X \cup \{0\}$ have the same polar in E', we may assume that X is closed, convex and contains 0. To apply Theorem 2 as in the proof of Theorem 3, let Y be a convex set in E' such that

$$1 < \inf_{y \in Y} \sup_{x \in X} \langle y,x \rangle . \tag{11}$$

Then (11) implies that $Y \cap U = \emptyset$. Since U has a nonempty $\tau(E',E)$-interior, by the Hahn-Banach theorem there exist a nonzero element $x \in E$ and a real number β such that

$$\langle y,x \rangle \geqq \beta \geqq \langle u,x \rangle \tag{12}$$

for all $y \in Y$ and $u \in U$. Since $0 \in U$, $\beta \geqq 0$. Suppose first that $\beta = 0$. By the bipolar theorem ([4], p.126) $x \in X$. But then X also contains all the positive multiples of x. Since $x \neq 0$, this contradicts the boundedness of X. Hence $\beta > 0$ and so we may assume that $\beta = 1$. Then again $x \in X$ and so (12) implies (9). This completes the proof.

As pointed out in [2], the $\tau(E,E')$-interior of the set U in (10) does not necessarily contain the origin of E', and so Theorem 4 strictly generalizes the Alaoglu-Bourbaki theorem. Our proofs show a dual relationship between Theorems 3 and 4.

Theorem 5. *Let X be a closed bounded set in E and let the closed convex hull C of X in E be* $\tau(E,E')$-*complete. Then the following statements are equivalent:*

(a) *X is* $\sigma(E,E')$-*compact;*

(b) *If* γ *is a real number and Z is the convex hull of an equi-*

continuous sequence $\{y_j\}$ in E' such that

$$\inf_{z\in Z}\ \sup_{x\in X}\ \langle z-u,x\rangle \geqq \gamma \tag{13}$$

for any $\sigma(E',E)$- cluster point u of the sequence $\{y_j\}$, then $\gamma \leqq 0$;

(c) If α, β are real numbers, $\{y_j\}$ is an equicontinuous sequence in E' and $\{x_i\}$ is a sequence in X such that

$$\left.\begin{aligned} \langle y_j,x_i\rangle &\geqq \alpha \qquad \text{for } j \leqq i \\ \langle y_j,x_i\rangle &\leqq \beta \qquad \text{for } j > i \end{aligned}\right\} \tag{14}$$

Then $\alpha \leqq \beta$;

(d) The inequality (1) holds for any equicontinuous (convex) set Y in E' (for which the left-hand side of (1) is positive).

Proof: Suppose that X is $\sigma(E,E')$-compact. (b) is a property of the vector space C(X) of real-valued continuous functions on X. For given a $\sigma(E',E)$-cluster point u of the sequence $\{y_j\}$ in E', there exists a subsequence of $\{y_j\}$ which converges pointwise on X to u (see [4], p.185). Without disturbing (13) we may assume that $\{y_j\}$ itself converges pointwise on X to u. Since $\{y_j\}$ is uniformly bounded in C(X), from Lebesgue's bounded convergence theorem, $\{y_j\}$ converges to u in the weak topology of C(X). Since the closures of Z in C(X) in the weak and the uniform topology coincide, u can be arbitrarily approximated uniformly on X by elements in Z. This shows that $\gamma \leqq 0$. To prove that (b) implies (c), let Z be the convex hull of the sequence $\{y_j\}$ given in (c). Then (14) implies that (13) holds for any $\sigma(E',E)$-cluster point u of $\{y_j\}$, where $\gamma = \alpha - \beta$. Hence $\alpha \leqq \beta$. Now we assume (c). Let Y be an equicontinuous set in E', for which we suppose that the left-hand side of (1) is not $-\infty$; the right-hand side of (1) is not $+\infty$. Let α, β be real numbers such that

$$\alpha < \inf_{\eta\in\mathscr{F}(Y)}\ \sup_{x\in X}\ \inf_{y\in\eta}\ \langle y,x\rangle \tag{15}$$

$$\beta > \sup_{\xi\in\mathscr{F}(X)}\ \inf_{y\in Y}\ \sup_{x\in\xi}\ \langle y,x\rangle \tag{16}$$

We choose arbitrarily an element $y_1 \in Y$, then by (15) there exists $x_1 \in X$ such that $\langle y_1,x_1\rangle > \alpha$. For the element $x_1 \in X$ by (16) there exists $y_2 \in Y$ such that $\langle y_2,x_1\rangle < \beta$. If $x_1,\dots,x_{n-1} \in X$ and $y_1,\dots,y_{n-1} \in Y$ have been chosen satisfying (14), then by (16) there exists $y_n \in Y$ such that $\langle y_n, x_i \rangle < \beta$ for all $1 \leqq i \leqq n-1$, and again by (15) there exists $x_n \in X$ such that $\langle y_j, x_n \rangle > \alpha$ for all $1 \leqq j \leqq n$. Continuing in this way we obtain an equicontinuous sequence $\{y_j\}$ in E' and a sequence $\{x_i\}$ in X satisfying (14). Hence $\alpha \leqq \beta$. This proves (d). In proving (d) implies (a), we shall make use of the assumption that the closed convex hull C of X is $\tau(E,E')$-complete. Suppose that X is not $\sigma(E,E')$-compact, then the closure of X in the algebraic dual E'^* of E' contains an element ϕ which is not in the $\tau(E,E')$-completion of E. By Grothendieck's completion theorem (see [4], p.148), there exists a balanced convex equicontinuous set U in E' such that the restriction of ϕ to U is not $\sigma(E',E)$-continuous at $0 \in U$. After suitably scaling U, the proof of Theorem 1 goes through for the set

$$Y = \{y \in U: \ \phi(y) \geqq 1\}$$

The proof is completed.

It is clear that if a bounded set X in E satisfies Theorem 5(b), so does also its closed convex hull C in E. Thus, we obtain a new proof of Krein's theorem (see [4], p.189), which says that if the closed convex hull C of a $\sigma(E,E')$-compact set X in E is $\tau(E,E')$-complete, then C is also $\sigma(E,E')$-compact. It is noteworthy that Theorem 5(b) was used by Pryce [3] as a starting point in his proof of James's theorem on weakly compact sets (see also [5]). Finally we remark that, if X is assumed to be convex, then Theorem 5(d) can be replaced by the condition that

(d') *The relation* (2) *holds for any equicontinuous convex set* Y *in* E' (*for which the left-hand side of* (2) *is positive*).

Thus, we obtain Simons's result Theorem 15 in [5].

References

1. L. DeBranges, Vectorial topology, J. Math. Anal. Appl., 69(1979), 443-454.

2. K. Fan, A generalization of the Alaoglu-Bourbaki theorem and its applications, Math. Z., 88(1965), 48-60.

3. J. D. Pryce, Weak compactness in locally convex spaces, Proc. Amer. Math. Soc., 17(1966), 148-155.

4. H. H. Schaefer, Topological vector spaces, Springer-Verlag, New York-Heidelberg-Berlin, 1970.

5. S. Simons, Maximinimax, minimax, and antiminimax theorems and a result of R. C. James, Pacific J. Math., 40(1972), 709-718.

Nonlinear Volterra Equations with Positive Kernels

NORIMICHI HIRANO Department of Mathematics, Yokohama National University, Yokohama, Japan

1. Introduction. Our purpose in this paper is to consider the existence of solutions to the nonlinear Volterra equation

$$(1.1) \qquad u(t) + \int_0^t a(t-s)Au(s)\,ds = f(t), \qquad 0 \le t \le T,$$

where $0 < T \le \infty$, $a(t)$ is a real valued function and A is a nonlinear operator from one space to another. In case where A is a maximal monotone operator on a real Hilbert space H, the existence of solutions of (1.1) has been studied by several authors(cf. [1],[6]). The setting in which A is a maximal monotone operator from a real Banach space $V \subset H$ to its dual V' has also been studied by Barbu[1], Crandall et al.[5], and Kiffe and Stecher[9]. In [7], the author considered the case where A is a pseudo-monotone operator from H into itself and gave existence theorems for the equation (1.1). From the point of view of applications to the case where A is a differential operator, this assumption on A is restrictive. In this paper, we extend the result in [7] to the case where A is a pseudo-monotone operator from a real Hilbert space $V \subset H$ into its dual V'. Assuming that $a(t)$ is a kernel of positive type, we show the existence of solutions of (1.1) for $f \in L^2(0,T;V)$.

2 Statement of the main result. Throughout this paper, V will denote a real Hilbert space, densely and continuously imbedded in a real Hilbert space H. Identifying H with its own

dual H', we have $V \subset H \subset V'$. Let (x,y) be the pairing between an element $x \in V'$ and $y \in V$. If $x, y \in H$, then (x,y) is the ordinary inner product in H. By $\|\cdot\|$, $|\cdot|$ and $\|\cdot\|_*$, we denote the norms of V, H and V', respectively. We denote by $\tilde{V}$, $\tilde{H}$, $\tilde{V}'$ the spaces $L^2(0,T;V)$, $L^2(0,T;H)$ and $L^2(0,T;V')$, respectively. We denote by $\langle\cdot,\cdot\rangle$ the pairing between $\tilde{V}'$ and $\tilde{V}$. Then for each $u, v \in \tilde{H}$, $\langle u, v\rangle$ is the inner product in $\tilde{H}$. The norms of $\tilde{V}$, $\tilde{H}$ and $\tilde{V}'$ are again denoted by $\|\cdot\|$, $|\cdot|$ and $\|\cdot\|_*$, respectively.

For a subset D of $\tilde{V}$, we denote by $\overline{co}^V D$ and $\overline{co}^H D$ the closed convex hull of D with respect to the topology of $\tilde{V}$ and $\tilde{H}$, respectively.

A nonlinear mapping A from V into V' is said to be monotone if $(y_1 - y_2, x_1 - x_2) \geq 0$ for all $y_i \in Ax_i$, $i = 1,2$. A is said to be maximal monotone if it has no proper monotone extension. A nonlinear (single-valued) mapping from V into V' is said to be pseudo-monotone if A satisfies the following conition:

(*) If a sequence $\{u_n\} \subset V$ satisfies that $u_n \to u$ (weak convergence) in V and $\limsup_{n\to\infty} (Au_n, u_n - u) \leq 0$, it follows that $(Au, u - v) \leq \liminf_{n\to\infty} (Au_n, u_n - v)$ for all $v \in V$, and Au_n converges weakly to Au in V'.

In the following, we will assume that A is a pseudo-monotone operator from V into V' with domain $D(A) = V$. In addition, we will assume that A satisfies the conditions

$$\|Au\|_* \leq C_1(1 + \|u\|) \qquad \text{for } u \in V, \tag{2.1}$$

$$C_2\|u\|^2 \leq C_3 + (Au, u) \qquad \text{for } u \in V, \tag{2.2}$$

where C_1, C_2 and C_3 are positive constants.

For each $a(t) \in L^1(0,T)$, L_a denotes the linear continuous operator defined by

$$(L_a f)(t) = \int_0^t a(t-s)f(s)\, ds \qquad \text{for } 0 \leq t \leq T,$$

for each $f \in L^2(0,T)$. Then the adjoint operator L_a^* is given by

$$(L_a^* f)(t) = \int_t^T a(s-t)f(s)\, ds \qquad \text{for } 0 \leq t \leq T.$$

We state the assumptions for the kernel function $a(t)$.

(i) $a(t) \in L^2(0,T)$ and is of positive type on $[0,T]$, i.e., for each $f \in L^2(0,T)$,

$$\int_0^t f(\tau) \int_0^\tau a(\tau-s)f(s)\, ds\, d\tau \geq 0, \quad \text{for } 0 \leq t \leq T.$$

(ii) L_a^* is injective, i.e., $L_a^* f = 0$ means $f = 0$.

Remark. the operator L_a and L_a^* are bounded linear on $\tilde{V}$, $\tilde{H}$ and $\tilde{V}'$. Also L_a and L_a^* are positive on these spaces if $a(t)$ satisfies the condition (i). Sufficient conditions for $a(t)$ to satisfy (i) are investigated in [11] and [12]. $a(t)$ satisfies (ii) if $a(0) \neq 0$ and $a'(t) \in L^1(0,T)$.

We now state our main result:

Theorem. Let $A:V \to V'$ be a pseudo-monotone operator satisfying (2.1) and (2.2). Let $a(t) \in L^2(0,T)$ be a function satisfying (i) and (ii). Then for each $f \in L^2(0,T;V)$, (1.1) has a solution in $L^2(0,T;V)$.

Remark. Since a monotone hemicontinuous operator is pseudo-monotone, our result is an extension of the results for monotone operators(cf. [1,9]). It is well known that the sum of monotone hemicontinuous operator and completely continuous operator is pseudo-monotone. Then we have

Corollary. Suppose that V is compactly imbedded in H. Let $A:V \to V'$ be a monotone single-valued hemicontinuous operator satisfying (2.1) and $B:H \to H$ be a continuous operator satisfying (2.1). Suppose that A + B satisfies (2.2) with A replaced by A + B. Let $a(t) \in L^2(0,T)$ be a function satisfying (i) and (ii). Then for each $f \in L^2(0,T;V)$, the equation

$$u(t) + \int_0^t a(t-s)(Au(s) + Bu(s))\,ds = f(t), \quad 0 \le t \le T, \tag{2.3}$$

has a solution in $L^2(0,T;V)$.

3. Proof of Theorem. We first state a well known result which is crucial for our argument.

Proposition A(cf. [3], [4]). Let K be a closed convex subset of $\tilde{V}$. Let $T:\tilde{V} \to \tilde{V}'$ be a monotone operator and $T_0:K \to \tilde{V}'$ be pseudo-monotone operator. Then there exists an element u of K such that

$$(g + T_0u, v - u) \ge 0 \quad \text{for all } g \in Tv, \quad v \in K.$$

The following lemma is also known and easy to verify.

Lemma A. Let u be an element of $\tilde{V}'$. Suppose that

$$\sup \{ (u, v) : v \in \tilde{V}, \ |v| \le 1 \} < \infty.$$

Then $u \in \tilde{H}$.

Throughout the rest of this section, we write L instead of L_a. We define an operator $\tilde{A}$ by $(\tilde{A}u)(t) = Au(t)$, $0 \leq t \leq T$, for each $u \in \tilde{V}$. Then it is easy to see, by the pseudo-monotonicity, that $\tilde{A}u$ is measurable for each $u \in \tilde{V}$. Also by (2.1), we have that $\tilde{A}u \in \tilde{V}'$ for each $u \in \tilde{V}$. Then $\tilde{A}$ is a operator from $\tilde{V}$ into $\tilde{V}'$. Moreover, it is easy to see that $\tilde{A}$ satisfies the following conditions:

(2.1)' $\quad \|\tilde{A}u\|_* \leq c_1(1 + \|u\|)$ $\qquad$ for $u \in \tilde{V}$,

(2.2)' $\quad c_2\|u\|^2 \leq c_3 + \langle \tilde{A}u, u\rangle$ $\qquad$ for $u \in \tilde{V}$,

where c_1, c_2 and c_3 are positive constants depending only on C_1, C_2, C_3 and T.

The following lemma is a direct consequence of the definition of the pseudo-monotonicity.

Lemma 1. Let $\{u_n\} \subset V$ be a sequence such that u_n converges weakly to u in V and $\liminf_{n\to\infty} (Au_n, u_n - u) \leq 0$. Then $\lim_{n\to\infty} (Au_n. u_n - u) = 0$.

Lemma 2. The equation (1.1) has a solution $u \in \tilde{V}$ if and only if the equation

(3.1) $\qquad v + \tilde{A}(Lv + f) = 0,$

has a solution $v \in \tilde{V}'$.

Proof. Let $v \in \tilde{V}'$ be a solution of (3.1). then we have that $Lv + L\tilde{A}(Lv + f) = 0$. Put $u = Lv + f$.(Note that $u \in \tilde{V}$). Then u is a solution of (1.1). Let u be a solution of (1.1). Then there exists $v \in \tilde{V}'$ such that $Lv = u - f$. Then it is easy to see that v is a solution of (3.1).

We denote by $\tilde{A}_f$ the operator

$$\tilde{A}_f(u) = \tilde{A}(u + f) \qquad \text{for each } u \in \tilde{V}',$$

i.e., $(\tilde{A}_f u)(t) = A(u(t) + f(t))$ for $t \in [0,T]$. Then it is easily verified that $\tilde{A}_f$ satisfies (2.1)' and (2.2)' for some c_1, c_2 and c_3. The equation (3.1) can be rewritten as

$$v + \tilde{A}_f Lv = 0.$$

Since L^* is injective, $v \in \tilde{V}$ is a solution of the equation above if and only if it is a solution of the equation

$$(3.1)' \qquad L^* v + L^* \tilde{A}_f Lv = 0.$$

So we will show the existence of a solution of the equation (3.1)'.

Proposition 1. Let $\{u_n\} \subset \tilde{V}'$ be a sequence such that u_n converges to u weakly in $\tilde{V}'$, $\{Lu_n\} \subset \tilde{V}$ and Lu_n converges to Lu weakly in $\tilde{V}$. Suppose that $\limsup_{n\to\infty} \langle \tilde{A}_f Lu_n, Lu_n - Lu\rangle \leq 0$. Then $\tilde{A}_f Lu_n$ converges to $\tilde{A}_f Lu$ weakly in $\tilde{V}'$ and $\langle \tilde{A}_f Lu, Lu - v\rangle \leq \liminf_{n\to\infty} \langle \tilde{A}_f Lu_n, Lu_n - v\rangle$, for each $v \in \tilde{V}$.

For the sake of simplicity of the proof, we prove Proposition 1 in case when $f = 0$. We need the following two lemmas to prove Proposition 1.

Lemma 3. Let $\{u_n\} \subset \tilde{V}'$ be a sequence such that u_n converges to u weakly in $\tilde{V}'$, $\{Lu_n\} \subset \tilde{V}$ and Lu_n converges to Lu weakly in $\tilde{V}$.

Suppose that $\limsup_{n\to\infty} \langle \tilde{A}Lu_n, Lu_n - Lu\rangle \le 0$. Then
$\liminf_{n\to\infty} (A(Lu_n)(t), (Lu_n)(t) - (Lu)(t)) = 0$ a.e. on $[0,T]$ and
$\lim_{n\to\infty} \langle \tilde{A}Lu_n, Lu_n - Lu\rangle = 0$.

Proof. Since $a(t) \in L^2(0,T)$ and $u_n \to u$ weakly in $\tilde{V}'$, we have that $(Lu_n)(t) \to (Lu)(t)$ weakly in V', for all $t \in [0,T]$. Then it follows that for each $t \in [0,T]$, there exists a subsequence $\{u_{n_i}\}$ of $\{u_n\}$ such that $(Lu_{n_i})(t)$ converges to $(Lu)(t)$ weakly in V, if and only if $\liminf_{n\to\infty} \|Lu_n\| < \infty$. In fact, any $\|\cdot\|$-bounded subsequence of $\{Lu_n(t)\}$ converges to $Lu(t)$ weakly in V. Fix $t \in [0,T]$ and suppose that

$$\liminf_{n\to\infty} (A(Lu_n)(t), (Lu_n)(t) - (Lu)(t)) \le 0. \tag{3.2}$$

By (2.1) and (2.2),

$$C_2\|(Lu_n)(t)\|^2 - C_3 - C_1(1 + \|(Lu_n)(t)\|)\|(Lu)(t)\| \le (A(Lu_n)(t), (Lu_n)(t) - (Lu)(t)). \tag{3.3}$$

Then from (3.2) and (3.3), we find that $\liminf_{n\to\infty} \|(Lu_n)(t)\| < \infty$. Thus we obtain that there exists a subsequence $\{u_{n_i}\}$ of $\{u_n\}$ such that

$$\lim_{i\to\infty} (A(Lu_{n_i})(t), (Lu_{n_i})(t) - (Lu)(t)) = \liminf_{n\to\infty} (A(Lu_n)(t), (Lu_n)(t) - (Lu)(t))$$

and $(Lu_{n_i})(t)$ converges to w weakly in V. Then since $(Lu_n)(t)$ converges weakly to $(Lu)(t)$ in V', we have that $(Lu_{n_i})(t)$ converges weakly to $(Lu)(t)$ in V. Then by Lemma 1, we have that $\lim (A(Lu_{n_i})(t), (Lu_{n_i})(t) - (Lu)(t)) = 0$. Therefore we obtain that

$$(3.4) \qquad \liminf_{n\to\infty} (A(Lu_n)(t), (Lu_n)(t) - (Lu)(t)) \geq 0$$

for $t \in [0, T]$. From (3.3), it follows that for each $n \geq 1$ and each $t \in [0,T]$,

$$(3.5) \qquad (A(Lu_n)(t),(Lu_n)(t) - (Lu)(t)) \geq - C_3 - (C_1\|(Lu)(t)\| - \frac{C_1^2}{4C_2}\|(Lu)(t)\|^2).$$

Then by using Fatou's lemma, we have that

$$0 \leq \int_0^T \liminf_{n\to\infty} (A(Lu_n)(t), (Lu_n)(t) - (Lu)(t))\, dt$$

$$\leq \liminf_{n\to\infty} \langle \tilde{A}Lu_n, Lu_n - Lu\rangle$$

$$\leq \limsup_{n\to\infty} \langle \tilde{A}Lu_n, Lu_n - Lu\rangle \leq 0.$$

Thus we obtain that $\limsup_{n\to\infty} \langle \tilde{A}Lu_n, Lu_n - Lu\rangle = 0$. Also by (3.4) and the inequality above, we have that

$$\liminf_{n\to\infty} (A(Lu_n)(t), (Lu_n)(t) - (Lu)(t)) = 0 \quad \text{a.e. on } [0,T].$$

This completes the proof.

Lemma 4. Let $\{u_n\} \subset \tilde{V}'$ be a sequence which satisfies the assumptions in Lemma 3. Then there exists a subsequence $\{u_{n_i}\}$ of $\{u_n\}$ such that $A(Lu_{n_i})(t)$ converges to $A(Lu)(t)$ weakly in V', a.e. on $[0,T]$ and $\limsup_{i\to\infty} (A(Lu_{n_i})(t), (Lu_{n_i})(t) - (Lu)(t)) \leq 0$, a.e. on $[0,T]$.

Proof. We put

$$f_n(t) = (A(Lu_n)(t), (Lu_n)(t) - (Lu)(t))$$

and

$$f_n^-(t) = \min(0, f_n(t)), \quad \text{for } t \in [0,T] \text{ and } n \geq 1.$$

Then $\int_0^T |f_n(t)|\,dt = \int_0^T f_n(t)\,dt - 2\int_0^T f_n^-(t)\,dt$ for $n \geq 1$. Also we note that $\liminf_{n\to\infty} f_n^-(t) = \min(0, \liminf_{n\to\infty} f_n(t))$ for $t \in [0,T]$. Then since $\liminf_{n\to\infty} f_n^-(t) = 0$, a.e. on $[0,T]$ by Lemma 3, we find that $\liminf_{n\to\infty} f_n^-(t) = 0$ a.e. on $[0,T]$. From the inequality (3.5), we can use Fatou's lemma. Thus we have that

$$0 \leq \int_0^T \liminf_{n\to\infty} f_n^-(t)\,dt \leq \liminf_{n\to\infty} \int_0^T f_n^-(t)\,dt$$
$$\leq \limsup_{n\to\infty} \int_0^T f_n^-(t)\,dt \leq 0.$$

On the other hand, $\lim_{n\to\infty} \int_0^T f_n(t)\,dt = 0$ by Lemma 3. Thus we find that $\lim_{n\to\infty} \int_0^T |f_n(t)|\,dt = 0$. Then there exists a subsequence $\{f_{n_i}\}$ of $\{f_n\}$ such that $f_{n_i}(t)$ converges to 0 a.e. on $[0,T]$, i.e, $\limsup (A(Lu_{n_i})(t), (Lu_{n_i})(t) - (Lu)(t)) \leq 0$ a.e. on $[0, T]$. Then By (3.3), it follows that $\{\|(Lu_{n_i})(t)\|\}$ is bounded a.e. on $[0, T]$. Then since $(Lu_{n_i})(t) \to (Lu)(t)$ weakly in V' on $[0, T]$, we have that any weakly convergent subsequence of $\{(Lu_{n_i})(t)\}$ converges to $(Lu)(t)$ weakly in V a.e. on $[0, T]$. Thus we obtain that $(Lu_{n_i})(t) \to (Lu)(t)$ weakly in V, a.e. on $[0, T]$. Hence by the pseudo-monotonicity of A, we find that $A(Lu_{n_i})(t) \to A(Lu)(t)$ weakly in V', a.e. on $[0, T]$.

Proof of Proposition 1. Let $v \in \hat{\mathcal{V}}$ and choose a subsequence $\{u_{n_i}\}$ of $\{u_n\}$ such that

$$\lim_{i\to\infty} \langle \tilde{A}Lu_{n_i}, Lu_{n_i} - v\rangle = \liminf_{n\to\infty} \langle \tilde{A}Lu_n, Lu_n - v\rangle.$$

By Lemma 4, we can choose a subsequence of $\{u_{n_i}\}$(denoted again by $\{u_{n_i}\}$) such that $\limsup_{i\to\infty} (A(Lu_{n_i})(t), (Lu_{n_i})(t) - (Lu)(t)) \leq 0$, a.e. on $[0,T]$ and $A(Lu_{n_i})(t) \to A(Lu)(t)$ weakly in V' a.e. on $[0,T]$. Also from the argument in the proof of Lemma 3, we find that $(Lu_{n_i})(t) \to (Lu)(t)$ weakly in V a.e. on $[0,T]$. Therefore by the pseudo-monotonicity of A, it follows that

$$(A(Lu)(t), (Lu)(t) - v(t))$$
$$\leq \liminf_{i\to\infty} (A(Lu_{n_i})(t), (Lu_{n_i})(t) - v(t)),$$

a.e. on $[0,T]$. Then since

$$(A(Lu_{n_i})(t), (Lu_{n_i})(t) - v(t))$$
$$\geq C_2\|(Lu_{n_i})(t)\|^2 - C_3 - C_1(1 + \|(Lu_{n_i})(t)\|)\|v(t)\|$$
$$\geq - C_3 - C_1\|v(t)\| - \frac{C_1^2}{4C_2}\|v(t)\|^2,$$

we have, by Fatou's lemma, that

$$\langle \tilde{A}Lu, Lu - v\rangle \tag{3.6}$$
$$\leq \int_0^T \liminf_{i\to\infty} (A(Lu_{n_i})(t), (Lu_{n_i})(t) - v(t))\, dt$$
$$\leq \liminf_{i\to\infty} \langle \tilde{A}Lu_{n_i}, Lu_{n_i} - v\rangle$$
$$= \liminf_{n\to\infty} \langle \tilde{A}Lu_n, Lu_n - v\rangle.$$

We next show that $\tilde{A}Lu_n$ converges weakly to $\tilde{A}Lu$ in $\hat{V}'$. Since $\{\tilde{A}Lu_n\}$ is bounded in $\hat{V}'$, it is sufficient to show that any weakly convergent subsequence of $\{\tilde{A}Lu_n\}$ converges to $\tilde{A}Lu$. Let $\{\tilde{A}Lu_{n_i}\}$ be a subsequence of $\{\tilde{A}Lu_n\}$ such that $\tilde{A}Lu_{n_i}$ converges to a point $w \in \hat{V}'$ weakly in $\hat{V}'$. Then noting that $\lim_{n\to\infty} \langle \tilde{A}Lu_n, Lu_n - Lu \rangle = 0$, we have that for each $v \in \hat{V}$,

$$\liminf_{i\to\infty} \langle \tilde{A}Lu_{n_i}, Lu_{n_i} - v \rangle$$

$$= \lim_{i\to\infty} \langle \tilde{A}Lu_{n_i}, Lu_{n_i} - Lu \rangle + \lim_{i\to\infty} \langle \tilde{A}Lu_{n_i}, Lu - v \rangle$$

$$= \langle w, Lu - v \rangle.$$

Therefore it follows by (3.6) that $\langle \tilde{A}Lu, Lu - v \rangle \leq \langle w, Lu - v \rangle$ for all $v \in \hat{V}$. This implies that $w = \tilde{A}Lu$ and therefore $\tilde{A}Lu_n$ converges to $\tilde{A}Lu$ weakly in $\hat{V}'$. This completes the proof.

Remark. It is easy to verify that the proofs above are valid, with minor changes, for $\tilde{A}_f$ with $f = 0$.

Proposition 1'. $L^*\tilde{A}_f L$ is a pseudo-monotone operator from $\hat{V}$ to $\hat{V}'$.

Proof. Let $\{u_n\} \subset \hat{V}$ be a sequence such that $u_n \to u$ weakly in $\hat{V}$ and $\limsup_{n\to\infty} \langle L^*\tilde{A}_f Lu_n, u_n - u \rangle \leq 0$. Then $Lu_n \to Lu$ weakly in $\hat{V}$ and $\limsup_{n\to\infty} \langle \tilde{A}_f Lu_n, Lu_n - Lu \rangle \leq 0$. Therefore by Proposition 1 we have that $\langle L^*\tilde{A}_f Lu, u - v \rangle \leq \liminf_{n\to\infty} \langle L^*\tilde{A}_f Lu, u - v \rangle$ for all $v \in \hat{V}$, and $\tilde{A}_f Lu_n$ converges weakly to $\tilde{A}_f Lu$ in V'. Since L^* is continuous on $\hat{V}'$, $L^*\tilde{A}_f Lu_n$ converges weakly to $L^*\tilde{A}_f Lu$ in $\hat{V}'$. This completes the proof.

Proposition 2. For $c > 0$, the equation

$$(3.7) \qquad L^*u + cu + L^*\tilde{A}_f Lu = 0$$

has a solution $u \in \hat{H}$ such that there exists a sequence $\{v_n\} \subset \hat{V}$

which satisfies that v_n converges to u strongly in $\tilde{H}$ and Lv_n converges to Lu strongly in $\tilde{V}$.

Proof. We put $K_n = \{v \in \tilde{V}: |v| \le n\}$ for each $n \ge 1$. Since $L^*\tilde{A}_f L$ is a pseudo-monotone operator from $\tilde{V}$ into $\tilde{V}'$, it follows that $cI + L^*\tilde{A}_f L$ is a pseudo-monotone operator from K_n to $\tilde{V}'$ for each $n \ge 1$. On the other hand, since L^* is a positive continuous operator from $\tilde{V}$ into $\tilde{V} \subset \tilde{V}'$, L^* is maximal monotone in $\tilde{V} \times \tilde{V}'$. Then we have, by Proposition A, that for each $n \ge 1$, there exists $u_n \in K_n$ such that

$$\langle L^*v + cu_n + L^*\tilde{A}_f Lu_n, v - u_n\rangle \ge 0 \quad \text{for all } v \in K_n. \tag{3.8}$$

By setting v = 0 in (3.8), we have

$$c\,|u_n|^2 + c_2\|Lu_n\|^2 \le c_3 \qquad \text{for all } n \ge 1. \tag{3.9}$$

Then (3.9) implies that $\sup_n |u_n| < \infty$ and $\sup_n \|Lu_n\| < \infty$. Then we may suppose without any loss of generality that $u_n \to u$ weakly in $\tilde{H}$ and $Lu_n \to Lu$ weakly in $\tilde{V}$. Noting that $u \in \overline{co}^H\{u_n : n \ge k\}$ and $Lu \in \overline{co}^V\{Lu_n : n \ge k\}$ for all $k \ge 1$, we find that there exists a sequence $\{v_n\}$ such that $\{v_n\}, \{Lv_n\} \subset \tilde{V}$, $v_n \to u$ strongly in $\tilde{H}$ and $Lv_n \to Lu$ strongly in $\tilde{V}$. We now show that u is a solution of (3.7). For given $\varepsilon > 0$, we choose $v_m \in \{v_n\}$ such that $|v_m - u| < \varepsilon$ and $\|Lv_m - Lu\| < \varepsilon$. Let n_0 be an integer such that $v_m \in K_{n_0}$. Then for each $n \ge n_0$, we have that

$$\langle L^*v_m + cu_n + L^*\tilde{A}_f Lu_n, v_m - u\rangle \ge 0. \tag{3.10}$$

Since $\sup_n |u_n| < M$, $\sup_n \|Lu_n\| < M$ and $\sup_n \|\tilde{A}_f Lu_n\|_* < M$, for some $M > 0$, it follows that

$$\langle u + \tilde{A}_f Lu_n, Lu - Lu_n\rangle + c\langle u_n, u - u_n\rangle \ge -\varepsilon M', \tag{3.11}$$

for $n \ge n_0$, where M' is a positive constant depending only on M and c. Then since ε is arbitrary, (3.11) implies that

$$\limsup_{n\to\infty} \langle \tilde{A}_f Lu_n, Lu_n - Lu\rangle \le 0,$$

because $\limsup_{n\to\infty} \langle u_n, u - u_n\rangle \le 0$ and $\lim_{n\to\infty} \langle u, Lu - Lu_n\rangle = 0$.

Hence by Proposition 1, we have that for each $v \in \tilde{V}$,

(3.12) $\langle \tilde{A}_f Lu, Lu - Lv\rangle \leq \liminf_{n\to\infty} \langle \tilde{A}_f Lu_n, Lu_n - Lv\rangle.$

Then combining (3.12) with (3.8), we have that for each $v \in \tilde{V}$,

$$\liminf_{n\to\infty} (\langle L^* v, v - u_n\rangle + c\langle u_n, v - u_n\rangle) \geq \langle \tilde{A}_f Lu, Lu - Lv\rangle.$$

Then since $\liminf_{n\to\infty} |u_n| \geq |u|$, it follows that

(3.13) $\langle L^* v, v - u\rangle + c\langle u, v - u\rangle \geq \langle \tilde{A}_f Lu, Lu - Lv\rangle,$ for $v \in \tilde{V}$.

Then

(3.14) $\langle L^*\tilde{A}_f Lu, -v\rangle \leq \langle L^* v, v - u\rangle + c\langle u, v - u\rangle - \langle \tilde{A}_f Lu, Lu\rangle$

$$\leq \|L\||v|^2 + (\|Lu\| + c)|u||v| - \langle \tilde{A}_f Lu, Lu\rangle,$$

where $\|L\|$ denotes the operator norm of L in $\tilde{H}$.
Then, (3.14) implies that there exists $R > 0$ such that for each $v \in V$ with $|v| \leq 1$, $\langle L^*\tilde{A}_f Lu, v\rangle \leq R$. Then by Lemma A, we obtain that $L^*\tilde{A}_f Lu \in \tilde{H}$. Then since

$$\langle L^*\tilde{A}_f Lu, u\rangle = \lim_{n\to\infty} \langle L^*\tilde{A}_f Lu, v_n\rangle = \lim_{n\to\infty} \langle \tilde{A}_f Lu, Lv_n\rangle = \langle \tilde{A}_f Lu, Lu\rangle,$$

it follows from (3.13) that

(3.15) $\langle L^* v - (-cu - L^*\tilde{A}_f Lu), v - u\rangle \geq 0$ for all $v \in \tilde{V}$.

Since $-(cu + L^*\tilde{A}_f Lu) \in \tilde{H}$ and L^* is continuous on $\tilde{H}$, (3.15) holds for all $v \in \tilde{H}$. Then since L^* is maximal monotone on $\tilde{H}$, we obtain that $-(cu + L^*\tilde{A}_f Lu) = L^* u$, and this completes the proof.

Proof of Theorem. By Proposition 2, we have that for each $n \geq 1$, the equation

(3.16) $L^* u_n + n^{-1} u_n + L^*\tilde{A}_f Lu_n = 0,$

has a solution $u_n \in \tilde{H}$. Fix $n \geq 1$. By Proposition 2, there exists $\{w_m\} \subset \tilde{V}$ such that $w_m \to u_n$ strongly in $\tilde{H}$ and $Lw_m \to Lu_n$ strongly in $\hat{V}$. Then multiply (3.16) by w_m and integrate. This gives

$$(3.17) \qquad \langle u_n + \tilde{A}_f Lu_n, Lw_m\rangle + n^{-1}\langle u_n, w_m\rangle = 0.$$

Since $w_m \to u_n$ strongly in $\hat{H}$ and $Lw_m \to Lu_n$ strongly in $\hat{V}$, we deduce from (3.17) that

$$(3.18) \qquad \langle u_n + \tilde{A}_f Lu_n, Lu_n\rangle + n^{-1}\langle u_n, u_n\rangle = 0.$$

By the condition (2.2)', it follows

$$(3.19) \qquad c_2\|Lu_n\|^2 + n^{-1}|u_n| \leq c_3 \quad .$$

The inequality (3.19) yeilds that for some $M > 0$, $\|Lu_n\| < M$ for $n \geq 1$. On the other hand, the equation (3.16) can be written

$$(3.20) \qquad u_n + n^{-1} z_n + \tilde{A}_f Lu_n = 0 , \quad \text{for } n \geq 1,$$

where $z_n \in \hat{V}'$ with $L^* z_n = u_n$, for $n \geq 1$.

Then we have

$$(3.21) \qquad \|u\|_*^2 + n^{-1} \langle\langle z_n, u_n\rangle\rangle \leq \|u\|_* \|A_f Lu_n\|_*$$

where $\langle\langle\cdot,\cdot\rangle\rangle$ denotes the inner product in the Hilbert space $\tilde{V}'$. Then since $\langle\langle z_n, u_n\rangle\rangle = \langle\langle z_n, L^* z_n\rangle\rangle \geq 0$, we obtain

$$\|u_n\|_* \leq c_1(\|Lu_n\| + 1) \leq c_1(M + 1), \quad \text{for all } n \geq 1.$$

Then we may assume without any loss of generality that $u_n \to u$ weakly in $\hat{V}'$ and $Lu_n \to Lu$ weakly in $\hat{V}$. This implies that there exists a sequence $\{z_m\} \subset \hat{H}$ such that $z_m \in \overline{co}^*\{u_n : n \geq m\}$ for $m \geq 1$, $z_m \to u$ strongly in $\hat{V}'$ and $Lz_m \to Lu$ strongly in $\hat{V}$. We recall that for each $n \geq 1$, u_n has an approximating sequence $\{v_m^{(n)}\} \subset \hat{V}$ such that $v_m^{(n)} \to u_n$ strongly in $\hat{H}$ and $v_n^{(n)} \to Lu_n$ strongly in $\hat{V}$. Therefore we find that there exists a sequence $\{v_n\} \subset \hat{V}$ such that $v_n \to u$ strongly in $\hat{V}'$ and $Lv_n \to Lu$ strongly in $\hat{V}$. We now show that

$\limsup_{n\to\infty} \langle \tilde{A}_f Lu_n, Lu_n - Lu \rangle \le 0$. For given $\varepsilon > 0$, choose $v_m \in \{v_n\}$ such that $\|v_m - u\|_* < \varepsilon$ and $\|Lv_m - Lu\| < \varepsilon$. Then multiply (3.16) by $u_n - v_m$ and integrate. This gives

$$(3.22)\quad \langle u_n + \tilde{A}_f Lu_n, Lu_n - Lv_m \rangle + n^{-1}\langle u_n, u_n - v_m \rangle = 0, \text{ for } n \ge 1.$$

Then since $\langle Lu_n - Lv_m, u_n - v_m \rangle \ge 0$ and $\langle \tilde{A}_f Lu_n, Lu_n - Lu \rangle = \langle \tilde{A}_f Lu_n, Lu_n - Lv_m \rangle + \langle \tilde{A}_f Lu_n, Lv_m - Lu \rangle$ for $n \ge 1$, we have by (2.1)' and (3.22) that

$$\begin{aligned} \langle \tilde{A}_f Lu_n, Lu_n - Lu \rangle + n^{-1}\langle u_n, u_n - v_m \rangle &\le \varepsilon \|\tilde{A}_f Lu_n\|_* + \langle v_m, Lu_n - Lv_m \rangle \\ &\le \varepsilon\, c_1(M + 1) + \langle v_m, Lu_n - Lv_m \rangle, \quad \text{for } n \ge 1. \end{aligned}$$

Noting that $\liminf_{n\to\infty} n^{-1}\langle u_n, u_n - v_m \rangle \ge 0$, it follows that

$$(3.23)\quad \begin{aligned} \limsup_{n\to\infty} \langle \tilde{A}_f Lu_n, Lu_n - Lu \rangle &\le \varepsilon\, c_1(M + 1) + \langle v_m, Lu - Lv_m \rangle \\ &\le \varepsilon\, c_1(M + 1) + \varepsilon(M + \varepsilon). \end{aligned}$$

Since ε is arbitrary, we obtain

$$\limsup_{n\to\infty} \langle \tilde{A}_f Lu_n, Lu_n - Lu \rangle \le 0.$$

Then by Proposition 1, we have that $\tilde{A}_f Lu_n \to \tilde{A}_f Lu$ weakly in $\tilde{V}'$. Therefore by (3.16), it follows that $L^* u + L^* \tilde{A}_f Lu = 0$. Since L^* is injective, we deduce that $u + \tilde{A}_f Lu = 0$. Then by Lemma 2, (1.1) has a solution.

REFERENCES

[1] V. Barbu, Nonlinear Volterra equations in Hilbert space, SIAM J. Math. Anal., 6(1975), 728-741.

[2] V. Barbu, Nonlinear semigroups and differential equations in Banach spaces, Editura Academiei, Bucaresti, Romania, and Noordhoff, Leyden, the Netherlands, 1976.

[3] F. E. Browder, Nonlinear operators and nonlinear equations of evolution in Banach spaces, Proc. Symp. Pure Math., XVIII, 2(1976).

[4] F. E. Browder, Pseudo-monotone operators and the direct method of the calculus of variations, Archiv Rat. Mech. Anal. 38(1970), 268-277.

[5] M. G. Crandall, S. -O. Londen, and J. A. Nohel, An abstract nonlinear Volterra integrodifferential equation, J. Math. Anal. Appl., 701-735(1978).

[6] G. Gripenberg, An existence result for a nonlinear Volterra integral equation in a Hilbert space, SIAM J. Math. Anal., 9(1978), 793-805.

[7] N. Hirano, Abstract nonlinear Volterra equations with positive kernels, SIAM J. Math. Anal., in press.

[8] T. Kiffe, A perturbation of an abstract Volterra equation, SIAM J. Math. Anal., 11(1980), 1036-1046.

[9] T. Kiffe and M. Stecher, An abstract Volterra integral equation in a reflexive Banach space, J. Differential equations 34(1979), 303-325.

[10] R. C. MacCamy and J. Wong, Stability theorems for some functional equations, Trans. Amer. Math. Soc., 164(1972), 1-37.

[11] J. A. Nohel and F. Shea, Frequency domain methods for Volterra equations, Advance in Math., 3(1976), 278-304.

[12] O. Staffans, Positive definite measures with applications to a Volterra equation, Trans. Amer. Math. Soc., 218(1976), 219-237.

Some Results on Multivalued Mappings and Inequalities Without Convexity

CHARLES D. HORVATH Champlain Regional College, St. Lambert, Quebec, Canada

INTRODUCTION. The classical KKM theorem can be stated thus:

Theorem A: *Let* $R_0,\dots,R_n$ *be closed subsets of the standard* n *dimensional simplex and let* $\{e_0,\dots,e_n\}$ *be the set of its vertices. If the convex hull of any subset* $\{e_{i_1},\dots,e_{i_k}\}$ *is contained in* $\bigcup_{j=0}^{k} R_{i_j}$ *then* $\bigcap_{i=0}^{n} R_i \neq \emptyset$.

This result was first extended to infinite dimensional vector spaces by Ky Fan [4] and then by Dugundji-Granas [3].
From Ky Fan's generalization one obtains Ky Fan fixed point theorem which was in turn generalized by Ben-El-Mechaiekh - Deguire - Granas [1] as follows:

Theorem B: *Let* X *be a compact convex set* $A,B : X \to X$ *two multivalued mappings such that*

(i) $\forall x \in X \quad B(x) \subset A(x)$,

(ii) $\forall x \in X \quad B(x) \neq \emptyset$ *and* $A(x)$ *is convex,*

(iii) $\forall x \in X \quad B^{-1}(x)$ *is open.*

Then $\exists x_0 \in X$ *such that* $x_0 \in A(x_0)$.

From theorem B follows the non linear alternative for two functions also given in [1].

Theorem C: *Let* X *be a convex compact set and* $f,g : X \times X \to \mathbb{R}$ *two real valued functions such that:*

(i) $\forall (x,y) \in X \times X \quad g(x,y) \leq f(x,y)$,

(ii) $\forall y \in X \quad x \to f(x,y)$ *is quasiconcave,*

(iii) $\forall x \in X \quad y \to g(x,y)$ *is l.s.c.,*

Then $\forall \lambda \in \mathbb{R}$ *one has the following alternative*

(a) $\exists y_0 \in X \quad \forall x \in X \quad g(x,y_0) \leq \lambda$ *or*

(b) $\exists x_0 \in X \quad f(x_0,x_0) > \lambda$.

In this paper convexity is replaced by contractibility. The KKM theorem, fixed point and coincidence theorem for multivalued mappings of Ky Fan and Ben-El-Mechaiekh - Deguire - Granas are thus generalized in part I.

In part II generalization of the non linear alternative and of Ky Fan's inequality are presented.

PART I

NOTATION. Let X and Y be two sets, capital letters $R : X \to Y$ will denote multivalued mappings while non capital letters $f : X \to Y$ will denote functions. If $R : X \to Y$ is a multivalued mapping then we can define two other multivalued mappings $R^{-1} : Y \to X$ and $R^* : Y \to X$ by: $x \in R^{-1}(y)$ if and only if $y \in R(x)$ and $R^*(y) = X \setminus R^{-1}(y)$, the complement of $R^{-1}(y)$ in X. It is clear that $(R^{-1})^{-1} = R$ and $(R^*)^* = R$ and $x \in R^*(y)$ if and only if $y \notin P(x)$.
Δ_n will represent the standard n dimensional simplex and $e_0, e_1, \ldots, e_n$ will be its vertices. If $J \subset \{1,\ldots,n\}$ is a non empty subset then $\Delta_J \subset \Delta_n$ will be the convex hull of the vertices $\{e_j : j \in J\}$.

Theorem 1. *Let* X *be a topological space and* $\{R_i : i = 0,\ldots,n\}$ *a family of closed subsets of* X *such that:*

(i) For any non empty subset J *of* $\{0,\ldots,n\}$ *there is a non empty contractible subset* F_J *of* X *such that* $F_J \subset \bigcup_{i \in J} R_i$,

(ii) If $J \subset J'$ *then* $F_J \subset F_{J'}$, .

Then $\bigcap_{i=0}^{n} R_i$ *is not empty and there is a continuous function* $f : \Delta_n \to X$ *such that* $f(\Delta_J) \subset \bigcup_{i \in J} R_i$ *for each non empty subset* J *of* $\{0,\ldots,n\}$.

PROOF. For each $i \in \{0,\ldots,n\}$ choose a point $x_i \in F_{\{i\}} \subset R_i$. Now let us assume that for each J of cardinality smaller or equal to k we have constructed a function $\sigma_J : \Delta_J \to X$ such that $\sigma_J(\Delta_J) \subset F_J$ and σ_J and $\sigma_{J'}$ coincide on $\Delta_J \cap \Delta_{J'}$ if $J \cap J'$ is not empty.
Now let $J = \{i_0,\ldots,i_k\}$ be a subset of cardinality $k+1$ of $\{0,\ldots,n\}$ and take $J_a = J \setminus \{i_a\}$. We have $J = J_0 \cup \ldots \cup J_k$, each Δ_{J_a} is a face of the simplex Δ_J and $\Delta_{J_0} \cup \ldots \cup \Delta_{J_k}$ is the boundary of the simplex Δ_J. For each a we also have a continuous function $\sigma_{J_a} : \Delta_{J_a} \to F_{J_a}$, since these functions coincide on the intersection of the faces we obtain a continuous function $\tilde{\sigma}_J : \Delta_{J_0} \cup \ldots \cup \Delta_{J_k} \to F_{J_0} \cup \ldots \cup F_{J_k} \subset F_J$. Since F_J is contractible $\tilde{\sigma}_J$ can be continuously extended to the simplex Δ_J.
For each subset J of cardinality $k+1$ we have built a continuous function $\sigma_J : \Delta_J \to F_J$ and it follows from the construction that $\Delta_J \cap \Delta_{J'} \neq \emptyset$ implies $\sigma_{J|\Delta_J \cap \Delta_{J'}} = \sigma_{J'|\Delta_J \cap \Delta_{J'}}$. After a finite number

of steps we obtain a continuous function $f : \Delta_n \to X$ such that $f(\Delta_J) \subset F_J$.
Now let $S_i = f^{-1}(R_i)$ and J be a non empty subset of $\{0,1,\ldots,n\}$. S_i is a closed subset of the simplex Δ_n and

$$\bigcup_{i\in J} S_i = f^{-1}(\bigcup_{i\in J} R_i) \supset f^{-1}(F_J) \supset \Delta_J.$$

Therefore $\bigcup_{i\in J} S_i \supset \mathrm{conv}\{e_i : i \in J\}$, by the Knaster-Kuratowski- Mazurkiewicz theorem,[10], [3], $\bigcup_{i=0}^{n} S_i$ is not empty. If $p_0 \in \bigcap_{i=0}^{n} S_i$ then obviously $f(p_0) \in \bigcap_{i=0}^{n} R_i$. The proof is complete. □

Remark. Notice that to construct a function $f : \Delta_n \to X$ such that $f(\Delta_J) \subset \bigcup_{i\in J} R_i$ the sets R_i do not have to be closed.

Corollary 1. *Let Y be a compact topological space and $R : X \to Y$ a closed valued mapping. If there exists a familly $\{F_A\}$ of non empty contractible subsets of Y, indexed by the finite subsets of X, such that*

(i) $A \subset B \Rightarrow F_A \subset F_B$,

(ii) $F_A \subset \bigcup_{x\in A} R(x)$.

Then $\bigcap_{x\in X} R(x)$ *is not empty.*

PROOF. Each $R(x)$ is closed, Y is compact and by theorem 1 the family $\{R(x) : x \in X\}$ has the finite intersection property. □

Theorem 2. *Let X be a compact topological space, $\{F_A\}$ a family of non empty contractible subsets of X indexed by the finite subsets of X such that $A \subset B \Rightarrow F_A \subset F_B$ and $F : X \to X$ and $G : X \to X$ two mappings such that:*

(i) $\forall\, x \in X$, $G(x)$ *is closed and* $F(x) \subset G(x)$,

(ii) $\forall\, x \in X$, $x \in F(x)$,

(iii) *If A is a finite subset of X such that $A \subset F^*(x)$ for some x in X then $F_A \subset F^*(x)$.*

Then $\bigcap_{x\in X} G(x) \neq \emptyset$.

PROOF. By theorem 1 it is sufficient to show that $F_A \subset \bigcup_{x\in A} G(x)$ for each finite subset A of X. Now assume that F_A is not in $\bigcup_{x\in A} G(x)$ for some finite subset A of X, then one can find $y \in F_A$ such that $\forall\, x \in A$, $y \notin G(x)$. In other words $A \subset G^*(y)$. Since $F(y) \subset G(y)$ is equivalent to $G^*(y) \subset F^*(y)$, by (iii) we conclude that $F_A \subset F^*(y)$, but since y was chosen in F_A we have $y \in F^*(y)$, which is equivalent to $y \notin F(y)$ and this contradicts (ii). □

Now, keeping the hypothesis on X and the family $\{F_A\}$ the same, theorem 2 has the following dual formulation:

Theorem 2'. *Let* $S : X \to X$ *and* $T : X \to X$ *be two mappings such that:*

(i) $\forall x \in X, S(x) \subset T(x)$,

(ii) $\forall x \in X, S^{-1}(x)$ *is open,*

(iii) If $A \subset T(x)$ *for some finite subset* A *of* X *then* $F_A \subset T(x)$.

Then either there is $x_0 \in X$ *such that* $S(x_0) = \emptyset$ *or there is* $x_0 \in X$ *such that* $x_0 \in T(x_0)$.

PROOF. Let $G(x) = S^*(x)$ and $F(x) = T^*(x)$, then hypothesis (i) and (iii) of the theorem 1 are verified. If $x_0 \notin F(x_0)$ then $x_0 \in T(x_0)$. If $x \in F(x)$ for each $x \in X$ then we conclude from theorem 1 that $\bigcap_{x \in X} S^*(x) \neq \emptyset$. If $\forall x \in X$, $x_0 \in S^*(x)$ then $\forall x \in X$, $x \notin S(x_0)$, in other words $S(x_0)$ is empty. This concludes the proof. □

As a last application of theorem 1 we give the following selection theorem.

Theorem 3. *Let* X *be a compact topological space,* Y *a contractible space and* $S,T : X \to Y$ *two multivalued mappings such that:*

(i) $\forall x \in X, S(x) \neq \emptyset$ *and* $\forall y \in Y, S^{-1}(y)$ *is open,*

(ii) $\forall x \in X, S(x) \subset T(x)$,

(iii) For any open set U *of* X, $\bigcap\{T(x) : x \in U\}$ *is empty or contractible.*

Then T *has a continuous selection.*

PROOF. Since X is compact, by (i) one can find a finite subset $\{y_0,\ldots,y_n\}$ of Y such that $X = S^{-1}(y_0) \cup \ldots \cup S^{-1}(y_n)$. Now for each non empty subset J of $\{0,\ldots,n\}$ define $F_J = \bigcap\{T(x) : x \in \bigcap_{i \in J} S^{-1}(y_i)\}$. If $x \in \bigcap_{i \in J} S^{-1}(y_i)$ then $\{y_i\}_{i \in J} \subset S(x)$, by (ii) if $x \in \bigcap_{i \in J} S^{-1}(y_i)$ then $\{y_i\}_{i \in J} \subset T(x)$, therefore F_J is not empty since $\{y_i\}_{i \in J} \subset F_J$, by (iii) F_J is therefore contractible. It is clear that if $J \subset J'$ then $F_J \subset F_{J'}$. By theorem 1 there is a continuous function $f : \Delta_n \to Y$ such that $f(\Delta_J) \subset F_J$, now using a partition of unity subordinate to the covering $\{S^{-1}(y_i)\}_{i=0}^{i=n}$ we obtain a continuous function $\psi : X \to \Delta_n$ such that $\psi(x) \in \Delta_{J(x)}$. So $f(\psi(x)) \in f(\Delta_{J(x)}) \subset F_{J(x)} \subset T(x)$, therefore $f \circ \psi$ is a continuous selection for T. □

From the proof of theorem 3 we have the following:

Corollary 2. *With the hypothesis of the previous theorem, for any continuous function* $g : Y \to X$ *there exists* $y_0 \in Y$ *such that* $y_0 \in T(g(y_0))$.

PROOF. The continuous function $\psi \circ g \circ f$ has a fixed point p_0 by the Brouwer fixed point theorem.

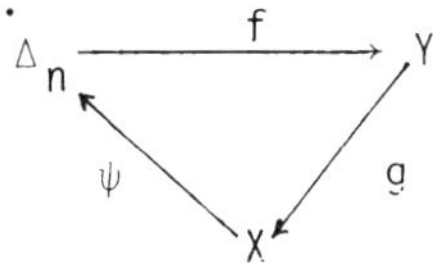

let $y_0 = f(p_0)$ then $f \circ \psi \circ g(y_0) = y_0$ and $f \circ \psi$ is a continuous selection of T. □

We have thus obtained the following generalization of Ky Fan's fixed point theorem:

Theorem 4. *Let* X *be a compact topological space,* Y *a subspace of* X *and* $A, B : Y \to X$ *two multivalued mappings such that:*

(i) $\forall\ x \in X,\ B^{-1}(x) \neq \emptyset$ *and* $\forall\ y \in Y,\ B(y)$ *is open,*

(ii) $\forall\ y \in Y,\ B(y) \subset A(y)$,

(iii) For any open set U *of* X, $\cap\{A^{-1}(x) : x \in U\}$ *is empty or contractible.*

Then A *has a fixed point.*

PROOF. Take $S = B^{-1}$, $T = A^{-1}$ and $g : Y \to X$ the inclusion.□

We finally give the following coincidence theorem:

Theorem 5. *Let* X, Y, A *and* B *be as in theorem 4 if a multivalued mapping* $R : Y \to X$ *has a continuous selection then there is* $y_0 \in Y$ *such that* $A(y_0) \cap R(y_0) \neq \emptyset$.

PROOF. Let $g : Y \to X$ be a continuous function such that $\forall\ y \in Y$, $g(y) \in R(y)$. By corollary 2 there exists $y_0 \in Y$ such that $g(y_0) \in A(y_0)$. □

PART II

We give here the analytic interpretations of the results from part I. In this section X will be a compact topological space and $s : X \times X \to \mathbb{R}$ a function. For $(x,r) \in X \times \mathbb{R}$ let $S(x,r) = \{y \in X : s(y,x) \leq r\}$ it will be assumed that the function s has the following property:

For any family $\{(x_i, r_i)\}_{i \in I} \subset X \times \mathbb{R}$ *with I non-empty.*

$\bigcap_{i \in I} S(x_i, r_i)$ *is either empty or contractible.*

The first result is a generalization of Granas [1], nonlnear alternative.

Proposition 1. *Let* $f : X \times X \to \mathbb{R}$ *and* $g : X \times X \to \mathbb{R}$ *be two functions such that:*

(i) $\forall\ (x,y) \in X \times X,\ g(x,y) \leq f(x,y)$,

(ii) $\forall\ x \in X$, $y \to g(x,y)$ *is lower semicontinuous,*

(iii) If $f(z,y) < f(x_i,y)$, $i = 1,\dots,k$, *then one can find* $w \in X$ *such that* $s(x_i,w) < s(z,w)$, $i = 1,\dots,k$.

Then for any $\lambda \in \mathbb{R}$ *either there existes* $y_0 \in X$ *such that for all* $x \in X$, $g(x,y_0) \le \lambda$ *or there exists* $y_0 \in X$ *such that* $f(y_0,y_0) > \lambda$.

PROOF. Let $G(x) = \{y \in X : g(x,y) \le \lambda\}$ and $F(x) = \{y \in X : f(x,y) \le \lambda\}$. From (i) we have $F(x) \subset G(x)$. Now if there is an $x_0 \in X$ such that $x_0 \notin F(x_0)$ then $f(x_0,x_0) > \lambda$ and we have the second part of the alternative. Now assume that $\forall\ x \in X$, $x \in F(x)$. Let A be a finite subset of X and $y \in X$ such that $A \cap F^{-1}(y) = \emptyset$, then $\forall\ a \in A$, $f(a,y) > \lambda$, now, if $z \in F^{-1}(y)$ we have $f(z,y) \le \lambda$, therefore if $A \cap F^{-1}(y) = \emptyset$ one has: $\forall\ z \in F^{-1}(y)$ $\forall\ a \in A$, $f(z,y) < f(a,y)$. A being finite, by (iii), one can find $w \in X$ such that $\forall\ a \in A$ $s(a,w) < s(z,w)$. Choose $r_0 \in \mathbb{R}$ such that $s(a,w) \le r_0 < s(z,w)$ for each $a \in A$, then $A \subset S(w,r_0)$ and $z \notin S(w,r_0)$. Let $F_A = \bigcap_{X\times\mathbb{R}} \{S(y,r) : A \subset S(y,r)\}$.

F_A is not empty since $A \subset F_A$ and furthermore F_A is contractible. The previous argument shows that if $A \cap F^{-1}(y) = \emptyset$ then $F_A \cap F^{-1}(y) = \emptyset$ but this is equivalent to saying that if $A \subset F^*(y)$ then $F_A \subset F^*(y)$. From theorem 2 we conclude that $\bigcap_{x\in X} G(x) \neq \emptyset$. If $x_0 \in \bigcap_{x\in X} G(x)$ then $\forall\ y \in X$, $g(x_0,y) \le \lambda$. This concludes the proof. □

Corollary 1. *With the hypothesis from proposition 1 one has*

$$\inf_{y\in X} \sup_{x\in X} g(x,y) \le \sup_{x\in X} f(x,x).$$

PROOF. Take $\lambda = \sup_{x\in X} f(x,x)$, ether $\lambda = \infty$ in which case the inequality is obvious or λ is finite in which case one apply proposition 1.□

Our next result is a generalization of Ky Fan's inequality [5].

Proposition 2. *Let* $\phi : X \times X \to \mathbb{R}$ *be a continuous function having the following property: If* $\phi(a_i,y) < \phi(z,y)$, $i = 1,\dots,k$, *then one can find* $w \in X$ *such that* $s(a_i,w) < s(z,w)$, $i = 1,\dots,k$. *Then* $\exists\ x_0 \in X$ *such that* $\forall\ x \in X$, $\phi(x_0,x_0) \le \phi(x,x_0)$.

PROOF. The result will follow from taking in proposition 1 $\lambda = 0$ and $g(x,y) = f(x,y) = \phi(y,y) - \phi(x,y)$. □

Corollary 2. *Let* $\phi : X \times X \to \mathbb{R}$ *be a continuous function such that:* $\forall\ \varepsilon > 0$, $\forall\ k \in \mathbb{N}$, $\forall\ (x_1,\dots,x_k,y) \in X^{k+1}$, $\exists\ w \in X$ *such that* $\phi(x_i,y) \le s(x_i,w) \le \phi(x_i,y) + \varepsilon$, $i = 1,\dots,k$. *Then* $\exists\ x_0 \in X$ *such that* $\forall\ x \in X$, $\phi(x_0,x_0) \le \phi(x,x_0)$.

PROOF. Let $a_1,\dots,a_{k-1},y,z$ be elements of X such that $\phi(a_i,y) < \phi(z,y)$, $i = 1,\dots,k-1$. Let $x_i = a_i$, $i = 1,\dots,k-1$ and $x_k = z$ and choose $\varepsilon > 0$ such that $\varepsilon < \min_i\{\phi(z,y) - \phi(a_i,y)\}$. Obviously we have $\phi(a_i,y) + \varepsilon < \phi(z,y)$, $i = 1,\dots,k-1$. Now, there is $w \in X$ such that $\phi(a_i,y) \le s(a_i,w) \le \phi(a_i,y) + \varepsilon$, $i = 1,\dots,k-1$ and $\phi(z,y) \le s(z,w) \le \phi(z,y) + \varepsilon$, from these inequalities we have $s(a_i,w) < s(z,w)$ for $i = 1,\dots,k-1$, and the conclusion follows from proposition 2. □

Corollary 3. *Let* $\phi : X \times X \to \mathbb{R}$ *be a continuous function such that:* $\forall \varepsilon > 0$, $\forall k \in \mathbb{N}$, $\forall (x_1,\dots,x_k,y) \in X^{k+1}$, $\exists w \in X$ *such that* $\phi(x_i,y) - \varepsilon \le s(x_i,w) \le \phi(x_i,y)$. *Then* $\exists x_0 \in X$ *such that* $\forall x \in X$, $\phi(x_0,x_0) \le \phi(x,x_0)$.

PROOF. It is similar to that of the previous corollary. If a_i, z,y, $i = 1,\dots,k-1$ are such that $\phi(a_i,y) < \phi(z,y)$, $i = 1,\dots,k-1$ then choose $\varepsilon > 0$ and $w \in X$ such that: $\phi(a_i,y) < \phi(z,y) - \varepsilon$ and $\phi(a_i,y) - \varepsilon \le s(a_i,w) \le \phi(a_i,y)$, $\phi(z,y) - \varepsilon \le s(z,w) \le \phi(z,y)$. It then follows that $s(a_i,w) \le s(z,w)$, $i = 1,\dots,k-1$ and proposition 2 can be used.□

The previous results assume a more elegant form if the function s is replaced by a function of a single variable.

Proposition 3. *Let* $a : X \to \mathbb{R}$ *and* $f,g : X \times X \to \mathbb{R}$ *be 3 functions such that:*

(i) $\forall r \in \mathbb{R}$: $\{y \in X : a(y) \le r\}$ *is empty or contractible,*

(ii) $\forall (x,y) \in X \times X : g(x,y) \le f(x,y)$,

(iii) $\forall x \in X : y \to g(x,y)$ *is lower semicontinuous,*

(iv) $f(z,y) < f(x,y) \Rightarrow a(x) < a(z)$.

Then $\forall \lambda \in \mathbb{R}$, $\exists y_0 \in X$ *such that* $\forall x \in X$, $g(x,y_0) \le \lambda$ *or* $\exists y_0 \in X$ *such that* $f(y_0,y_0) > \lambda$.

PROOF. Taking $s(x,y) = a(x)$ we have $S(x,r) = \{y \in X : a(y) \le r\}$ and $\bigcap_{i\in I} S(x_i,r_i) = \{y \in X : a(y) \le \bar{r}\}$ where $\bar{r} = \inf_{i\in I} r_i$ and proposition 1 can be applied. □

Corollary 4. *With the hypothesis of proposition 3 we have*

$$\inf_{y\in X} \sup_{x\in X} g(x,y) \le \sup_{x\in X} f(x,x).$$

Proposition 4. *Let* $a : X \to \mathbb{R}$ *and* $\phi : X \times X \to \mathbb{R}$ *be two functions such that:*

(i) $\forall r \in \mathbb{R}$: $\{y \in X : a(y) \le r\}$ *is either empty or contractible,*

(ii) ϕ *is continuous,*

(iii) $\phi(x,y) < \phi(z,y) \Rightarrow a(x) < a(z)$.

Then $\exists x_0 \in X$ *such that* $\forall x \in X$, $\phi(x_0,x_0) \le \phi(x,x_0)$.

PROOF. Use proposition 3 with $g(x,y) = f(x,y) = \phi(y,y) - \phi(x,y)$.□

Remarks. (a) If X is a compact convex subset of a topological vector space and if $f : X \times X \to \mathbb{R}$ is a quasi-concave function, (i.e. $\forall\ y \in X : \{x \in X : f(x,y) \geq r\}$ is convex) then in proposition 1 one can take $s(x,y) = -f(x,y)$ and we obtain Granas non-linear alternative, [1].

(b) If X is a compact convex subset of a Hilbert space and if $s(x,y) = \|x - y\|$ then the set F_A constructed in the proof of proposition 1 is the convex hull of A, this is not true any more in a Banach space, in that case F_A will also be convex but might be larger then the convex hull of A as can be seen by taking in $\mathbb{R}^2$ $\|x\| = \max\{|x_1|,|x_2|\}$. Now in a Hilbert space the convex hull of a finite set of points is the intersection of all the closed, (or open), balls containing these points and this is not true in Banach spaces, this leads us to the following question: Given a Banach space $(E,\|\cdot\|_1)$ is it possible to find a norm $\|\cdot\|_2$ equivalent to $\|\cdot\|_1$ such that the convex hull of any finite set of points is exactly the intersection of the closed balls containing these points ?

REFERENCES.

[1] Ben-El-Mechaiekh, Deguire P., Granas A., Pointes fixes et coïncidences pour les fonctions multivoques. (Application de type Φ et Φ^*). C. R. Acad. Sci. Paris, 292, Série I, (1982).

[2] Browder F.E., The fixed point theory of multi valued mappings in topological vector spaces. Math. Ann., 177 (1968).

[3] Dugundji J., Granas A., KKM-maps and variational inequalities. Ann. Scuola. Norm. Sup. Pisa (4) 5 (1978), 679-682.

[4] Fan K., A generalization of Tychonoff's fixed point theorem. Math. Ann. 142 (1961).

[5] Fan K., Sur un théorème Minimax. C. R. Acad. Sc. Paris 259 (1964).

[6] Fan K., A minimax inequality and applications. in O. Shisha ed. Inequalities III, Academic Press, 1972.

[7] Granas A., KKM-maps and their applications to nonlinear problems. The Scottish Book, (Mathematics from Scottish Café). (Edited by R.D. Mauldin), Birkhäuser, Boston 1982.

[8] Horvath C., Points fixes et coïncidences pour les applications multivoques sans convexité. C. R. Acad. Sci. Paris 296 (1983).

[9] Horvath C., Points fixes et coïncidences dans les espaces topologiques compacts contractibles. C. R. Acad. Sci. Paris 299 (1984).

[10] Knaster B., Kuratowski K., Mazurkiewicz S., Ein Beweis des Fixpunktsatzes für n-Dimensionale Simplexe. Fund. Math. 14 (1929).

[11] McClendon J.F., Minimax and variational inequalities for compact spaces. Proceedings of the A. M. S., V. 89 (4) (1983).

[12] Yen C.L., A minimax inequality and its applications to variational inequalities. Pacific J. Math. 97 (1981).

Strong Equilibria

TATSURO ICHIISHI* Department of Economics, The University of Iowa, Iowa City, Iowa

1. INTRODUCTION

Given a finite set of players N, a game in normal form is defined as a list of specified data $G := \{X^j, u^j\}_{j\in N}$, where X^j is the set of pure strategies and $u^j: \prod_{i\in N} X^i \to \mathbf{R}$ is a utility function of player j. This paper surveys several results on various types of cooperative behavior in game G.

Cooperative behavior is most clearly described within the setup of a game in characteristic function form without side-payments (or simply, a non-side-payment game), defined as a correspondence (set-valued map) $V: \mathcal{N} \to \mathbf{R}^N$, where $\mathcal{N}$ is the family of nonempty coalitions $2^N\setminus\{\phi\}$. Here, for each $S \in \mathcal{N}$, the subset V(S) of $\mathbf{R}^N$ (or rather, its projection to $\mathbf{R}^S$) specifies the set of utility allocations attainable by coalition S. A typical cooperative solution concept for game V is the core C(V), the set of all utility allocations that are feasible (via the grand coalition N) and stable (in the sense that no coalition can improve upon them).

Three types of cooperative behavior in a normal-form game G are summarized by three non-side-payment games constructed from G, so one can associate with G three core-like solution concepts. One type is based on the pessimistic view held by each coalition S as to achieving a utility allocation for the members of S in relation to strategy-choice of the outsiders N\S; it is summarized by the α-characteristic function V_α^p, and the associated solution concept for G is the α-core $C_\alpha^p(G)$. (The superscript p stands for pure strategies.) Another type is based on the optimistic view, summarized by the β-characteristic

*Current affiliation: Ohio State University, Columbus, Ohio

function V^p_β, and the associated solution concept is the β-core $C^p_\beta(G)$. The third type is based on the passive view, and the associated solution concept is the strong equilibrium: denote by $SE^p(G)$ the set of all strong equilibrium utility allocations. Note that a strong equilibrium is a Pareto optimal Nash equilibrium.

The first result reported here is a theorem for nonemptiness of $SE^p(G)$. Ky Fan's (1969, 1972) coincidence theorem, and the works of Scarf (1967) and Shapely (1973) were essential for establishing this theorem.

There are situations in which player j is allowed to use a mixed strategy, a probability on X^j. The coalitional analogue of a mixed strategy is a correlated strategy for coalition S, a probability on $X^S := \prod_{j \in S} X^j$. The product measure of mixed strategies is a correlated strategy, but the converse is not true (unless $\#S = 1$). One may consider cooperative behavior based on the pessimistic (optimistic, resp.) view when coalitions can use correlated strategies, and define the associated α-core $C^c_\alpha(G)$ (β-core $C^c_\beta(G)$, resp.).

Randomized strategies specify how correlated strategies are generated. A probability space (Ω, $\mathcal{A}$, p) is given a priori, and a randomized strategy of coalition S is defined as a measurable function $f^S: \Omega \to X^S$. Notice that its distribution $p \circ (f^S)^{-1}$ is a correlated strategy of S. The three core-like solution concepts when coalitions can use randomized strategies are denoted by $C^r_\alpha(G)$, $C^r_\beta(G)$, and $SE^r(G)$, respectively.

The second result reported here establishes relationships among the above eight core-like solution concepts under mild conditions; in particular, $SE^p(G)$ is shown to be the smallest set. As a corollary to the first and the second results, one obtains conditions under which $C^c_\beta(G)$ is nonempty.

A repeated game G* is constructed from a normal-form game G as the repetition of G over infinitely many periods. Roughly stated, the Aumann Proposition asserts $C^c_\beta(G) = SE^r(G^*)$.

The third result reported here establishes a version of the Aumann Proposition in a general setup which is consistent with the conditions for nonemptiness of $C^{c}_{\beta}(G)$ (as pointed out as a corollary to the first two results), so that this version is non-vacuous.

2. PRELIMINARIES

2.1. Nash Equilibrium of a Game in Normal Form. Let N be a finite set of players, fixed throughout this paper. A game in normal form is a list of specified data, $G := \{X^j, u^j\}_{j \in N}$, where X^j is a pure-strategy set of player j, and $u^j : \prod_{i \in N} X^i \to \mathbf{R}$ is a utility function of player j. Here, player i influences player j through j's utility function u^j, and vice versa. Set $X := \prod_{i \in N} X^i$.

A Nash equilibrium of game G is a strategy bundle $x^* := \{x^{j*}\}_{j \in N} \in X$ such that no single player has an incentive to change his strategy given the others' strategy choice; formally it is a strategy bundle x^* such that

$$\neg \; \exists \, j \in N : \quad \exists \, x^j \in X^j : \quad u^j(x^j, x^{N \setminus \{j\}*}) > u^j(x^*). \qquad (1)$$

It is a solution concept for the situations in which each player behaves noncooperatively and passively. Cournot [1838] proposed this equilibrium concept for $N = \{1, 2\}$ as an equilibrium concept for the duopoly, where X^j ($\subset \mathbf{R}_+$) is interpreted as the set of possible output levels and u^j is the profit function for firm $j = 1, 2$. Nash [1950] proposed this concept (almost) in its full generality as an extension of von Neumann's [1928] minimax concept. Indeed, for a zero-sum two-person game ($N = \{1, 2\}$, $u^1 + u^2 = 0$), the Nash equilibria are precisely the optimal solutions for the minimax equality.

A Nash equilibrium can be re-formulated in terms of the following notion of reaction correspondence (set-valued map) $\Phi: X \to X$. Define the individual reaction correspondence $\Phi^j: X \to X^j$ by

$$\Phi^j(x) := \{\xi^{j*} \in X^j \mid \forall \xi^j \in X^j : u^j(\xi^{j*}, x^{N\setminus\{j\}}) \geq u^j(\xi^j, x^{N\setminus\{j\}})\},$$

and set $\Phi(x) := \prod_{j\in N} \Phi^j(x)$. Then, the Nash equilibria are the fixed-points of the correspondence Φ. A Nash equilibrium existence theorem can be proved, therefore, by Kakutani's [1941] fixed-point theorem when X^j is a compact convex subset of a Euclidean space (a technique pioneered by Nash [1950]), and by the Fan-Glicksberg fixed-point theorem (Fan [1952], Glicksberg [1952]) when X^j is a compact convex subset of a Hausdorff, locally convex topological vector space. See, e.g., Khan [1985] for a survey of the recent existence literature on the Nash equilibrium and modified or generalized Nash equilibria for the second case (i.e., in the context of locally convex topological vector spaces). Indeed, there are several generalized game-theoretical models and the corresponding generalized Nash equilibrium concepts. Prakash and Sertel [1974] developed one extension; their logic for the existence theorem goes beyond fixed-point theorems for topological vector spaces.

2.2. Core of a Non-Side-Payment Game. Given a finite set of players N, and hence given the family of nonempty coalitions $\mathcal{N} := 2^N\setminus\{\phi\}$, a game in characteristic function form without side-payments (or simply, a non-side-payment game) is defined as a correspondence (set-valued map), $V\colon \mathcal{N} \to \mathbf{R}^N$, such that for every $S \in \mathcal{N}$, $V(S)$ is a cylindar based on a subset of $\mathbf{R}^S$, viz., $[u, v \in \mathbf{R}^N.\ \forall j \in S\colon u_j = v_j]$ implies $[u \in V(S)$ if and only if $v \in V(S)]$. Here, $V(S)$ (or rather, its projection to $\mathbf{R}^S$) is interpreted as the set of utility allocations attainable in coalition S.

The core of a non-side-payment game V is the set $C(V)$ of all utility allocations $u^* \in V(N)$ such that no coalition can improve upon them; formally it is the set of all $u^* \in V(N)$ such that

$$\neg \exists S \in \mathcal{N} : \exists u \in V(S): \forall j \in S: u_j > u_j^*. \qquad (2)$$

It is a cooperative solution concept which characterizes feasibility (via the grand coalition N) and coalitional stability. (Compare (2) with (1).) Given a general equilibrium model of pure exchange $\mathcal{E}$, one can define the non-side-payment game associated with $\mathcal{E}$ (see, e.g., Scarf [1967, pp. 51-55]). In this economic setup, Edgeworth [1881] proposed the core concept. A non-side-payment game V is called a <u>side-payment game</u>, if there exists a function $v : \mathcal{N} \to \mathbf{R}$ such that $V(S) = \{u \in \mathbf{R}^N | \sum_{i \in S} u_i \leq v(S)\}$. The core of a side-payment game was discovered by Gillies [1959] and Shapley during their investigation of the von Neumann-Morgenstern solution in early 1950's. The present model of non-side-payment game and the definition of core are due to Aumann and Peleg [1960]. Peleg [1985] established a characterization of the core correspondence, $V \mapsto C(V)$.

A subfamily $\mathcal{B}$ of $\mathcal{N}$ is called <u>balanced</u> if there exist associated balancing coefficients, i.e., if there exists a set $\{\lambda_S | S \in \mathcal{B}\}$ of nonnegative real numbers such that

$$\forall j \in N: \quad \sum_{S \in \mathcal{B} : S \ni j} \lambda_S = 1.$$

A non-side-payment game V is called <u>balanced</u>, if for every balanced subfamily $\mathcal{B}$ of $\mathcal{N}$,

$$\bigcap_{S \in \mathcal{B}} V(S) \subset V(N).$$

If a given pure exchange economy $\mathcal{E}$ satisfies the standard assumptions (like convexity of the consumption sets and quasi-concavity of the utility functions), the non-side-payment game associated with $\mathcal{E}$ is proved to be balanced.

Using an elaborate path-following technique of Lemke and Howson [1964], Scarf [1967] established that a balanced non-side-payment game satisfying mild regularity conditions has a nonempty core. For alternative proofs of the Scarf theorem, see Shapley [1973] and Ichiishi [1983, pp. 82-84] (the latter simplied the major step of the former by applying Ky Fan's [1969, 1972] coincidence theorem).

3. CORE-LIKE SOLUTIONS IN PURE STRATEGIES

3.1. Strong Equilibria in Pure Strategies. Given a game in normal form $G := \{X^j, u^j\}_{j \in N}$, construct (parameterized) non-side-payment games, V_x, $x \in X$, by

$$V_x(S) := \bigcup_{\xi^S \in X^S} \{u \in \mathbf{R}^N | \forall\, j \in S: \;\; u_j \leq u^j(\xi^S, x^{N \setminus S})\}.$$

Here, a coalition forms based on the passive cooperative behavior of its members; members of coalition S jointly choose ξ^S, taking the outsiders' strategies $x^{N \setminus S}$ as given. One can consider the core of game V_x, but a strategy bundle which gives rise to a member of the core may not be the same as the given parameter x. A strong equilibrium in pure strategies of game G is a strategy bundle $x^* \in X$ such that $\{u^j(x^*)\}_{j \in N}$ is in the core of V_{x^*}, or equivalently such that

$$\neg \exists\, S \in \mathcal{N} : \exists\, \xi^S \in X^S: \;\; \forall\, j \in S: \quad u^j(\xi^S, x^{N \setminus S *}) > u^j(x^*). \tag{3}$$

Compare (3) with (1) and (2). A strong equilibrium concept was proposed by Aumann [1959]; it is a Pareto optimal Nash equilibrium. Denote by $SE^p(G)$ the set of all strong equilibrium utility allocations of G with pure strategies.

A strong equilibrium is a fixed-point of the core-strategy correspondence,

$$x \mapsto \{x' \in X \mid \{u^j(x')\}_{j \in N} \in C(V_x)\}.$$

Since the core C(V) is in general a disconnected set, however, one cannot apply any fixed-point theorem to this correspondence in order to establish a theorem for nonemptiness of $SE^p(G)$. See Ichiishi [1983, Corollary 5.8.2, p. 101] for the following Theorem 1. (Actually, Theorem 1 is a special case of the main result of Ichiishi [1981]. An exposition of Theorem 1 in the finite dimensional setup is given in Ichiishi [1983].) Again, its proof ultimately boils down to Shapley's [1973] contribution and Ky Fan's [1969, 1972] coincidence theorem. See also contributions of Border [1984].

THEOREM 1. *Let* $G := \{X^j, u^j\}_{j \in N}$ *be a game in normal form. The set* $SE^p(G)$ *is nonempty, if*

(i') X^j *is a nonempty, convex, compact subset of a Hausdorff, locally convex topological vector space, for every* j;

(ii') u^j *is continuous and quasi-concave in* X, *for every* j; *and*

(iii) *the non-side-payment game* V_x *is balanced, for every* $x \in X$.

3.2. *α-Core and β-Core in Pure Strategies.* Let $G := \{X^j, u^j\}_{j \in N}$ be a game in normal form. The *α-characteristic function* of game G is a correspondence, $V^p_\alpha : \mathcal{N} \to \mathbf{R}^N$, defined by

$$V^p_\alpha(S) := \bigcup_{x^S \in X^S} \; \bigcap_{y^{N \setminus S} \in X^{N \setminus S}} \{u \in \mathbf{R}^N \mid \forall j \in S: \; u_j \leq u^j(x^S, y^{N \setminus S})\}.$$

According to the non-side-payment game V^p_α, utility allocation u is attainable in coalition S if the members of S find one of their strategy bundles such that as long as they hang on to it, every member j in S can enjoy a utility level no less than u_j regardless of strategy choice of the outsiders. The α-characteristic function is, therefore, convenient to analyze situations in which a coalition is formed based on the *prudent* or pessimistic cooperative behavior of its members. The α-core of G with pure strategies is the core of game V^p_α, and is denoted by $C^p_\alpha(G)$.

The β-characteristic function of game G is a correspondence, V^p_β: $\mathcal{N} \to \mathbf{R}^N$, defined by

$$V^p_\beta(S) := \bigcap_{y^{N\setminus S} \in X^{N\setminus S}} \bigcup_{x^S \in X^S} \{u \in \mathbf{R}^N | \forall j \in S: u_j \leq u^j(x^S, y^{N\setminus S})\}.$$

According to the non-side-payment game V^p_β, utility allocation u is attainable in coalition S if for any strategy choice of the outsiders, the members of S can counteract it so as to make member j's utility level to be no less than u_j for every $j \in S$. The β-characteristic function is, therefore, convenient to analyze situations in which a coalition is formed based on the optimistic cooperative behavior of its members. The β-core of G with pure strategies is the core of game V^p_β, and is denoted by $C^p_\beta(G)$.

Notice that there is a revival of the classical concepts of "min" and "max" (in von Neumann's [1928] minimax theorem) by means of the operations " $\cap$ " and " $\cup$ ".

PROPOSITION. Let G be a game in normal form. Then,

$$SE^p(G) \subset C^p_\beta(G) \subset C^p_\alpha(G).$$

Aumann [1959] introduced the β-core concept; he called it the acceptable payoff vectors at that time. The terms α-core and β-core are due to Aumann and Peleg [1960]. The α-concepts and β-concepts were proposed independently also by Jentzsch [1964]. In the above Proposition, the larger a set, the more likely that it is nonempty. Indeed, Scarf [1971] established that for any normal-form game which satisfies Assumptions (i') and (ii') of Theorem 1, its α-characteristic function is balanced, so its α-core is nonempty.

4. GAME WITH CORRELATED STRATEGIES

Given a topological space Y, denote by $\mathcal{M}(Y)$ the space of all probability measures defined on the σ-algebra of Borel subsets of Y.

Given a game G with topological spaces of pure strategies and continuous utility functions, a game with correlated strategies is now constructed. A mixed strategy of player j is a member of $\mathcal{M}(X^j)$. A correlated strategy of coalition S is a member of $\mathcal{M}(X^S)$. One can imbed the mixed strategy bundles $\prod_{j\varepsilon S} \mathcal{M}(X^j)$ into the correlated strategy bundles $\mathcal{M}(X^S)$ naturally by a function, $(\mu^j)_{j\varepsilon S} \mapsto \otimes_{j\varepsilon S} \mu^j$, but this function is not onto (unless #S = 1). When correlated strategies $\mu \varepsilon \mathcal{M}(X^S)$ and $\nu \varepsilon \mathcal{M}(X^{N\setminus S})$ are chosen, the expected utility of player j is given by

$$U^j(\mu\otimes\nu) := \int_X u^j(x)(\mu\otimes\nu)(dx).$$

The game with correlated strategies associated with G is the list $(\{\mathcal{M}(X^S)\}_{S\varepsilon\mathcal{N}}, \{U^j\}_{j\varepsilon N})$. Two types of non-side-payment games are readily constructed.

The α-characteristic function of game $(\{\mathcal{M}(X^S)\}_{S\varepsilon\mathcal{N}}, \{U^j\}_{j\varepsilon N})$ is a non-side-payment game, V_α^c: $\mathcal{N} \to \mathbf{R}^N$, defined by

$$V_\alpha^c(S) := \bigcup_{\mu\varepsilon\mathcal{M}(X^S)} \bigcap_{\nu\varepsilon\mathcal{M}(X^{N\setminus S})} \{u \varepsilon \mathbf{R}^N | \forall\ j \varepsilon S: \ u_j \leq U^j(\mu\otimes\nu)\}.$$

The α-core of G with correlated strategies is the core of game V_α^c, and is denoted by $C_\alpha^c(G)$.

The β-characteristic function of game $(\{\mathcal{M}(X^S)\}_{S\varepsilon\mathcal{N}}, \{U^j\}_{j\varepsilon N})$ is a non-side-payment game, V_β^c: $\mathcal{N} \to \mathbf{R}^N$, defined by

$$V_\beta^c(S) := \bigcap_{\nu\varepsilon\mathcal{M}(X^{N\setminus S})} \bigcup_{\mu\varepsilon\mathcal{M}(X^S)} \{u \varepsilon \mathbf{R}^N | \forall\ j \varepsilon S: \ u_j \leq U^j(\mu\otimes\nu)\}.$$

The β-core of G with correlated strategies is the core of game V_β^c, and is denoted by $C_\beta^c(G)$.

It is also straightforward that $C_\beta^c(G) \subset C_\alpha^c(G)$.

It should be pointed out that a strong equilibrium cannot be defined here without an additional structure. To see this, suppose that the grand coalition N has chosen a correlated strategy $\nu \in \mathcal{M}(X)$, and coalition S is going to form in spite of that. In accordance with the behavioral principle underlying the strong equilibrium concept, members of S passively take the outsiders' strategies as given. But what is the part of ν that players N\S alone are responsible for? The required additional structure (e.g., chance devices which correlate pure strategies--see Section 5) clarifies the role of players N\S in correlated strategy ν.

5. GAME WITH RANDOMIZED STRATEGIES

Given a probability space $(\Omega, \mathcal{A}, p)$ and a Hausdorff locally convex topological vector space E, the Pettis integrability of a weakly p-measurable function $f : \Omega \to E$ is well-defined as in Diestel and Uhl [1977, Definition 2, pp. 52-53]. Let X be a subset of E, and denote by $L_1(p; X)$ the set of all (equivalence classes modulo p-null sets of) Pettis integrable functions $f : \Omega \to E$ such that $f(\omega) \in X$, p-a.e. in Ω. Given this setup the following Jensen's inequality holds true: Let X be a compact, convex subset of E, u be a continuous, concave function from X to $\mathbf{R}$, and $f \in L_1(p; X)$. Then:

$$\int_\Omega u(f(\omega))p(d\omega) \leq u\left(\int_\Omega f(\omega)p(d\omega)\right).$$

Notice that under the same assumptions the numerical function $u \circ f$ is integrable.

To model a chance device, let $(\Omega_S, \mathcal{A}_S, p_S)$ be a probability space, $S \in \mathcal{N}$, and denote by $(\Omega, \mathcal{A}, p)$ the product space of $\{(\Omega_S, \mathcal{A}_S, p_S) \mid S \in \mathcal{N}\}$. By abuse of notation denote the family $\{[\mathrm{proj}_S]^{-1}(A) \mid A \in \mathcal{A}_S\}$ also by $\mathcal{A}_S$, where $\mathrm{proj}_S: \Omega \to \Omega_S$ is the natural projection. The restriction of p to $\mathcal{A}_S$ (as a sub-σ-algebra of $\mathcal{A}$) is denoted by p_S. No confusion arises by identifying $(\Omega, \mathcal{A}_S, p_S)$ and $(\Omega_S, \mathcal{A}_S, p_S)$. The set Ω is the <u>states of the world</u>, and the σ-algebra $\mathcal{A}_S$ is the <u>randomizability structure</u> of coalition S.

Given a game with pure strategies G, for which each strategy space X^j is a compact, convex subset of a Hausdorff locally convex topological vector space and each utility function u^j is continuous in X, a game with randomized strategies is now constructed. A <u>randomized strategy</u> of player j is a member of $L_1(p; X^j)$. The product of the strategy spaces, $\prod_{j\in N} L_1(p; X^j)$, is identified with the space $L_1(p; X)$. When a randomized strategy bundle $f \in L_1(p; X)$ is chosen, the <u>expected utility</u> of player j is given by

$$Eu^j(f) := \int_\Omega u^j(f(\omega))\, p(d\omega).$$

Each coalition S actually has access only to its own chance device $(\Omega_S, \mathcal{A}_S, p_S)$. The <u>feasible</u> randomized strategy space of coalition S is, therefore, $L_1(p_S; X^S)$. [One might use $L_1(\otimes_{T\subset S} p_T; X^S)$ as such a space, which will allow for departmentalization within coalition S. All the following arguments are applicable to this modified case.] The <u>game with randomized strategies associated with</u> G is the list $(\{L_1(p; X^j), Eu^j\}_{j\in N}, \{L_1(p_S; X^S)\}_{S\in\mathcal{N}})$. The concept of randomized strategy is due to Aumann [1974].

Given a randomized strategy $f^S \in L_1(p_S; X^S)$, its distribution $p_S\circ[f^S]^{-1}$ is a correlated strategy on X^S. Notice that if p_S concentrates only on a finite subset D of Ω_S (i.e., $p_S(D) = 1$), then one can obtain only correlated strategies each of whose support has a cardinality no greater than that of D. In order to obtain a rich family of correlated strategies, the probability space $(\Omega_S, \mathcal{A}_S, p_S)$ is often chosen to be non-atomic.

The α-<u>characteristic function</u> and the β-<u>characteristic function</u> of the game with randomized strategies are given by

$$V^r_\alpha(S) := \bigcup_{f^S} \bigcap_{g^{N\setminus S}} \{u \in \mathbf{R}^N | \forall\ j \in S:\ u_j \leq Eu_j(f^S, g^{N\setminus S})\}, \text{ and}$$

$$V^r_\beta(S) := \bigcap_{g^{N\setminus S}} \bigcup_{f^S} \{u \in \mathbf{R}^N | \forall\ j \in S:\ u_j \leq Eu^j(f^S, g^{N\setminus S})\},$$

respectively, where f^S runs through $L_1(p_S;\ X^S)$, and $g^{N\setminus S}$ runs through $L_1(p_{N\setminus S};\ X^{N\setminus S})$. The α-core of G with randomized strategies is the core of game V^r_α and is denoted by $C^r_\alpha(G)$. The β-core of G with randomized strategies is the core of game V^r_β, and is denoted by $C^r_\beta(G)$.

Given any $f \in L_1(p;\ X)$, a non-side-payment game V^r_f is defined by

$$V^r_f(S) := \bigcup_{\xi^S} \{u \in \mathbf{R}^N | \forall\ j \in S:\ u_j \leq Eu^j(\xi^S, f^{N\setminus S})\},$$

where ξ^S runs through $L_1(p_S;\ X^S)$. A strong equilibrium of G with randomized strategies is a randomized strategy bundle $f^* \in L_1(p_N;\ X)$ such that $\{Eu^j(f^*)_{j\in N}\}$ is a point in the core of game $V^r_{f^*}$. Define:

$$SE^r(G) := \{\{Eu^j(f^*)\}_{j\in N} \in \mathbf{R}^N | f^* \text{ is a strong equilibrium of G with randomized strategies}\}.$$

The argument made in Ichiishi [1985] for the finite dimensional setup can easily be generalized to establish:

THEOREM 2. Let $G := \{X^j, u^j\}_{j\in N}$ be a game in normal form. Assume

(i") X^j is a convex, compact, metrizable subset of a Hausdorff locally convex topological vector space, for every j;

(ii") u^j is continuous and concave in X, for every j; and

(iv) For each coalition S, its chance device is a non-atomic probability space.

Then the following inclusions hold true:

$$\begin{array}{ccccc}
SE^{p}(G) & \subset & C_{\beta}^{p}(G) & \subset & C_{\alpha}^{p}(G) \\
\cap & & \cap & & \| \\
SE^{r}(G) & & C_{\beta}^{r}(G) & \subset & C_{\alpha}^{r}(G) \\
 & \subset & \cap & & \| \\
 & & C_{\beta}^{c}(G) & \subset & C_{\alpha}^{c}(G)
\end{array}$$

The inclusions, $SE^p(G) \subset SE^r(G)$, $C_\beta^p(G) \subset C_\beta^r(G)$, $C_\alpha^p(G) \subset C_\alpha^r(G)$, are consequences of Jensen's inequality. The inclusion $C_\alpha^r(G) \subset C_\alpha^p(G)$ is proved by the compactness argument. The inclusions, $SE^r(G) \subset C_\beta^c(G)$, $C_\beta^r(G) \subset C_\beta^c(G)$, $C_\alpha^r(G) \subset C_\alpha^c(G)$ may be proved by the w*-denseness of the probabilities with a finite support and Liapunov's theorem. It remains to show $C_\alpha^c(G) \subset C_\alpha^r(G)$ for at least one chance device, e.g., for the unit interval with the Lebesgue measure as $(\Omega_S, \mathcal{A}_S, p_S)$, which is true due to Skorokhod [1965, Lemma, p. 10]. It should be emphasized that a randomized strategy for S and a randomized strategy for T, $S \neq T$, are p-independent due to the fact that $(\Omega, \mathcal{A}, p)$ is the product space of $\{(\Omega_S, \mathcal{A}_S, p_S) | S \in \mathcal{N}\}$, and that this p-independence is crucial for establishing Theorem 2. Here, $C_\beta^r(G) = C_\beta^c(G)$ for the standard case. Each of the other inclusions (unless represented by "=") may be strict.

COROLLARY. *Let* G *be a game in normal form with nonempty pure strategy spaces*. *The* β-*core with correlated strategies*, $C_\beta^c(G)$, *is nonempty if game* G *satisfies Assumptions (i"), (ii"), and (iii)*.

6. REPEATED GAME WITH RANDOMIZED STRATEGIES

Given a game in normal form (or its variant) $G := \{X^j, u^j\}_{j \in N}$, the associated repeated game G* is defined as repetition of G over infinitely many periods, where each player j chooses his repeated-game-strategy $f_t^j(x_1, \ldots, x_{t-1})$ of period t conditioned on realized strategy bundles of periods up to t-1, $(x_1, \ldots, x_{t-1})$; and where the repeated-game-utility of player j is given by

$$u^{j*}(f) := \underline{\lim}_{T\to\infty} \frac{1}{T} \sum_{t=1}^{T} u^j(\sigma_t(f)),$$

i.e., the Cèsaro limit of the averages of the utility stream made by the realized strategy sequence $\{\sigma_t(f)\}_{t=1}^{\infty}$ when each player i chooses his repeated-game-strategy $f^i = \{f_1^i, f_2^i, \ldots, f_t^i, \ldots\}$, $i \in N$ ($\sigma_t(f) \in X$ is defined inductively by $\sigma_t(f) := f_t(\sigma_1(f), \ldots, \sigma_{t-1}(f))$). In contrast with this repeated game G*, the original game G is sometimes called a one-shot game. Player j's repeated-game-strategy of period t amounts to his statement at the outset of the repeated game that he will <u>react</u> in period t to the preceding history $(x_1, \ldots, x_{t-1})$ by choosing $f_t^j(x_1, \ldots, x_{t-1})$. This statement can serve sometimes as a commitment, and sometimes as a threat to the other players if the others are trying to realize the history $(x_1, \ldots, x_{t-1})$. This committing/threatening nature of the repeated-game-strategies is a basic logic underlying the Folk Theorem (whose authorship is not known) in game theory: the set of feasible, individually rational utility allocations of a one-shot game G is precisely the set of Nash equilibrium utility allocations of the associated repeated game G*. One can thus explain cooperative phenomena in a static game as a noncooperative equilibrium outcome of the associated dynamic game. Aumann [1959] introduced the strong equilibrium concept, and asserted the Aumann Proposition that the β-core of a one-shot game <u>with correlated strategies</u> is characterized as the set of strong equilibrium utility allocations of the associated repeated game -- a coalitional analogue of the Folk Theorem. He analyzed correlated strategies in the repeated game, and the paper is consequently difficult to read.

Subsequent analyses on (variants of) the Aumann Proposition were made within the framework of the repeated game <u>with pure strategies</u> by Yanovskaya [1971/72] and Rubinstein [1980]. Yanovskaya's strong equilibrium concept is based on (group) preference relations that are quite different from the type of preference relations used in Aumann [1959]. As a result, the set of outcomes of G she characterized there is not the β-core with correlated strategies, but rather a certain set which under mild assumptions turns out to be the β-core of G with pure strategies. Rubinstein's strong equilibrium concept is based on the same type of preference relations (at least in spirit) as in Aumann [1959]. Due to his restriction to the pure strategies in the repeated game, however, Rubinstein actually characterized (what he called) the desired payoffs of a one-shot game.[1] Roughly stated, a desired payoff is a solution concept based on the players' optimistic behavioral principle such that while the members of S choose a correlated strategy, they believe that the outsiders choose only pure strategies.

The Aumann Proposition within the framework of the repeated game with probabilistic choice of strategies was re-considered by Aumann [1976]. There, he proposed new concepts of lower strong equilibrium and upper strong equilibrium of the repeated game with randomized strategies, and studied their relationships with the β-core of the one-shot game with correlated strategies. When randomized strategies are introduced, the realized utilities become random variables. The difference between lower and upper strong equilibria lies in the criteria based on which coalitions may form. According to the former, coalition S will form if each member of S can improve upon the given utility level almost surely. According to the latter, coalition S will form if there exists an event with positive probability on which each member of S can improve upon the given level. In Aumann's [1976] work, each pure-strategy set of the one-shot game was assumed to be finite (therefore, non-convex).

The use of randomized strategies in a repeated game has advantage over the use of correlated strategies in that: (1)

---

[1]Rubinstein [1980] characterized the desired payoffs of a one-shot game also as the strong perfect equilibrium utility allocations of the associated repeated game.

strong equilibrium concepts are well-defined in the former; and (2) one can pin down how the grand coalition identifies the defectors.[2] Once there are differences in the information structures (σ-algebras on Ω) available to coalitions, to incorporate in a model the distinction between what players know about the others' behavior and what players can observe about the others' behavior becomes a formidable task.

While the β-core of a two-person game with correlated strategies is always nonempty, one has to impose certain conditions in order to assert its nonemptiness for a general n-person game. The only available general theorem for nonemptiness of the β-core with correlated strategies is the Corollary (Section 5 of this paper), which assumes, among others, convexity of the pure-strategy sets X^j. Thus, for the Aumann Proposition to be non-vacuous one has to generalize it to an infinite pure-strategy set. Ichiishi [1984] extended Aumann's [1976] argument to the situation in which the problem about differences in information structures is settled and in which pure-strategy sets are merely assumed to be compact metric spaces. The main result there is:

THEOREM 3. *Let* G *be a game in normal form. Assume*:
(i) X^j *is a compact subset of a metric space, for every* j;
(ii) u^j *is continuous in* X, *for every* j;

---

[2]Suppose player j commits to the others that he will choose strategy x (strategy y, resp.) with probability 1/100 (with probability (99/100, resp.) in period t. This is a correlated strategy. If he actually chooses x in period t, and if he tells the others that the event of probability of 1/100 has somehow occurred, nobody can say whether player j has fulfilled his commitment. Suppose everybody can observe event A which occurs with probability 1/100, and suppose player j commits to the others that he will choose strategy x (strategy y, resp.) if event A occurs (if event A^c occurs, resp.) in period t. This is a randomized strategy. Since the others can observe whether A occurs or not in period t, they can also detect whether player j fulfilled his commitment.

(v) For each player j and period t, probability space (Ω_t, $\mathcal{A}_t^j$, p_t) is non-atomic, and $\mathcal{A}_t^S$ and $\mathcal{A}_t^T$ ($S \cap T = \phi$) are p_t-independent.

Choose any $u^* \in \mathbf{R}^N$. Then the following three statements are equivalent:

(a) $u^* \in C_\beta^c(G)$;

(b) u^* is an upper strong equilibrium utility allocation of the repeated game with randomized strategies associated with G;

(c) u^* is a lower strong equilibrium utility allocation of the repeated game with randomized strategies associated with G.

The core-like solutions can be refined once one introduces "credibility" of threats by others; this concept is the most clearly formulated in the context of sequential games. Chakrabarti [1985] re-considered the Aumann Proposition, taking explictly into account his "credibility" concept.

REFERENCES

Aumann, R. J., "Acceptable Points in General Cooperative n-Person Games," Paper 16 in: A. W. Tucker and R. D. Luce, eds., Contributions to the Theory of Games, Vol. IV, 287-324. Princeton: Princeton University Press, 1959.

Aumann, R. J., "Subjectivity and Correlation in Randomized Strategies," Journal of Mathematical Economics, **1**, 67-96, 1974.

Aumann, R. J., Lectures in Game Theory, Lecture Notes, Stanford University, 1976.

Aumann, R. J., and Peleg, B., "Von Neumann-Morgenstern Solutions to Cooperative Games Without Side Payments," Bulletin of the American Mathematical Society, **66**, 173-179, 1960.

Border, K., "A Core Existence Theorem for Games Without Ordered Preferences," Econometrica, **52**, 1537-1542, 1984.

Chakrabarti, S. K., "Refinements of the β-Core and the Strong Equilibrium and the Aumann Proposition," Working Paper No. 8502, Committee for Mathematical Analysis in the Social Sciences, The University of Iowa, March 1985.

Cournot, A. A., Recherches sur les principes mathématiques de la théorie des richesse, Paris: Libraire des Sciences politiques et sociales, M. Riviere & cie., 1838. (English translation: Researches into the Mathematical Principles of the Theory of Wealth, New York: Macmillan, 1897.)

Diestel, J., and Uhl, J.J., Jr., Vector Measures, Providence: American Mathematical Society, 1977.

Edgeworth, F. Y., Mathematical Psychics, London: Kagan Paul, 1881.

Fan, K., "Fixed-Point and Minimax Theorems in Locally Convex Topological Linear Spaces," Proceedings of the National Academy of Sciences of the U.S.A., **38**, 121-126, 1952.

Fan, K., "Extensions of Two Fixed Point Theorems of F. E. Browder," Mathematisches Zeitschrift, **112**, 234-240, 1969.

Fan, K., "A Minimax Inequality and Applications," in: O. Shisha, ed., Inequalities, III, 103-113. New York: Academic Press, 1972.

Gillies, D. B., "Solutions to General Non-Zero-Sum Games," Paper 3 in: A. W. Tucker and R. D. Luce, eds., Contributions to the Theory of Games, Vol. IV, 47-85. Princeton: Princeton University Press, 1959.

Glicksberg, I. L., "A Further Generalization of the Kakutani Fixed Point Theorem with Application to Nash Equilibrium Points," Proceedings of the American Mathematical Society, **3**, 170-174, 1952.

Ichiishi, T., "A Social Coalitional Equilibrium Existence Lemma," Econometrica, **49**, 369-377, 1981.

Ichiishi, T., Game Theory for Economic Analysis, New York: Academic Press, 1983.

Ichiishi, T., "Strong Equilibria of a Repeated Game with Randomized Strategies," Working Paper No. 8412, Committee for Mathematical Analysis in the Social Sciences, The University of Iowa, May 1984. (To appear in Mathematical Social Sciences)

Ichiishi, T., "Core-like Solutions for Games with Probabilistic Choice of Strategies," mimeo, February 1985. (A revised version to appear in Mathematical Social Sciences)

Jentzsch, G., "Some Thoughts on the Theory of Cooperative Games," Paper 20 in: M. Dresher, L.S. Shapley and A.W. Tucker, eds., Advances in Game Theory, 407-442. Princeton: Princeton University Press, 1964.

Kakutani, S., "A Generalization of Brouwer's Fixed-Point Theorem," Duke Mathematical Journal, **8**, 457-459, 1941.

Khan, M. A., "On Extensions of the Cournot-Nash Theorem," in: C.D. Aliprantis, O. Burkinshaw, and N. J. Rothman, eds., Advances in Equilibrium Theory, 79-106. Berlin/Heidelberg: Springer-Verlag, 1985.

Lemke, S. E., and Howson, J. T., "Equilibrium Points of Bi-Matrix Games," SIAM Journal of Applied Mathematics, **12**, 413-423, 1964.

Nash, J. F., Jr., "Equilibrium Points in n-Person Games," Proceedings of the National Academy of Sciences of the U.S.A., **36**, 48-49, 1950.

Peleg, B., "An Axiomatization of the Core of Cooperative Games Without Side Payments," Journal of Mathematical Economics, **14**, 203-214, 1985.

Prakash, P., and Sertel, M. R., "Existence of Noncooperative Equilibria in Social Systems," Discussion Paper No. 92, The Center for Mathematical Studies in Economics and Management Science, Northwestern University, May 1974. Revised: November 1974.

Rubinstein, A., "Strong Perfect Equilibrium in Supergames," International Journal of Game Theory, **9**, 1-12, 1980.

Scarf, H., "The Core of an N Person Game," Econometrica, **35**, 50-69, 1967.

Scarf, H., "On the Existence of a Cooperative Solution for a General Class of N-Person Games," Journal of Economic Theory, **3**, 169-181, 1971.

Shapley, L. S., "On Balanced Games Without Side Payments," in: T. C. Hu and S. M. Robinson, eds., Mathematical Programming, 267-290. New York: Academic Press, 1973.

Skorokhod, A. V., Studies in the Theory of Random Processes, (English translation), Reading: Addison-Wesley, 1965.

von Neumann, J., "Zur Theorie der Gesellschaftsspiele," Mathematische Annalen, **100**, 295-320, 1928. (English translation: "On the Theory of Games of Strategy," Paper 1 in: A. W. Tucker and R. D. Luce, eds., Contributions to the Theory of Games, Vol. IV, 13-42. Princeton: Princeton University Press, 1959.)

Yanovskaya, E. B., "Core in Noncooperative Games," International Journal of Game Theory, **1**, 209-215, 1971/72.

A Variational Principle Application to the Nonlinear Complementarity Problem

GEORGE ISAC* Department of Mathematics, Royal Military College, St.-Jean, Quebec, Canada

M. THERA Department of Mathematics, University of Limoges, Limoges, France

Dedicated to Professor Ky Fan on the occasion of his retirement.

值樊教授光荣退休献敬詞

ABSTRACT : We state a variational principle for a locally compact convex cone K by means of the concept of a K-uniformly sublinear mapping. Then, we apply this principle to the nonlinear complementarity problem N.C.P(T,K), where T is the difference of two nonlinear operators, subject to suitable assumptions.

Our results are originated from the theory of the buckling problem of a thin elastic plate subjected to unilateral conditions.

I INTRODUCTION.

Throughout E will designate a Banach space, E^* its topological dual, $\langle\cdot,\cdot\rangle$ the bilinear canonical pairing over $E \times E^*$. Let K denote a closed nonempty convex cone in E. K^* stands for the dual cone of K and is given by

$$K^* = \{x^* \in E^* : \langle x,x^*\rangle \geq 0, \text{ for all } x \in K\} .$$

If T is a mapping from K into E^*, we recall that the complementarity problem relative to T and K is denoted by C.P(T,K) and is defined by the conditions :

$$\text{C.P.(T,K) : find } x_0 \in K \text{ such that } T(x_0) \in K^* \text{ and } \langle x_0,T(x_0)\rangle = 0 .$$

*Research partly carried out while the author was visiting the University of Limoges (France).

Many complementarity problems such as for example, the complementarity problem used in the elasticity theory, have the property that T is the Fréchet derivative of a potential functional Φ defined on E, and as is well known, every optimal solution of the minimization problem,

$$(\mathbb{P}) \qquad \text{minimize } \Phi(x) \text{ over all } x \in K,$$

solves C.P(T,K).

Our aim in this paper, is to improve this result when T is the Gâteaux derivative of a nonconvex functional Φ. This is done via a variational principle for closed locally (weakly) compact convex cones, that is stated by using the concept of a K-uniformly sublinear mapping.

2 PRELIMINARY RESULTS.

We recall that $K \subset E$ is said to be a *convex cone* if

$$(1) \qquad K+K \subset K$$

and

$$(2) \qquad \lambda K \subset K \text{ for each } \lambda \in \mathbb{R}_+ .$$

If furthermore,

$$(3) \qquad K \cap (-K) = \{0\}$$

K is said to be *pointed*.

We say that a subset $B \subset K$ is a *base* for K, if B generates K, that is, $K = \bigcup_{\lambda \geq 0} \lambda B$ and if $0 \notin \overline{co}\, B$ (here, as usually, $\overline{co}$ stands for the closed convex hull of B).

If B is a base of a closed convex cone K, the relation $K \subset \bigcup_{\lambda \geq 0} \overline{co}\, B \subset \overline{K} = K$, shows that B can always be chosen closed and convex.

It results also from a separation argument, that in any locally convex topological vector space (ℓ.c.t.v.s. in short), for every closed convex base B, we may find $f \in E^*$ such that $B := \{x \in K : f(x)=1\}$ is another base which satisfies the uniqueness decomposition property :

(4) for each $x \in K \setminus \{0\}$, there exists a unique $\lambda > 0$ such that $\lambda^{-1} x \in B$.

Hence, every closed convex cone with a base is necessarily pointed.

Proposition 2.1 (V. Klee [31])

Let K *be a closed convex pointed cone in* E*. The following are equivalent :*

(1) K *is locally (weakly) compact ;*

(2) K *admits a (weakly) compact base.*

When K is not pointed, an useful property initially remarked by V. Klee [31] (see also B. Anger and J. Lembcke [1]) and pointed out to the authors by J.M. Borwein [8]) will be a key tool for what follows.

Proposition 2.2.

Let K *be a convex closed cone in* E. *The following are equivalent :*

(1) K *is locally (weakly) compact ;*

(2) K *is the direct sum of a locally (weakly) compact convex pointed subcone* S *with a finite dimensional subspace* L *;*

(3) there exists a continuous sublinear functional g *such that* $B := \{x \in K : g(x)=1\}$ *is (weakly) compact, generates* K. *Furthermore, there exists* $c > 0$ *such that* $g(x) \geq c\|x\|$ *for each* $x \in K$.

Proof : (1)$\Rightarrow$(2) As easily seen, the set $L := K \cap (-K)$ is a subspace which is finite dimensional since K is locally (weakly) compact. Hence, if we denote by M a topological complement of L and by P (resp. Q) the continuous linear projection onto L (resp. onto M), the set $S := K \cap M$ is a locally (weakly) compact convex subcone of K which is pointed since $S \cap (-S) = (K\cap(-K)) \cap M = L \cap M = \{0\}$.

Since for each $x \in K$, $Q(x) = x-P(x) \in K-L \subset K+K \subset K$ and $Q(x) \in M$, we deduce that $Q(K) \subset K \cap M = S$, and therefore $K \subset P(K) + S \subset L+S \subset K$. Hence, $S+L = K$ as desired.

(2)$\Rightarrow$ (3) Since $S+L = K$ and S is a locally (weakly) compact convex cone, the standard Klee's result, provides a compact convex base B for S. If $e^* \in E^*$ separates B from the origin, then $x^* := e^* \circ Q \in K^* := \{x^* \in E^* : x^*(x) \geq 0$ for all $x \in K\}$ and $x^*(x) > 0$ for each $x \in K \setminus \{0\}$.

For any equivalent norm in L, the functional g given by

$$g(x) := \|P(x)\| + x^*(x)$$

is sublinear and continuous. As easily seen, $B := \{x \in K : g(x)=1\}$ generates K and is (weakly) compact. Hence, B is bounded and there exists $c > 0$ such that $g(x) \geq c\|x\|$ for all $x \in K$.

(3)$\Longrightarrow$ (1) Let $a < 1$ and set $U := \{x \in E : g(x) \leq a\}$. Then U is a convex closed neighbourhood of the origin such that $U \cap K \subset [0,a] B$ which is (weakly) compact. Hence $U \cap K$ is (weakly) compact and therefore K is locally (weakly) compact, as desired. □

In the locally convex setting, we say that $g : E \to \mathbb{R}$ is K-uniformly positive if for a given saturated family of semi-norms $\{p_i\}_{i \in I}$ that generates the topology on E, there exists $c > 0$ such that $g(x) \geq c\, p_i(x)$, for all $i \in I$ and each $x \in K$, $x \neq 0$.

We observe that if we only require the existence of a sublinear continuous functional, which is K-uniformly positive, we only get a convex open neighbourhood U of the origin such that $U \cap K$ is bounded. Then we say that K is *locally bounded* [22]. In fact we have :

Proposition 2.3

Let K *be a cone in* E *which is supposed to be locally convex. The following are equivalent :*

(1) There exists a sublinear functional g *which is* K-*uniformly positive and continuous ;*

(2) K *is locally bounded.*

Proof : We only have to prove (2)$\Longrightarrow$(1). Choose a neighbourhood U of the origin, which is closed, convex and circled. Then, the gauge p_U of U given by $p_U(x) := \inf\{t > 0 : t^{-1}x \in U\}$ is sublinear and continuous. Furthermore, if $p_U(x) = 0$ for some $x \in K \setminus \{0\}$ we can select a net $(x_i)_{i \in I}$ with $t_i > 0$, $\lim t_i = 0$ and $t_i^{-1} x \in U$. Since K is a cone, $t_i^{-1} x \in U \cap K$ and therefore $x = t_i(t_i^{-1} x)$ belongs to the *asymptotic cone* $T_\infty(U \cap K)$ of $U \cap K$, i.e., $x \in \{d \in E : \exists (t_i)_{i \in I} \to 0,\ t_i > 0,\ \exists (d_i)_{i \in I},\ d_i \in E,\ d = \lim t_i d_i\}$. Since $U \cap K$ is bounded, $T_\infty(U \cap K)$ reduces to $\{0\}$, so that $x = 0$ and we get a contradiction. Hence, P_U is a convenient sublinear mapping. □

REMARK : We observe that in Proposition 2.2, whenever K is a pointed convex locally (weakly) compact cone, L reduces to $\{0\}$ and we recapture Klee's result and its well-known consequence : E^* admits a K-uniformly continuous linear functional which is necessarily strictly positive.

3 A VARIATIONAL PRINCIPLE.

Theorem 3.1

Let $K \subset E$ *be a cone in a normed space and let* T_1, T_2 *be functionals defined on* K *that satisfy :*

(1) T_1 *is positively homogeneous of order* p *and* $T_1(x) > 0$ *for each* $x \in K$, $x \neq 0$ *:*

(2) $\limsup\limits_{\substack{\|x\| \to +\infty \\ x \in K}} \dfrac{T_2(x)}{\|x\|^p} \leq 0.$

If K *is closed, locally (weakly) compact and convex, and* T_1 *is (weakly) lower semi-continuous, then* $\{x \in E : T_1(x) - T_2(x) \leq \lambda\}$ *is (weakly) relatively compact for all* $\lambda \in \mathbb{R}$.

Proof : Since K is a closed locally (weakly) compact convex cone, by Proposition 2.2(3) we get $c > 0$ and a sublinear continuous mapping g such that

$$g(x) \geq c^{1/p} \|x\| \quad \text{for all} \quad x \in K.$$

Note $L_\lambda := \{x \in K : T_1(x) - T_2(x) \leq \lambda\}$, the λ-*level set* of $T_1 - T_2$. Since g is weakly continuous, it suffices to show that $\sup\limits_{x \in L_\lambda} g(x)$ is finite.

If not, pick $x_n \in L_\lambda$ such that $g(x_n) \geq n$. Then $y_n := \dfrac{x_n}{g(x_n)}$ belongs to $g^{-1}(\{1\}) \cap K$ which is (weakly) compact. Thus, there exists a subsequence (x_{n_k}), which tends (weakly) to $\bar{y}$ with $g(\bar{y}) = 1$. Hence,

$$T_1(y_{n_k}) = \frac{T_1(x_{n_k})}{[g(x_{n_k})]^p} \leq \frac{\lambda + T_2(x_{n_k})}{[g(x_{n_k})]^p} \leq \frac{\lambda + T_2(x_{n_k})}{c\|x_{n_k}\|^p}$$

so that,

$$T_1(\bar{y}) \leq \frac{1}{c} \liminf_{k \to +\infty} \frac{T_2(x_{n_k})}{\|x_{n_k}\|^p} \leq 0 .$$

Thus, $T_1(\bar{y}) \leq 0$, $\bar{y} \in K$ and $g(\bar{y}) = 1$, which is impossible. □

Corollary 3.2

Let T *be a* p*-positively homogeneous functional defined on a cone* K *given in a normed space* E*, which satisfies*

$$T(x) > 0 \quad \textit{for } x \in K,\ x \neq 0.$$

If K is closed, locally (weakly) compact and convex and T is (weakly) lower semi-continuous, then each level set of T is (weakly) compact.

REMARK : A similar result was proved by Janos [28 : Thm. 1.2] for a positively continuous functional.

The next result, extends the Weierstrass variational principle.

Theorem 3.3

Under the assumptions of Proposition 3.1, if $T_1 - T_2$ *is (weakly) lower semi-continuous then* $T_1 - T_2$ *achieves its minimum on* K *and the set of minimal points of* $T_1 - T_2$ *is (weakly) compact.*

Proof : Every level set of $T_1 - T_2$ is (weakly) relatively compact, hence (weakly) compact, since (weakly) closed by (weak) lower semi-continuity of $T_1 - T_2$. Therefore, $m := \inf_{x \in K} \{T_1(x) - T_2(x)\}$ is finite and the finite intersection property applied to the family $(F_\varepsilon)_{\varepsilon > 0}$ given by $F_\varepsilon := \{x \in K : T_1(x) - T_2(x) \leq m + \varepsilon\}$ provides an element $\bar{x} \in \bigcap_{\varepsilon > 0} F_\varepsilon$ that satisfies :

$$T_1(\bar{x}) - T_2(\bar{x}) = \min_{x \in K} \{T_1(x) - T_2(x)\} . \qquad \square$$

Corollary 3.4

Under the assumptions of Corollary 3.2, every p*-positively homogeneous functional* T *that satisfies*

$$T(x) > 0 \quad \text{for } x \in K,\ x \neq 0 ,$$

achieves its minimum on K *and moreover the set of minimal points of* T *is (weakly) compact.*

Proof : Set $T_2 = 0$ in the preceding Theorem. □

See also J.P. Penot [43] for a more general criterion using the concept of asymptotically compact set.

REMARK : We notice that assumption (2) of Theorem 2.1 applies with $p > 1$, for a *quasi-bounded* mapping i.e, a mapping such that

$$q(f) := \limsup_{\substack{\|x\| \to +\infty \\ x \in K}} \frac{|f(x)|}{\|x\|} < +\infty .$$

As observed in [21], this occurs whenever f is *asymptotically linear*, that is, whenever we may find $D_\infty f \in E^*$ such that

$$\lim_{\substack{\|x\| \to +\infty \\ x \in K}} \frac{|f(x) - D_\infty f(x)|}{\|x\|} = 0 .$$

The reader interested by these concepts should consult for instance [21], [33].

4 APPLICATION TO THE NONLINEAR COMPLEMENTARITY PROBLEM.

Let T map a closed convex cone $K \subset E$ into E^* and consider the *nonlinear complementarity* problem relative to T and K

N.C.P.(T,K) : Find $\bar{x} \in K$ such that $T(\bar{x}) \in K^*$ and $\langle T(\bar{x}), \bar{x} \rangle = 0$.

Since K is a cone, as easily observed, N.C.P.(T,K) is equivalent to the *variational inequality* V.I(T,K) defined by

V.I(T,K) : Find $\bar{x} \in K$, such that $\langle T(\bar{x}), y - \bar{x} \rangle \geq 0$ for each $y \in K$.

which can also be rewritten as,

V.I(T,K) : Find $\bar{x} \in K$ such that $-T(\bar{x}) \in \partial\psi_K(\bar{x})$,

where $\partial\psi_K(\bar{x})$ stands for the convex subdifferential of the indicator function of K defined by,

$$\psi_K(x) = 0 \text{ if } x \in K \quad \text{and} \quad \psi_K(x) = +\infty, \text{ else.}$$

The complementarity problem was studied in infinite dimensional spaces in [2] [4] [7] [24] [25] [26] [29] and variational inequalities for potential operators in [44].

Let K be a closed convex set of a Banach space E. We recall that a continuous operator P defined on E is a projection onto K if $P(E) = K$ and $P(x) = x$ for each $x \in K$.

We obtain the existence of continuous projections using different approaches. When K is a closed convex cone, the following holds true :

Theorem 4.1 [34]

Let K *be a closed convex cone in a Banach space* E. *For each* $\alpha > 0$, *there exists a projection* P_α *onto* K *such that*

$$\|x-P_\alpha(x)\| \leq (1+\alpha)d(x,K) \qquad \textit{for all} \quad x \in E$$

where as usual $$d(x,K) := \inf_{y\in K} \|x-y\| .$$

We recall that a norm $\| \ \|$ defined on E is *strictly convex* if it satisfies the property $\|x+y\| = \|x\| + \|y\|$ implies $x = t\,y$ for some $t \geq 0$ or else $y = 0$.
The corresponding normed space is then called strictly convex. The class of strictly convex Banach spaces introduced by Clarkson [13] includes all Hilbert spaces, as well all $L^p(\Omega,\tau,\mu)$ for $1 < p < +\infty$. Moreover, every separable Banach space admits a strictly convex equivalent renorm [36].

When E is a strictly convex Banach space and $C \subset E$ a closed locally compact convex subset, we always may define the *nearest-point mapping* P_C which maps each $x \in E$ into its uniquely determined nearest point $P_C(x)$ in C, i.e. $\|x-P_C(x)\| = \min_{y\in C}\|x-y\|$. Since C is locally compact, P_C is continuous with respect to the norm topology and therefore P_C is a projection onto C.

We say that a sequence $\{K_n\}_{n\in \mathbb{N}}$ of closed convex cones in a Banach space E is a *Galerkin approximation* of a given convex cone $K \subset E$, if the following are satisfied for all $n \in \mathbb{N}$:

(1) K_n is a locally compact convex subcone of K ;

(2) for each $n \in \mathbb{N}$, there exists a continuous projection P_n onto K_n such that $\{P_n(x)\}$ norm-converges to x for each $x \in K$.

This definition is less demanding than the one used by Cimetière [12].

Proposition 4.2.

Let K *be a locally compact convex cone in a Banach space* E. *If there exists an increasing sequence* $\{K_n\}$ *of closed convex subcones of* K *satisfying* $K = \overline{\bigcup_{n\in \mathbb{N}} K_n}$, *then* $\{K_n\}$ *defines a Galerkin approximation of* K.

Proof : By virtue of Theorem 4.1, we get for each n a continuous projection P_n into K_n such that $||x-P_n(x)|| \leq 2d(x,K_n)$, for each $x \in E$. In particular, for each $x \in K$, and each $\varepsilon > 0$, there exists some $n \in \mathbb{N}$ with $||x_n - x|| < \varepsilon$. Hence, for $m \geq n$ we get $||x-P_m(x)|| \leq 2d(x,K_m) \leq 2d(x,K_n) \leq 2||x-x_n|| < 2\varepsilon$ and $\lim_{n\infty} P_n(x) = x$. □

We say that a convex cone K has a *Schauder* basis, if there exists a sequence $\{x_n\} \subset K$ such that every $x \in K$ has the decomposition :

$$x = \sum_{n=1}^{+\infty} a_n x_n, \quad a_n \geq 0 \text{ for each } n \in \mathbb{N}.$$

It is easily seen that if $\{x_n\}$ is a Schauder basis for K, then $K_n := \operatorname{cone}\left(\bigcup_{0 \leq i \leq n} \{x_i\}\right) = \left\{\sum_{i=1}^{n} \lambda_i x_i \;;\; \lambda_i \geq 0\right\}$ defines a Galerkin approximation of K. Also we now give a result which provides, under suitable assumptions, the existence of a Schauder basis for the positive cone of a Banach lattice.

We recall that a point x_o is an *extreme point* for a convex set $A \subset E$, if $x_o \in A$ and if $A \setminus \{x_o\}$ is convex. This in turn is equivalent to the condition :

$$x \in A,\ y \in A,\ tx + (1-t)y = x_o \text{ for some } t \in (0,1)$$

then $x = y = x_o$.

In what follows, ext(B) will stand for the set of extreme points of B.

A Banach space E is said to have the *Krein-Milman* property (K.M.P. in short), if each closed convex bounded subset of E is the norm-closed convex hull of its extreme points. As shown by Lindenstrauss [15, Prop. 1] this occurs if in particular, each (non empty) closed convex bounded subset of E has an extreme point.

The class of Banach spaces with the K.M.P. is large since it contains all dual spaces that are subspaces of weakly compactly generated Banach spaces, as does in particular every dual separable Banach space as well as any reflexive Banach space. For more details the reader is refered to the book by Diestel [15].

Let E be a vector lattice. Denote $x \vee y := \sup(x,y)$ and $x \wedge y := \inf(x,y)$. $x^+ := x \vee 0$, $x^- := (-x) \vee 0$, $|x| = x \vee (-x)$ are the *positive part, the negative part, the absolute value* of x, respectively. E is called a *normed vector lattice,* if the topology and the order are related by

$$|x| \leqslant y| \Longrightarrow \|x\| \leqslant \|y\|.$$

If, furthermore E is complete we say that E is a Banach lattice. For example, all $L_p(\mu)$ and $C(K)$-spaces with their usual order are Banach lattices, while the class of separable Banach lattices includes all Banach spaces having inconditional basis $\{x_n\}$ for the order given by $x = \sum_{n=1}^{+\infty} a_n x_n \geqslant 0$, if and only if $a_n \geqslant 0$ for each n.

The following Proposition gives existence of a Schauder basis for the positive cone of a separable Banach lattice :

Proposition 4.3. -

Let E be an infinite-dimensional separable Banach lattice, with the K.M.P. If $E_+ := \{x \in E : x \geqslant 0\}$ has a bounded convex base then E_+ has a Schauder basis.

Proof : Let B be a bounded base for E_+. Since E is a Banach lattice, E_+ is closed and as observed before, B can always be chosen convex closed with the uniqueness decomposition property.

Since $0 \notin B$, pick $a > 0$, $b > 0$ such that

(1) $$a \leqslant \|x\| \leqslant b \quad \text{for each } x \in B.$$

Let x be an extreme point of B. For each y in E_+ such that $0 < y \leqslant x$, we claim that $y = \lambda x$ for some $\lambda \in]0,1]$. Indeed, $x = y+z$ with $z \in E_+$ and we may find $\lambda_1 > 0$, $\lambda_2 > 0$, b_1, b_2 in B such that $x = (\lambda_1+\lambda_2)\left\{\frac{\lambda_1}{\lambda_1+\lambda_2} b_1 + \frac{\lambda_2}{\lambda_1+\lambda_2} b_2\right\}$. By virtue of the uniqueness of the decomposition of x we necessarily get that x is a convex combination of b_1 and b_2. Hence, since $x \in \text{ext}(B)$, $x = b_1 = b_2$ and the claim is proved. In particular, for $x,y \in \text{ext}(B)$, with $x \wedge y \neq 0$, the relations $0 < x \wedge y \leqslant x$ and $0 < x \wedge y \leqslant y$, and the preceding observation yield $x \wedge y = \lambda x = \beta y$ for some $\lambda \in]0,1]$, $\beta \in]0,1]$. Hence, $x = \frac{\lambda}{\beta} y$ and by (1) $x = y$. Thus, the extreme points of B are pairwise disjoint and according to the relations $x + y = x \vee y + x \wedge y$ and $2(x \wedge y) = x+y - |x-y|$, (which are valid in any vector lattice), for $x,y \in \text{ext}(B)$ we necessarily get

(2) $$0 < x \leqslant x+y = |x-y| .$$

This forces the extreme points of B to be discrete, since by virtue of (1), (2) yields :

(3) $$a \leq \|x\| \leq \|x-y\| \quad \text{for all} \quad x,y \in \mathrm{ext}(B).$$

Since E is separable and normed, it satisfies the second axiom of countability as do also its subspace $\mathrm{ext}(B)$. Being discrete, $\mathrm{ext}(B)$ is then necessarily at most countable. Also, we may assume there exists $\{b_n\}_{n\in\mathbb{N}}$ such that $\mathrm{ext}(B) = \{b_n\}_{n\in\mathbb{N}}$. For each $x \in B$, since $0 \leq x \wedge b_n \leq b_n$ and $b_n \in \mathrm{ext}(B)$, there exists $x(n)$ such that $x \wedge b_n = x(n)b_n$. Since $x(n)b_n \leq b_n$, we get $x(n)\|b_n\| \leq \|b_n\|$, which implies that necessarily $x(n) \in [0,1]$. For each $n \in \mathbb{N}$ we have

$$0 \leq \sum_{i=1}^{n} x(i)b_i = (x(1)b_1 \vee \dots \vee x(n)b_n) +$$

$$(x(1)b_1 \wedge \dots \wedge x(n)b_n) = x(1)b_1 \vee \dots \vee x(n)b_n$$

since the b_i's are pairwise disjoint.
Hence we get,

$$0 \leq \sum_{i=1}^{n} x(i)b_i \leq x ,$$

relation which implies that $\sum_{i=1}^{+\infty} x(i)b_i = \bar{x}$ exists and $0 \leq \bar{x} \leq x$.

By assumption, $B = \overline{\mathrm{co}}\ \mathrm{ext}(B)$, since E has the Krein-Milman property. Hence, $x = \lim_{k\to+\infty} (x_k)$ with $x_k \in \mathrm{co}(\{b_i\}_{i\in\mathbb{N}})$, i.e $x_k = \sum_{i\in I(k)} \lambda_i(k)b_i$ where $I(k)$ is a finite subset of $\mathbb{N}$, $\lambda_i(k) \in [0,1]$ and $\sum_{i\, I(k)} \lambda_i(k) = 1$.
We observe that,

$$x \wedge x_k = x \wedge \sum_{i\in I(k)} \lambda_i(k)b_i \leq \sum_{i\in I(k)} x \wedge \lambda_i(k)b_i$$

$$\leq \sum_{i\in I(k)} x \wedge b_i = \sum_{i\in I(k)} x(i)b_i \leq \bar{x} .$$

By continuity of the operator $(u,v) \longmapsto u \wedge v$ we derive

$$x = \lim_k (x \wedge x_k) \leq \bar{x} \quad \text{and therefore} \quad x = \bar{x} = \sum_{i=1}^{+\infty} x(i)b_i$$

and the proof is complete. □

Proposition 4.4

Let E be a Banach space and let $K \subset E$ be a closed convex cone.

Consider a Galerkin approximation $(K_n)_{n\in\mathbb{N}}$ of K. Suppose that for $T : K \to E^*$ the following assumption are satisfied :

(a) if (x_n) is weakly convergent to $\bar{x}$ and if (y_n) is strongly convergent to $\bar{y}$, then

$$\limsup_{n\to+\infty} \langle T(x_n),y_n\rangle \leq \langle T(\bar{x}),\bar{y}\rangle \ ;$$

(b) $x \mapsto \langle T(x),x\rangle$ *is sequentially weakly lower semi-continuous.*

Then, if x_n *solves* N.C.P.(T,K_n), *each weak sequential limit point of* (x_n) *solves* N.C.P.(T,K). *Moreover, if* K *is locally compact or* E *is reflexive, it suffices that* (x_n) *be bounded.*

Proof : Let $\bar{x}$ be a weak sequential limit point of (x_n), that is, $\bar{x} = (w) - \lim_{k\to+\infty} x_{n_k}$. Since x_{n_k} solves N.C.P.(T,K_{n_k}), assumption (b) yields

$$\langle T(\bar{x}),\bar{x}\rangle \leq \liminf_{k\to+\infty} \langle T(x_{n_k}),x_{n_k}\rangle = 0$$

Letting $\tilde{x}_{n_k} := P_{n_k}(\bar{x})$ and $z_{n_k} := \frac{1}{n_k}\tilde{x}_{n_k} + (1-\frac{1}{n_k})x_{n_k}$ and using the convexity of K_{n_k} we have $z_{n_k} \in K_{n_k}$. Furthermore, since x_{n_k} solves N.C.P.(T,K_n), x_{n_k} also solves V.I(T,K_n) so that,

$$0 \leq \langle T(x_{n_k}),z_{n_k}-x_{n_k}\rangle = \frac{1}{n_k}\langle T(x_{n_k}),\tilde{x}_{n_k}\rangle - \frac{1}{n_k}\langle T(x_{n_k}),x_{n_k}\rangle .$$

Hence, $\langle T(x_{n_k}),\tilde{x}_{n_k}\rangle \geq 0$ and therefore, since $(\tilde{x}_{n_k})$ norm-converges to $\bar{x}$, assumption (a) yields $\langle T(\bar{x}),\bar{x}\rangle \geq 0$. Hence $\langle T(\bar{x}),\bar{x}\rangle = 0$.

Let $y \in K$ be arbitrary and set $y_n := P_n(y) \in K_n$. Since x_{n_k} solves N.C.P.(T,K_{n_k}), we have $\langle T(x_{n_k}),y_{n_k}\rangle \geq 0$, and using again assumption (a), we obtain, $\langle T(\bar{x}),y\rangle \geq 0$ for every $y \in K$. Hence $T(\bar{x}) \in K^*$ and $\bar{x}$ solves N.C.P(T,K), as desired. □

REMARK (1) : Assumption (a) is fulfilled if T is sequentially weakly continuous. In this case, if (x_n) weakly converges to $\bar{x}$ and (y_n) norm-converges to $\bar{y}$, since

$$\langle T(x_n),y_n\rangle - \langle T(\bar{x}),\bar{y}\rangle = \langle T(x_n)-T(\bar{x}),\bar{y}\rangle - \langle T(x_n),\bar{y}-y_n\rangle$$

and $(T(x_n))$ is strongly bounded, we have $\lim_{n\to+\infty} \langle T(x_n),y_n\rangle = \langle T(\bar{x}),\bar{y}\rangle$.

REMARK (2) : If T is *completely continuous*, i.e, T maps weakly convergent sequences to $\bar{x}$ into strongly convergent sequences to $T(\bar{x})$, as do most of the operators used in the deformation equations of Elasticity theory, then $x \mapsto \langle T(x),x\rangle$ is sequentially weakly continuous and Proposition 4.4 is trivial.

REMARK (3) Suppose E has the Dunford-Pettis property (D.P.P in short), i.e. for each sequence $\{x_n\} \subset E$ (resp. $\{x_n^*\} \subset E^*$) that tends weakly to zero in E (resp. in E^*) then $\lim_{n \infty} x_n^*(x_n) = 0$.

It is well-known that for each compact set T, $\mathfrak{C}(T)$ admits the D.P.P, as well each $L^1(\mu)$ for each probability measure.

If E is Dunford-Pettis then for each operator T continuous from E equipped with the weak topology into E^* equipped with the weak topology $x \mapsto \langle T(x),x\rangle$ is weakly continuous and Proposition 4.4 is again trivial.

Lemma 4.5. -

Let T *be the Gâteaux derivative of a convex functional* ϕ. *The following implication holds true :*

$$\limsup_{||x|| \to +\infty} \frac{\langle T(x),x\rangle}{||x||^p} \leq 0 \Rightarrow \limsup_{||x|| \to +\infty} \frac{\phi(x)}{||x||^p} \leq 0.$$

Proof : Because of the convexity of ϕ we have

$$\phi(x) - \phi(0) \leq \phi'(x)(x) = \langle T(x),x\rangle$$

from which we derive the desired result. □

Theorem 4.6. -

Let K *be a closed locally compact convex cone in a Banach space. Let* $T := T_1 - T_2$ *where* T_i $(i=1,2)$ *maps* K *into* E^* *and suppose the following assumptions hold true :*

(1) If $\{x_n\}$ *is weakly convergent to* $\overline{x}$ *and if* $\{y_n\}$ *norm-converges to* $\overline{y}$, *then* $\limsup_{n \infty} \langle T(x_n),y_n\rangle \leq \langle T(\overline{x}),\overline{y}\rangle$;

(2) $x \mapsto \langle T_1(x),x\rangle$ *is sequentially weakly lower semicontinuous while* $x \mapsto \langle T_2(x),x\rangle$ *is sequentially weakly upper semicontinuous ;*

(3) T_i *is the Gâteaux derivative of* ϕ_i *and* $\phi := \phi_1 - \phi_2$ *is lower semicontinuous ;*

(4) ϕ_1 *is positively homogeneous of order* p, *lower semicontinuous and satisfies* $\phi_1(x) > 0$ *for each* $x \in K \setminus \{0\}$;

(5) ϕ_2 *is convex and* $\displaystyle\limsup_{\substack{||x|| \to +\infty \\ x \in K}} \frac{\langle T_2(x),x\rangle}{||x||^{p+1}} \leq 0.$

Then, N.C.P(T,K) is solvable.

Proof : Let $\{K_n\}$ be a Galerkin approximation of K. Using Lemma 4.5, conditions (3), (4) and (5), Theorem 3.3 applied to ϕ provides $x_n \in K_n$

such that $\phi(x_n) = \min_{x \in K_n} \phi(x)$. By virtue of (3), for each $x \in K_n$, $\langle T(x_n), x-x_n \rangle = \phi'(x_n)(x-x_n)$ and therefore $\langle T(x_n), x-x_n \rangle \geq 0$, so that x_n solves N.C.P(T,K_n).

We claim that $\{x_n\}$ is bounded. If $\{x_n\}$ fails to be bounded, Proposition 2.2 provides a sublinear continuous mapping g which is K-uniformly positive so that $\lim_{n\to\infty} g(x_n) = +\infty$. Then consider $y_n := \frac{x_n}{g(x_n)}$ which belongs to the weakly compact set $g^{-1}(\{1\}) \cap K$. If $\bar{y}$ is a weak cluster point of $\{y_n\}$, then necessarily $\bar{y} \in K \setminus \{0\}$ since $\bar{y}$ is the weak limit of a subsequence $\{y_{n_k}\}$ and $g(\bar{y}) = 1$.

Since $\langle T_1(y),\bar{y}\rangle = \lim_{t\to 0^+} \frac{\phi_1(\bar{y}+t\bar{y})-\phi_1(\bar{y})}{t} = \lim_{t\to 0^+} \frac{(1+t)^p-1}{t}\, \phi_1(\bar{y}) = p\, \phi_1(\bar{y})$,

condition (4) yields $\langle T_1(\bar{y}),\bar{y}\rangle > 0$.

From the relation,

$$\langle T_1(y_{n_k}),y_{n_k}\rangle = \frac{\langle T_1(x_{n_k}),x_{n_k}\rangle}{[g(x_{n_k})]^{p+1}} =$$

$$\frac{\langle T_2(x_{n_k}),x_{n_k}\rangle}{[g(x_{n_k})]^{p+1}} \leq \frac{1}{c}\, \frac{\langle T_2(x_{n_k}),x_{n_k}\rangle}{\| x_{n_k}\|^{p+1}}$$

and using (2) and (5) we derive

$$0 < \langle T_1(\bar{y}),\bar{y}\rangle \leq \liminf_{k\ \infty} \langle T_1(y_{n_k}),y_{n_k}\rangle \leq \frac{1}{c} \liminf_{k} \frac{\langle T_2(x_{n_k}),x_{n_k}\rangle}{\| x_{n_k}\|^{p+1}} \leq 0,$$

and a contradiction. Hence $\{x_n\}$ is bounded as desired, and so admits a weak cluster point $\bar{x}$. Proposition 4.4 and assumptions (1) and (2) achieve the proof. □

To conclude the paper, we now give a last result in the lines of a previous paper [26] which is originated from the buckling problem of a thin elastic plate subjected to unilateral conditions.

<u>Theorem 4.7</u>

Let K *be a closed convex cone in a reflexive Banach space. Let* $\{K_n\}_{n\in \mathbb{N}}$ *be a Galerkin approximation of* K *and suppose given the mappings,* $A_i : K \to E^*$ $(i = 1,2)$; $f : K \to \mathbb{R}$ *and* $T_i : K \to \mathbb{R}$ $(i = 1,2)$.

If the following conditions are satisfied :

(a) $A := A_1 - A_2$ *fulfils the assumptions (a), (b) of Proposition 4.4 ;*

(b) A *is the Gâteaux derivative of* $f : K \to \mathbb{R}$ *;*

(c) f *is lower semi-continuous ;*

(d) $f(x) \geq T_1(x) - T_2(x)$ *for each* $x \in K$ *;*

(e) T_1 *and* T_2 *satisfy the assumptions (1), (2) of Theorem 3.1 ;*

(f) there exists a subset $D \subset K$ *that contains every solution of* N.C.P.(A, K_n) *for every* $n \in \mathbb{N}$ *and* $\lim_{\substack{\|x\| \to +\infty \\ x \in D}} \langle A(x), x\rangle = +\infty$.

Then the problem N.C.P.(A, K) *is solvable.*

Proof : Since each K_n is locally compact and all assumptions of Theorem 3.3 are satisfied, we may find $x_n \in K_n$ such that, $f(x_n) = \min_{x \in K_n} f(x)$.

From assumptions (b), and since

$$\langle x - x_n, A(x_n)\rangle = \left[\frac{d}{dt} f(x_n + t(x - x_n))\right]_{|t=0}$$

for every $x \in K_n$, we deduce that x_n solves V.I.(A, K_n) and therefore x_n solves also N.C.P.(A, K_n).

Since $\{x_n\}_{n \in \mathbb{N}}$ is bounded according to assumption (f) and E is reflexive, $\{x_n\}$ admits a weak cluster point $\bar{x} \in K$ and we obtain that N.C.P.(A,K) is solvable by Proposition 4.4. □

REMARK (1) As in our paper [26], if $\{x_n\}_{n \in \mathbb{N}}$ has a strongly convergent subsequence and there exists an element $z \in K$ such that $f(z) < f(0)$ then problem N.C.P.(A, K) has a non-zero solution.

REMARK (2) If we use the *universal coercivity condition* defined by Minty [40], we obtain that problem N.C.P.(A, K) has a solution if we suppose that for A and K there exists a closed convex cone $K_o \subset \mathbb{K}$ such that for A_1, A_2 and K_o the assumptions of Theorem 4.3 are satisfied and K_o and K are connected by the geometrical condition (M) of our paper [26].

Acknowledgment

Thanks are due to Professor J.M. Borwein (Dalhousie University) for his suggestions especially regarding the local compactness conditions.

REFERENCES

[1] Anger, B. ; Lembcke, J. (1980) Extension of linear forms with strict domination on locally compact cones, Math. Scand. 47, n°2 : 251-265.

[2] Allen, G. (1977) Variational inequalities, complementarity problems and duality theorems, J. Math. Anal. and Appl. Vol. 58 : 1-10.

[3] Baiocchi, C. ; Capelo, A. (1984) Variational and Quasi variational Inequalities. Applications to free-boundary problems, John Wiley.

[4] Bazaraa, M.S., Goode, J.J. and Nashed, M.Z. (1972) A nonlinear complementarity problem in mathematical programming in Banach space, Proc. Amer. Math. Soc. Vol. 35 Nr. 1 : 165-170.

[5] Berger, M.S. (1977) Nonlinearity and functional analysis, Academic Press, New York.

[6] Berger, M.S., and Schechter, M. (1977) On the solvability of semi-linear gradient operator equations, Advances in Math. Vol. 25 : 97-132.

[7] Borwein, J.M. (1984) Generalized linear complementarity problem treated without fixed point theory, J. Opt. Theory and Appl., Vol. 43 Nr. 3: 343-355.

[8] Borwein, J.M. (1985) Alternative theorems for general complementarity problems, Proceeding of the Cambridge Conference on Infinite Dimensional Programming, Springer-Verlag, New York, in press.

[9] Bourbaki, N. (1981) Espaces vectoriels topologiques, Chap. 2, Hermann Paris.

[10] Cain, Jr. G.L., ; Nashed, M.Z. (1971) Fixed points and stability for a sum of two operators in locally convex spaces.

[11] Ciarlet, Ph. G. ; Rabier, P. (1980) Les équations de Von Kármán Lecture Notes in Math., Nr. 826, Springer-Verlag.

[12] Cimetière, A. (1980) Un problème de flambement unilatéral en théorie des plaques, J. de Mécanique, Vol. 19 Nr. 1 : 183-202.

[13] Clarkson, J.A. (1936) Uniformly convex spaces, Trans. Amer. Math. Soc. 40 : 396-414.

[14] Diestel, J. (1980) A survey of results related to the Dunford-Pettis property, Contemporary Mathematics Vol. 2 : 5-60.

[15] Diestel, J. (1975) Geometry of Banach spaces. Selected Topics, Lecture Notes in Math., Springer-Verlag n° 485.

[16] Do, C. (1976) The Buckling of a thin elastic plate subjected to unilateral conditions. In : Applications of methods of functional analysis to problems in mechanics. Lecture Notes in Math. Springer-Verlag Nr. 50 : 307-316.

[17] Do, C. (1975) Problèmes de valeurs propres pour une inéquation variationnelle sur un cône et application au flambement unilatéral d'une plaque mince, C.R. Acad. Sci. Paris A. 280 : 45-48.

[18] Do, C. (1977) Bifurcation theory of elastic plates subjected to unilateral conditions, Math. Anal. Appl. Vol. 60 Nr. 2 : 435-448.

[19] Duvaut, G. ; Lions, J.L (1974) Problèmes unilatéraux dans la flexion forte des plaques. (Le cas stationnaire), J. de Mécanique, vol. 13 N° 1 : 51-74.

[20] Duvaut, G. ; Lions, J.L. (1972) Les Inéquations en Mécanique et en Physique, Dunod Paris.

[21] Granas, A. (1962) The theory of compact vector fields and some of its applications to topology of functional spaces (I), Rosprawy Matematyezne 30 : 1-93.

[22] Isac, G. (1980) Cônes localement bornés et cônes complètement réguliers. Applications à l'analyse non linéaire, Université de Sherbrooke.

[23] Isac, G. (1982) Opérateurs asymptotiquement linéaires sur des espaces localement convexes, Colloquium Math. Vol. XLVI Fasc. 1 : 67-73

[24] Isac, G. (1985) Nonlinear complementarity problem and Galerkin method, J. of Math. Anal. and Appl. Vol. 108 Nr. 2 : 563-574.

[25] Isac, G. (in press) On the implicit complementarity problem in Hilbert spaces, Bull. of the Australian Math. Soc.

[26] Isac, G., ; Théra, M. (Preprint 1985) Complementarity problem and post-critical equilibrium state of thin elastic plates, Dépt. de Math. Univ. de Limoges.

[27] Jameson, G.J.O. (1970) Ordered linear spaces, Lecture Notes in Math., Vol. 141 Springer-Verlag.

[28] Janos, L. (1960) Homogeneous functional on locally compact cones, (Zech. Math. J. Vol. 10(85) : 380 : 399. (In russian).

[29] Karamardian, S. (1971) Generalized complementarity problem, J. Opt. Theory and Appl. Vol. 8 Nr. 3 : 161-16.

[30] Kinderlehrer, D. ; Stampacchia, G. (1980) An introduction to Variational Inequalities and their Applications, Academic Press, New York.

[31] Klee, V.L. (1955) Separation properties of convex cones, Proc. Amer. Math. Soc. 6 : 313-318.

[32] Köthe, G. (1969) Topological vector spaces I, Springer-Verlag, New York.

[33] Krasnoselskii, M.A. (1964) Positive solutions of operator equations Groningen, P. Noordhoff.

[34] Krasnoselskii, M.A. and Zabreiko, P.P. (1984) Geometrical methods of nonlinear analysis, Springer-Verlag.

[35] Kubrusly, C. ; Odean, J.T. (1981) Nonlinear eigenvalue problems characterized by variational inequalities with applications to the postbuckling analysis of unilaterally supported. Nonlinear Anal. The. Math. and Appl. Vol. 5 Nr. 12 : 1265-1284.

[36] Lindenstrauss, J., ; Tzafriri, L., (1973) Classical Banach spaces Lecture Notes in Math. Springer-Verlag n° 338.

[37] Lions, J.L. ; Stampacchia, G. (1967) Variational inequalities, Comm. Pure Appl. Math. 20 : 493-519.

[38] Lipkin, L.J. (1981) Weak continuity and compactness of nonlinear mapping. Nonlinear Anal. Th. Meth. and Appl. Vol. 5 Nr. 11 : 1257-1263.

[39] Luna, G. (1975) A remark on the nonlinear complementarity problem. Proc. Amer. Math. Soc. Vol. 48 Nr. 1 : 132-134.

[40] Minty, G.J. (1978) On variational inequalities for monotone operators, Advances in Math. Vol. 30 : 1-7.

[41] Moré, J.J. (1974) Coercivity conditions in nonlinear complementarity problems, Siam Review 16 : 1-16.

[42] Panagiotopoulos, P.D. (1985) Inequality problems in Mechanics and Applications. Convex and Nonconvex Energy Functions, Birkauser.

[43] Penot, J.P. Private communication.

[44] Szulkin, A. (1984) On a class of variational inequalities involving gradient operators, J. Math. Anal. and Appl. Vol. 100 : 486-499.

[45] Vainberg, M.M. (1984) Variational methods for the study of nonlinear operators, Holden-Day Inc. San Francisco, London.

On a Best Approximation Theorem

HIDETOSHI KOMIYA College of Commerce, Nihon University, Tokyo, Japan

1. The purpose of the present note is to give a refinement of Simons's best approximation theorem [6, Theorem 5.1] and to derive best approximation theorems over ordered normed vector spaces from this refinement.

Fan [2, Theorem 1] extended Kakutani's fixed point theorem for multi-valued mappings to the infinite dimensional case:

Let X be a compact convex subset of a locally convex topological vector space. Let A be an upper semicontinuous and nonempty closed convex-valued mapping of X into X. Then there exists a point x of X such that $x \in Ax$.

The following coincidence theorem is a corollary of Fan's fixed point theorem (cf. [3, Proof of Theorem 3])

Let X and Y be compact convex subsets of locally convex topological vector spaces. Let A be an upper semicontinuous and nonempty closed convex-valued mapping of X into Y, and let B be an upper semicontinuous and nonempty closed convex-valued mapping of Y into X. Then there exist a point x of X and a point y of Y such that $y \in Ax$ and $x \in By$.

We shall prove a refinement of Simons's result by making use of the coincidence theorem.

2. A multi-valued mapping A of a topological space X into a topological vector space E is called continuous if A is both upper semicontinuous and lower semicontinuous. In this note, upper semicontinuity of A means that for any $x \in X$ and any 0-neighborhood U of E, there exists a neighborhood V of x such that $Ay \subset Ax + U$ for all $y \in V$. Note that this upper semicontinuity is not necessarily the same as the usual upper semicontinuity if A is not compact-valued. Our upper semicontinuity is a weaker notion than the usual one and the sum of upper semicontinuous multi-valued mappings in our sense is also upper semicontinuous in our sense. Note that the functional $x \mapsto \inf \phi(Ax)$ on X is both upper semicontinuous and lower semicontinuous if ϕ is a continuous sublinear

functional on E and A is a continuous multi-valued mapping of X into E.

<u>Theorem 1</u> <u>Let N be a normed vector space, X a compact convex subset of N, and ϕ a continuous sublinear functional on N. Let F be a continuous multi-valued mapping of X into N with nonempty values Fx such that for any u, v ε Fx, there exists w ε Fx with $\phi(w - (u + v)/2) \leq 0$. Then there exist x ε X and f ε N' such that</u>

$$f \leq \phi, \qquad f(x) = \max f(X)$$

<u>and</u>

$$\inf f(Fx) = \inf \phi(Fx).$$

<u>Proof</u> The dual space N' of N will have the weak* topology. If $Y = \{f \,\varepsilon\, N': f \leq \phi\}$, then Y is compact. Define a multi-valued mapping A of X into Y by

$$Ax = \{f \,\varepsilon\, Y: \inf f(Fx) = \inf \phi(Fx)\}.$$

Ax is nonempty by [4, Basic Theorem] and it is easy to see Ax is convex. Let $\{(x_\alpha, f_\alpha)\}$ be a net in the graph Gr(A) of A converging to $(x,f) \,\varepsilon\, X \times Y$. Take a point x' of Fx. Since the functional $y \longmapsto \inf \phi(Fy - x')$ is upper semicontinuous, for any $\varepsilon > 0$, $\inf \phi(Fx_\alpha - x') < \varepsilon$ eventually. Hence there exist $x'_\alpha \,\varepsilon\, Fx_\alpha$ with $\phi(x'_\alpha - x') < \varepsilon$ eventually. Then we have

$$\begin{aligned} f_\alpha(x') &= f_\alpha(x'_\alpha) - f_\alpha(x'_\alpha - x') \\ &\geq f_\alpha(x'_\alpha) - \phi(x'_\alpha - x') \\ &> \inf f_\alpha(Fx_\alpha) - \varepsilon \\ &= \inf \phi(Fx_\alpha) - \varepsilon. \end{aligned}$$

Hence we have $f(x') \geq \inf \phi(Fx)$, since the functional $y \longmapsto \inf \phi(Fy)$ is lower semicontinuous, and hence $\inf f(Fx) \geq \inf \phi(Fx)$. The reverse inequality is easy, hence we have $(x,f) \,\varepsilon\, Gr(A)$. We have proved that Gr(A) is closed. Since Y is compact, A is upper semicontinuous and has closed values.

Define a multi-valued mapping B of Y into X by

$$Bf = \{x \,\varepsilon\, X: f(x) = \max f(X)\}.$$

It is easily seen that Bf is nonempty and convex. If $\{(f_\alpha, x_\alpha)\}$ is a net in Gr(B) converging to $(f,x) \in Y \times X$, then $f(x) = \lim_\alpha f_\alpha(x_\alpha)$. Hence for any $x' \in X$,

$$f(x') = \lim_\alpha f_\alpha(x') \leq \lim_\alpha f_\alpha(x_\alpha) = f(x).$$

Therefore we have $(f,x) \in Gr(B)$. We have proved that Gr(B) is closed. Since X is compact, B is upper semicontinuous and has closed values.

By the coincidence theorem, there exist $x \in X$ and $f \in Y$ such that $f \in Ax$ and $x \in Bf$. These x and f are the desired ones.

Remark (1) The conditions for the multi-valued mapping F of Theorem 1 are satisfied if F is a continuous multi-valued mapping of X into N with nonempty convex values.

(2) We obtain [6,Theorem 5.1] by taking $Fx = Px - Qx$. We see from Theorem 1 that the assumption that P and Q have closed values in [6, Theorem 5.1] is unnecessary.

(3) Theorem 1 holds even if the normed vector space N is replaced by a Macky space, but it is not known whether Theorem 1 holds or not when N is replaced by a general locally convex topological vector space.

3. A normed vector space N with a vector order $\geq$ is called an ordered normed vector space. The functional Φ on N defined by

$$\Phi(x) = \inf \{\|y\|: y \geq x\}$$

is called the canonical half-norm for the given vector order (cf. [1]). The functional Φ is continuous and sublinear. A point f of N' is said to be poisitve if $f(x) \geq 0$ for all $x \geq 0$, and we write $f \geq 0$ if f is positive. When the positive cone is closed, $x \leq 0$ if and only if $\Phi(x) = 0$, and $f \leq \Phi$ if and only if $f \geq 0$ and $\|f\| \leq 1$. Now we have the following thorem.

Theorem 2 Let N be an ordered normed vector space with closed positive cone and X be a compact convex subset of N. Let F be a continuous multi-valued mapping of X into N with nonempty values Fx such that for any $u, v \in Fx$, there exists $w \in Fx$ with $w \leq (u + v)/2$. Then there exist $x \in X$ and $f \in N'$ such that

$$f \geq 0, \quad \|f\| \leq 1, \quad f(x) = \max f(X)$$

and

$$\inf f(Fx) = \inf \Phi(Fx).$$

It is known that if N is a normed vector lattice (cf. [5]), then the canonical half-norm is the norm of the positive part, that is, $\Phi(x) = \|x^+\|$. Hence we have the following corollary of Theorem 2.

Corollary 3 Let L be a normed vector lattice and X be a compact convex subset of L. Let F be a continuous multi-valued mapping of X into L with nonempty values Fx such that for any u, v ε Fx, there exists w ε Fx with $w \leq (u + v)/2$. Then there exist x ε X and f ε N' such that

$$f \geq 0, \quad \|f\| \leq 1, \quad f(x) = \max f(X),$$

and

$$\inf f(Fx) = \inf \{\|y^+\| : y \in Fx\}.$$

REFERENCES

[1] Arendt W., Chernoff P.R. and Kato T., "A Generalization of Dissipativity and Positive semigroup", J. Optim. Theory, vol. 8, pp. 167-180, 1983.

[2] Fan K.,"Fixed-point and Minimax Theorems in Locally Convex Topological Linear Spaces", Proc. Nat. Acad. Sci. U.S.A., vol. 38, pp. 121-125, 1952.

[3] Kakutani S., "A Generalization of Brouwer's Fixed Point Theorem", Duke Math. J., vol. 8, pp. 457-459, 1941.

[4] König H., "On Certain Applications of the Hahn-Banach and Minimax Theorems", Arch. Math., vol. 21, pp. 583-591, 1970.

[5] Schaefer H.H., "Banach Lattices and Positive Operators", Berlin/Heidelberg/New York: Springer, 1974.

[6] Simons S., "An Existence Theorem for Quasiconcave Functions with Applications", an expanded version of the talk given at AMS meeting in Eugene, Oregon in August 1984.

A Vector-Minimization Problem in a Stochastic Continuous-Time n-Person Game

HANG-CHIN LAI Institute of Mathematics, National Tsing Hua University, Hsinchu, Taiwan, Republic of China

KENSUKE TANAKA Department of Mathematics, Niigata University, Niigata, Japan

1. Introduction

The authors investigated a cooperative n-person game with a discount factor in the case of discrete time space and a countable state space. (See Lai and Tanaka [10]). We now study the game system with a side-payment in case of the continuous time space. In this case we would apply the Kolmogorov forward differential equation to settle the transition probabilities which plays an important role in our game system. From the transition probability, we can define a one parameter contraction semigroup of operators on a space of all bounded real valued functions on the countable state space so that it has a fixed point. Then it will reduce to an optimal multistrategy on which the collective total expected multiloss will make a collective stability in the dynamic game system for a side-payment. This vector valued optimization problem in game theory is pretty different from usual multiobjective mathematical problems which one could refer to Yu [16], Tanino and Sawaragi [15], Lai and Ho [11] and their references. The reasons come from that the total expected multiloss is depending on each component which is steered by each player upon his chosen in the strategy space. The cooperative discounted Markov game considered in this paper has many connection with the noncooperative cases which are developed by the authors in [7], [8], [9] and [14].

We organize this paper into five sections. Section 2 is the formulation of our game system which will be discussed in this paper. In section 3, we give some necessary assumptions and the preliminaries on this game system in which we introduce the Kolomogorov forward differential equation which will derive a one paprmeter semigroup of operators on the space of bounded real valued functions of the state space S. Section 4 will give a cone dominated structure in which we would describe an optimal multi-strategy existed in the game so that the total expected multiloss makes a collective stability under a side-payment, and finally we summarized the main results of our present game system that under some conditions, there is a side-payment forced a multistrategy to be optimum, and the related properties, for example, a convex multiplier, lower support function and super-differential are discussed in the context.

2. Formulation of an n-person discounted dynamic game

Consider a game system as the following $2n+3$ objects:

$$(S,\ A^1,\ A^2, \cdots, A^n,\ q,\ r^1,\ r^2, \cdots, r^n, \alpha). \quad (2.1)$$

Here

(1) $S = \{1, 2, \cdots,\}$ is a countable set, namely, the state space;

(2) A^i is the action space of player i. We assume that each A^i is a compact metric space, $i \in N \equiv \{1, 2, \cdots, n\}$ is the player set;

(3) $q = q(\cdot|s, \bar{a})$ is a bounded function, namely a transition rate function, for any $(s, \bar{a}) \in S \times A$,

where $\bar{a} = (a^1, a^2, \cdots, a^n) \in A \equiv \prod_{i=1}^{n} A^i$.

(4) $r^i(s, \bar{a})$ is a real valued function on $S \times A$, namely a loss rate function of player $i \in N$.

(5) α is a positive number, as a discount factor.

In order to know easily our game system, we give some interpretation as follows.

For any time $t \in [0,\infty)$, all players observe the state of the game process and classify it to one possible state $s_t \in S$ which is determined by a stochastic process $\{X(t)\}$ at time t. Then all players choose their actions $\bar{a}_t \in A$ under some probability $\bar{\mu}_{s_t}$ such that each player $i \in N$ will loss $r^i(s_t,\bar{a}_t)$. After this the game process moves to a new state $s_{t'}$ according to the transition function $q(s_{t'}|s_t, \bar{a}_t)$ associated with a Markov process.

A strategy $\pi^i = \pi^i(t)$ of the player $i \in N$ chosen at time $t \in [0, \infty)$ is described as a function which maps the information of the game process from past history up to present state into the action space carried by i th player. Especially if $\pi^i = \pi^i(t)$ is independent of the time t and depends only on the present state, the strategy is said to be stationary. In this case, for any $t \in [0, \infty)$, there is mapping $\mu^i_t = \mu^i : S \to P(A^i)$, that is, $\mu^i \in [P(A^i)]^S$, for simplicity, we write $P(A^i)$ instead of $[P(A^i)]^S$ as the stationary strategy space of the player $i \in N$. Here $P(A^i)$ is the set of all probability measures on the Borel measurable space $(A^i, \beta(A^i))$ where $\beta(A^i)$ is the Borel field of the metric space A^i for each $i \in N$. Since A^i is assumed to be compact so $C(A^i)$, the space of all real-valued continuous functions on A^i, is a separable Banach space in the supremum norm topology. It is know that $P(A^i)$ is a weak* closed unit sphere in $M(A^i)$, the dual space of $C(A^i)$. Hence $P(A^i)$ is a metrizable compact convex subset of $M(A^i)$. (Cf. Lai and Tanaka [8]). Throughout this paper, we assume that each player uses only the stationary strategy.

In this game systems, as the multistrategy $\bar{\mu}$ is chosen by all players, the i th player's total expected discounted loss under an initial state $s \in S$ is defined by

$$
\begin{aligned}
\psi^i(\bar{\mu})(s) &= \int_0^\infty e^{-\alpha t} E_{\bar{\mu}}[r^i(s_t, \bar{\mu}) | s_0 = s] dt \\
&= \int_0^\infty e^{-\alpha t} \Big(\sum_{s' \in S} r^i(s', \bar{\mu}) p_t(s' | s, \bar{\mu}) \Big) dt \\
&= \int_0^\infty e^{-\alpha t} \Big(\sum_{s' \in S} \int_A r^i(s', \bar{a}) \bar{\mu}(d\bar{a} | s') p_t(s' | s, \bar{\mu}) \Big) dt
\end{aligned} \tag{2.2}
$$

$i = 1, 2, \cdots, n$. We denote a vector

$$\psi(\bar{\mu})(s) = (\psi^1(\bar{\mu})(s), \psi^2(\bar{\mu})(s), \cdots, \psi^n(\bar{\mu})(s)) \tag{2.3}$$

and ask whether there exists an optimal multistrategy $\bar{\mu}^*$ such that $\psi(\bar{\mu})(s)$ makes a collective stability rather than the individual optimal strategies of players under some convex dominated structure. In this paper, we will use a side-payment $d = (d^1, d^2, \cdots, d^n) \in \mathbb{R}^n$ with

$\sum_{i=1}^{n} |d^i| = 1$ to make a collective loss function

$$\langle d, \psi(\bar{\mu})(s)\rangle = \sum_{i=1}^{n} d^i \psi^i(\bar{\mu})(s) \tag{2.4}$$

and find an optimal multistrategy $\bar{\mu}^*$ to minimize the above equation. To this end, we need some assumptions.

3. Some assumptions and Preliminaries

For convenience, we let $\hat{i} = \{j \in N | j \neq i\}$, then for each $i \in N$, the strategy $\mu^j \in P(A^j)$ distinguishes

$$\mu^{\hat{i}} = (\mu^1, \cdots, \mu^{i-1}, \mu^{i+1}, \cdots, \mu^n) \in \prod_{j\neq i} P(A^j)$$

which is employed by all players except the i th player so that the multistrategy can be decomposed by

$$\bar{\mu} = (\mu^{\hat{i}}, \mu^i) \in \prod_{j\neq i} P(A^j) \times P(A^i) = P(A)$$

and if $\sigma^i \in P(A^i)$ then

$$(\mu^{\hat{i}}, \sigma^i) = (\mu^1, \cdots, \mu^{i-1}, \sigma^i, \mu^{i+1}, \cdots, \mu^n) \in P(A).$$

In the game system (2.1), we need the following assumptions upon the functions $q(\cdot|\cdot,\cdot)$ and $r(\cdot,\cdot)$ (cf. Lai and Tanaka [7]).

(A1) $q(s'|s,\bar{a})$ is a continuous function on $\bar{a} \in A$ for every $(s',s) \in S \times S$, moreover, for any $\bar{a} \in A$

$$q(s'|s,\bar{a}) \geqq 0 \quad \text{if} \quad s' \neq s$$

$$\sum_{s'\in S} q(s'|s, \bar{a}) = 0$$

$$|q(s'|s, \bar{a})| \leqq M.$$

(A2) For each $i \in N$, $r^i(s, \bar{a})$ is continuous on A for a given $s \in S$ and bounded on $S \times A$, that is, there is a constant $K > 0$ such that

$$|r^i(s, \bar{a})| \leqq K \quad \text{for all} \quad (s, \bar{a}) \in S \times A.$$

Under the assumptions (A1) and (A2), for a multistrategy $\bar{\mu} = (\mu^1, \cdots, \mu^n) \in P(A)$, we have

$$r^i(s, \bar{\mu}) = \int_A r^i(s, \bar{a})\bar{\mu}(d\bar{a}|s)$$

and

$$q(s'|s, \bar{\mu}) = \int_A q(s'|s, \bar{a})\bar{\mu}(d\bar{a}|s).$$

Then, the assumption (A1) can be derived to

(A1') $q(s'|s, \bar{\mu}) \geqq 0$ if $s' \neq s$

$\sum_{s' \in S} q(s's, \bar{\mu}) = 0$

$|q(s'|s, \bar{\mu})| \leqq M$, for any $s \in S$ and $\bar{\mu} \in P(A)$.

Now we introduce the transition rate matrix by

$$Q(\bar{\mu}) = \{q(s'|s, \bar{\mu})|s, s' \in S\}. \tag{3.1}$$

Under (A1') it is know (cf. Kakumanu [5]) that there exists a unique stationary transition probability matrix

$$F(t, \bar{\mu}) = \{p_t(s'|s, \bar{\mu})|s', s \in S\}$$

associated with $Q(\bar{\mu})$ satisfying the Kolomogorov forward differential equation:

$$\frac{d}{dt}F(t, \bar{\mu}) = F(t, \bar{\mu})Q(\bar{\mu}) \tag{3.2}$$

with an initial condition

$$F(0, \bar{\mu}) = I, \text{ the identity operator.}$$

Let $B(S)$ be the set of all bounded real valued functions on S. Then for any $u \in B(S)$ we define

$$T_t(\bar{\mu})u(s) = \sum_{s' \in S} u(s')p_t(s'|s, \bar{\mu}),\ t \in [0,\infty),$$

and one find easily that

(T1) $\lim_{t \to 0^+} T_t(\bar{\mu})u(s) = T_0(\bar{\mu})u(s) = u(s),$

(T2) $T_{t_1}(\bar{\mu})T_{t_2}(\bar{\mu})u(s) = T_{t_1+t_2}(\bar{\mu})u(s).$

Thus $\{T_t(\bar{\mu})\}_{t \in [0,\infty)}$ is a one parameter semigroup of operator on $B(S)$. Applying the Kolomogorov forward differential equations, it would have

$$\frac{d}{dt} T_t(\bar{\mu})u(s) = T_t(\bar{\mu})Q(\bar{\mu})u(s) \tag{3.3}$$

with an initial condition

$$T_0(\bar{\mu}) = I.$$

From (3.3) we see that $Q(\bar{\mu})$ plays an infinitesimal generator of the semigroup $\{T_t(\bar{\mu});\ t \in [0, \infty)\}$. We then obtain

Lemma 3.1. For $\bar{\mu} \in P(A)$ and $u \in B(S)$, we have

$$u(s) = \int_0^\infty e^{-\alpha t} T_t(\bar{\mu})(I-Q(\bar{\mu}))u(s)dt \tag{3.4}$$

and

$$T_t(\bar{\mu})u(s) = e^{tQ(\bar{\mu})}u(s) \qquad \text{for } t \in [0, \infty) \tag{3.5}$$

where α is the discount factor in the game system (2.1) and I is the identity operator on $B(S)$.

Proof. The proof is the same as [8, Lemma 2]. The formula (3.4) follows directly from the elementary calculation

$$\begin{aligned}
&\int_0^\infty e^{-\alpha t} T_t(\bar{\mu})(\alpha I-Q(\bar{\mu}))u(s)dt \\
&= \int_0^\infty \{\alpha e^{-\alpha t}\ T_t(\bar{\mu})u(s) - e^{-\alpha t}\frac{d}{dt}\ T_t(\bar{\mu})u(s)\}dt \\
&= -\int_0^\infty \frac{d}{dt}[e^{-\alpha t}T_t(\bar{\mu})u(s)]dt \\
&= T_0(\bar{\mu})u(s) \\
&= u(s).
\end{aligned}$$

Formula (3.5) follows directly from the equation (3.3). Thus the proof is complete. #

4. Cone convexity and dominated structure

In order to discuss the vector minimization of $\psi(\bar{\mu})(s)$, we need a convex dominated structure. Thus we consider a subset $L \subset \mathbb{R}^n$ such that

(i) $L \not\ni 0$

(ii) $L \cup \{0\} \equiv D$ is a convex cone with vertex at the origin 0

(iii) $L^+ \equiv \{y \in \mathbb{R}^n | \langle x, y\rangle > 0$ for all $x \in L\} \neq \phi$.

Denote by $L_1^+ \equiv \{y \in \mathbb{R}^n |\ y \in L^+,\ |y| = \sum_{i=1}^n |y_i| = 1\}$. We use the notation L, L^+, L_1^+ and D throughout.

Definition 1. A multistrategy $\bar{\mu}^*$ is said to be a D-convex solution for the minimization problem of $\psi(\bar{\mu})(s)$ (see (2.3)) for an intial state $s \in S$ if there is no other multistrategy $\bar{\mu}$ such that

$$\psi(\bar{\mu}^*)(s) \in \psi(\bar{\mu})(s) + L. \tag{4.1}$$

Let $E_s \equiv \{\psi(\bar{\mu})(s) |$ for all $\bar{\mu} \in P(A)\}$ and Ext $[E_s | D]$ be the set of all total expected discounted multiloss corresponding to D-convex solutions for a given initial state $s \in S$.

Lemma 4.1. Let $d \in L_1^+$ be a side-payment and

$$\langle d, \psi(\bar{\mu}_d^*)(s)\rangle \leqq \langle d, \psi(\bar{\mu})(s) \quad \text{for all } \bar{\mu} \in P(A).$$

Then $\bar{\mu}_d^*$ is a D-convex solution for an initial state s.

Proof. It is the scalarization of the usual vector minimization and the proof of Lemma follows from the fact that a feasible solution $\bar{\mu}^* \in P(A)$ of the vector minimization problem:

$$\underset{\bar{\mu}\in P(A)}{\text{minimize}}\ \psi(\bar{\mu})(s)$$

is an optimal (D-convex) solution of the game if it is an optimal solution of the scalar minimization problem:

$$\underset{\bar{\mu}\in P(A)}{\text{minimize}}\ \langle d, \psi(\bar{\mu})(s)\rangle \quad \text{for a given } d \in L_1^+. \qquad \#$$

Note that the inner product

$$\begin{aligned}\langle d, \psi(\bar{\mu})(s)\rangle &= \sum_{i=1}^{n} d^i \psi^i(\bar{\mu})(s) \\ &= \int_0^\infty e^{-\alpha t} \sum_{n=1}^{n} d^i E_{\bar{\mu}}[r^i(s_t, \bar{\mu}) | s_0 = s] dt \\ &= \int_0^\infty e^{-\alpha t} E_{\bar{\mu}}[\langle d, r(s_t, \bar{\mu})\rangle | s_0 = s] dt\end{aligned}$$

where $r(s, \bar{\mu}) = (r^1(s, \bar{\mu}), \cdots, r^n(s, \bar{\mu}))$.

Now for $\bar{\mu} \in P(A)$ we define a new one step transition probability matrix $P(\bar{\mu})$: $B(S) \to B(S)$ by

$$P(\bar{\mu}) = I + \frac{1}{M} Q(\bar{\mu}) \tag{4.2}$$

where M is the positive constant appeared in (A1). Evidently an element of $P(\bar{\mu})$ is defined by

$$p(s'|s, \bar{\mu}) = \delta_{s',s} + \frac{1}{M} q(s'|s, \bar{\mu}) \tag{4.3}$$

where

$$\delta_{s',s} = 0 \quad \text{if} \quad s' \neq s \quad \text{and} \quad \delta_{s,s} = 1.$$

Since $P(A)$ is compact and $r^i(s, \bar{\mu})$ and $p(s'|s, \bar{\mu})$ are continuous on $P(A)$ for any states $s', s \in S$, the supremum of these functions are surely attainable. Hence we can define an operator $T_d^i: B(A) \to B(S)$ depending on $\mu^{\hat{i}} \in \prod_{j \neq i} P(A^j)$ by

$$T_d^i u(s) = \min_{\sigma^i} \left[\frac{\langle d, r(s; \mu^{\hat{i}}, \sigma^i)\rangle}{M + \alpha} + \frac{M}{M+\alpha} P(\mu^{\hat{i}}, \sigma^i) u(s)\right]. \tag{4.4}$$

Since $0 < \frac{M}{M+\alpha} < 1$, it is easy to see that T_d^i is a contraction mapping on the Banach space $B(S)$ for the uniform norm. Hence there exists a fixed point u^* associated with a multistrategy $\bar{\mu}^* = (\mu_*^{\hat{i}}, \mu_*^i)$ such that

$$u^*(s) = T_d^i(\mu_*^{\hat{i}}) u^*(s). \tag{4.5}$$

Actually this fixed point u^* depends on $\mu_*^{\hat{i}}$, so we write

$$u^*(s) = u(\mu_*^{\hat{i}})(s). \tag{4.6}$$

Evidently from (4.4), (4.5) and (4.6), we obtain

$$u(\mu_*^{\hat{i}})(s) = \min_{\sigma^i \in P(A^i)} \left[\frac{\langle d, r(s; \mu_*^{\hat{i}}, \sigma^i)\rangle}{M + \alpha} + \frac{M}{M+\alpha} P(\mu_*^{\hat{i}}, \sigma^i) u(\mu_*^{\hat{i}})(s)\right]$$

$$\leqq \frac{\langle d, r(s; \mu_*^{\hat{i}}, \sigma^i)\rangle}{M + \alpha} + \frac{M}{M+\alpha} P(\mu_*^{\hat{i}}, \sigma^i) u(\mu_*^{\hat{i}})(s)$$

for all $\sigma^i \in P(A^i)$. Then from (4.5), the above inequality can be rewritten as

$$(\alpha I - Q(\hat{\mu}^i_*, \sigma^i))u(\hat{\mu}^i_*)(s) \leqq \langle d, r(s, \hat{\mu}^i_*, \sigma^i). \tag{4.7}$$

Substituting (4.7) into (3.4) (Lemma 1), we get

$$u(\hat{\mu}^i_*)(s) \leqq \int_0^\infty e^{-\alpha t} T_t(\hat{\mu}^i_*, \sigma^i)\langle d, r(s; \hat{\mu}^i_*, \sigma^i)\rangle dt$$

$$= \langle d, \int_0^\infty e^{-\alpha t} \sum_{s' \in S} r(s'; \hat{\mu}^i_*, \sigma^i) p_t(s' | s; \hat{\mu}^i_*, \sigma^i)\rangle dt$$

$$= \langle d, \psi(\hat{\mu}^i_*, \sigma^i)(s)\rangle. \tag{4.8}$$

On the other hand, for any $\bar{\mu} \in P(A)$, define

$$T_d(\bar{\mu})u(s) = \frac{\langle d, r(s, \bar{\mu})\rangle}{M + \alpha} + \frac{M}{M + \alpha} P(\bar{\mu})u(s), \quad u \in B(S).$$

Then by the same argument as $T^i_\alpha(\hat{\mu}^i)$ has done, one sees easily that $T_d(\bar{\mu})$ has a unique fixed point v^* which associates with the multi-strategy $\bar{\mu}^* = (\hat{\mu}^i_*, \mu^i_*)$. Thus we write

$$v^*(s) = u(\bar{\mu}^*)(s)$$

and

$$u(\bar{\mu}^*)(s) = T_d(\bar{\mu}^*)u(\bar{\mu}^*)(s).$$

This identity reduces to

$$u(\bar{\mu}^*)(s) = \langle d, \psi(\bar{\mu}^*)(s)\rangle. \tag{4.9}$$

As the Lemma 3 of [8] has shown that $u(\bar{\mu})(s)$ is continuous on $\bar{\mu} \in P(A)$. We want to show that as $\sigma^i \to \mu^i_*$

$$\langle d, \psi(\hat{\mu}^i_*, \sigma^i)(s)\rangle \longrightarrow \langle d, \psi(\bar{\mu}^*)(s)\rangle$$

and then the inequality

$$\langle d, \psi(\bar{\mu}^*)(s)\rangle \leqq \langle d, \psi(\hat{\mu}^i_*, \sigma^i)(s)\rangle \tag{4.10}$$

holds for any $\sigma^i \in P(A^i)$. To do this, the following lemma is essential. At first the expression (4.5) is reduced to

$$\alpha u(\hat{\mu}^i_*)(s) = \min_{\sigma^i} \{\langle d, r(s; \hat{\mu}^i_*, \sigma^i)\rangle + Q(\hat{\mu}^i_*, \sigma^i)u(\hat{\mu}^i_*)(s)\} .$$

For simplicity, we let

$$L^i(\hat{\mu}^i_*, \sigma^i)u(\hat{\mu}^i_*)(s) \equiv \langle d, r(s; \hat{\mu}^i_*, \sigma^i)\rangle + Q(\hat{\mu}^i_*, \sigma^i)u(\hat{\mu}^i_*)(s).$$

Then we have the following

Lemma 4.2. For each $i \in N$, there exists a multistrategy $\bar{\mu}^* \in P(A)$ such that

$$L^i(\bar{\mu}^*)u(\bar{\mu}^*)(s) = \min_{\sigma^i} L^i(\hat{\mu}^i_*, \sigma^i)u(\hat{\mu}^i_*)(s).$$

The proof of this lemma does mainly deal with Fan [1]. More detail is in Lai and Tanaka [8, Lemma 4]. As this lemma is established, the inequality (4.10) holds and we have the main results as next section.

5. Main results of n-person discounted game with a side-payment

The inequality (4.10) proves the following

Theorem 5.1. Let the game system (2.1) satisfying the assumptions (A1) and (A2). Then for a given side-payment $d \subset L_1^+ \subset \mathbb{R}^N$, there exists a cooperative optimal multistrategy $\bar{\mu}^*_d$ in the game (2.1) for an initial state $s \in S$. That is, $\bar{\mu}^*_d$ is a D-convex solution of the game (2.1), or equivalently

$$\psi(\bar{\mu}^*_d)(s) \in \text{Ext } [E_s|D].$$

Proof. The result follows from Lemma 4.2 and the inequality (4.10) which is derived in previous Section 4. #

Note that in Theorem 5.1, we suppose that a side-payment $d \in L_1^+ \subset \mathbb{R}^n$ is given so that it corresponds an optimal multistrategy for the collective stability in the game system (2.1). Next we would ask under what condition there will exist such a side-payment. In order to answer, we give the following.

Definition 2. A set $E \subset \mathbb{R}^n$ is said to be D-convex if $E + D$ is convex in $\mathbb{R}^n$.

As the authors had shown in [10] for the game in case of discrete-time space, we have a converse version of Theorem 5.1 in our present game system (2.1) which we stat as following

Theorem 5.2. For a multistrategy $\bar{\mu}^* \in P(A)$, if the set E_s (see Section 4) is a D-convex set in $\mathbb{R}^n$ for an initial state $s \in S$ and satisfying the condition

$$c\ell[E_s + D - \psi(\bar{\mu}^*)(s)] \cap (-c\ell\ D) = 0 , \tag{5.1}$$

then there exists a side-payment $\tilde{d} \in L_1^+$ such that $\bar{\mu}^*$ will minimize a collective loss function $\langle\tilde{d}, \psi(\bar{\mu})(s)\rangle$. That is,

$$\psi(\bar{\mu}^*)(s) \in \text{Ext}\ [E_s | D] . \tag{5.2}$$

Proof. The proof is similar to Theorem 4.2 of Lai and Tanaka [10]. For completeness we resketch the proof as follows. We would show that there is a nonzero

$$\hat{d} \in L^+ \cap (E_s + D - \psi(\bar{\mu}^*)(s))_0^+$$

where F_0^+ denotes the nonnegative polar cone of the set F. We suppose contrary that

$$L^+ \cap (E_s + D - \psi(\bar{\mu}^*)(s))_0^+ = \emptyset. \tag{5.3}$$

and write

$$W_s(\bar{\mu}^*) \equiv E_s + D - \psi(\bar{\mu}^*)(s) .$$

Then since L^+ and $W_s(\bar{\mu}^*)_0^+$ are convex disjoint and $\text{int}(L^+) \neq \emptyset$, by the separation theorem on the Euclidean space, we see that there exists a nonzero continuous linear functional $\langle d, \cdot\rangle$ on $\mathbb{R}^n$ such that

$$\sup_{x \in L^+} \langle d, x\rangle \leqq \inf_{y \in W_s(\bar{\mu}^*)_0^+} \langle d, y\rangle .$$

But $0 \in W_s(\bar{\mu}^*)_0^+$ we find that for $x \in L^+$, $\langle d, x\rangle \leqq 0$, and hence

$$d \in -(L^+)_0^+ = -c\ell\ D. \tag{5.4}$$

On the orher hand, $D = L \cup \{0\}$ is a convex cone with vertex at 0, thus as x tends to zero in L^+, the continuity of d will imply $\langle d, y\rangle \geqq 0$ for any $y \in W(\bar{\mu}^*)_0^+$. Then

$$d \in W_s(\bar{\mu}^*)_0^{++} = c\ell\ W_s(\bar{\mu}^*) \ . \tag{5.5}$$

From (5.4) and (5.5) we see that

$$d \in c\ell[W_s(^-*)] \cap (-c\ell\ D).$$

This contradicts to (5.1). So the identity (5.3) is not true. Hence there exists $d \neq 0$ in $L \cap (E_s+D-\psi(\bar{\mu}^*)(s))_0^+$ such that for any $\psi(\bar{\mu})(s) \in E_s$ and $x \in D$,

$$\langle d, \psi(\bar{\mu})(s) + x - \psi(\bar{\mu}^*)(s)\rangle \geqq 0,$$

that is,

$$\langle d, \psi(\bar{\mu}^*)(s)\rangle \leqq \langle \hat{d}, \psi(\bar{\mu})(s)+x\rangle$$

holds for any $\bar{\mu} \in P(A)$. Since $\langle d, x\rangle \geqq 0$ for any $x \in D$, thus by letting $\tilde{d} = d/|\hat{d}|$, we have $\tilde{d} \in L_1^+$ and

$$\langle \tilde{d}, \psi(\bar{\mu}^*)(s)\rangle \leqq \langle \tilde{d}, \psi(\bar{\mu})(s)\rangle. \tag{5.6}$$

That is, $\psi(\bar{\mu}^*)(s) \in \text{Ext}\ [E_s|D]$. #

Definition 3. A side-payment $d \in L_1^+$ satisfying (5.6) is called a D-multiplier of $\bar{\mu}^*$ for an initial state s, and denote $D_s(\bar{\mu}^*)$ as the set of all D-multipliers of $\bar{\mu}^*$ for an initial state $s \in S$.

From this definition we have the following

Theorem 5.3. Let $d_1 \in D_s(\bar{\mu}_1)$ and $d_2 \in D_s(\bar{\mu}_2)$. Then

$$\langle d_1-d_2, \psi(\bar{\mu}_1)(s)-\psi(\bar{\mu}_2)(s)\rangle \leqq 0. \tag{5.7}$$

Proof. The proof follows directly by the definition of D-multipliers. #

Let $L_0^+ = L^+ \cup \{0\}$. We define a lower support function K_s on L_0^+ by

$$K_s(d) = \inf_{\bar{\mu}\in P(A)} \langle d, \psi(\bar{\mu})(s)\rangle \quad \text{for } d \in L_0^+ \ . \tag{5.8}$$

A vector $x \in \mathbb{R}^n$ is said to be a <u>super-differential</u> of K_s at $\hat{d} \in L_0^+$ if

$$K_s(d) - K_s(\hat{d}) \leqq \langle d-\hat{d},\ x \rangle. \tag{5.9}$$

We denote the set of all super-differentials of K_s at $\hat{d}$ by $\tilde{\partial}K_s(d)$. Then the following theorem holds immediately.

<u>Theorem 5.4</u>. A multistrategy $\bar{\mu}^*$ is a D-convex solution associated with a D-multiplier $d \in D_s(\bar{\mu}^*)$ if and only if $\psi(\bar{\mu}^*)(s)$ is a super-gradient of the lower support function K_s at $d \in L_1^+$, that is,

$$\psi(\bar{\mu}^*)(s) \in \tilde{\partial}K_s(d). \tag{5.10}$$

It is remarkable that the optimization problem of the game system with side-payment is first developed by the authors in [10] for discrete state space and discrete time space. This paper is studied in the case of a stochastic continuous-time model with n players which is pretty different from the discrete-time case because in the continuous-time case, the stationary transition probabilities will satisfy the Kolomogorov forward differential equation for our game system and so the vector minimization problem of this pesudo dynamic game system is treated with a different method from the previous paper [10]. In this paper, the state space is assumed to be countable. It is natural to ask about this pseudo dynamic game with a side-payment in which the state space is assumed to be usual metric space. To treat this optimization problem, it seems necessary to apply the integro-differential equation given in Feller [4]. One can refer to Lai and Tanaka [8]. We hope the further results concerning this game system can appear elsewhere.

References

[1] K. Fan, Fixed points and minimax theorems in locally convex topological linear space, <u>Proc. Nat. Acad. Sci.</u> U.S.A., 39(1078), 452-471.

[2] K. Fan, Minimax theorems, Proc. Nat. Acad. Sci. U.S.A., 39(1953), 42-47.

[3] A. Federgruen, On N-person stochastic games with denumerable state space, Adv. Appl. Probab. 10(1078), 452-471.

[4] W. Feller, On the integro-differential equations of purely discontinuous Markov processes, Trans. Amer. Math. Soc., 48 (1940), 488-515.

[5] P. Kakumanu, "Continuous Time Markov Decision Models with Applications to Optimization Problems," Tech. Rep. 63, Dept. O.R. Cornell University 1969.

[6] P. Kakumanu, Continuous discounted Markov decision model with countable state and action space, Ann. Math. Statist., 42(1971), 919-926.

[7] H.C. Lai and K. Tanaka, Noncooperative n-person game with a stopped set, J. Math. Anal. Appl., 85(1982), 153-171.

[8] H.C. Lai and K. Tanaka, On an N-person noncooperative Markov game with a metric state space, J. Math. Anal. Appl., 101(1984), 78-96.

[9] H.C. Lai and K. Tanaka, A noncooperative n-person semi-Markov game with a separable metric state space, Appl. Math. Optim., 11 (1984), 23-42.

[10] H.C. Lai and K. Tanaka, On a D-convex solution of a cooperative m-person discounted Markov game, J. Math. Anal. Appl., to appear.

[11] H.C. Lai and C.P. Ho, Duality theorem and nondifferentiable convex multiobjective programming, J. Optim. Theory Appl, to appear.

[12] A Maitra and T. Parthasarathy, On stochastic games, J. Optim. Theory Appl., 5(1970), 289-300.

[13] K. Tanaka, On some vector valued Markov game, Japan J. Appl. Math., to appear.

[14] K. Tanaka and H.C. Lai, A two-person zero-sum Markov game with a stopped set, J. Math. Anal. Appl., 86(1982), 54-68.

[15] T. Tanino and Y. Sawaragi, Duality theory in multiobjective programming, J. Optim. Theory Appl., 27(1979), 509-529.

[16] P.L. Yu, Cone convexity, cone extreme points and nondominated solutions in decision problem with multiobjectives, J. Optim. Theory Appl., 14(1974), 319-377.

Calculation of the λ-Function for Several Classes of Normed Linear Spaces

ROBERT H. LOHMAN Department of Mathematics and Sciences, Kent State University, Kent, Ohio

THADDEUS J. SHURA Department of Mathematics and Sciences, Kent State University at Salem, Salem, Ohio

Let X be a normed space and let $x \in B_X$. A triple (e, y, λ) is said to be amenable to x in case $e \in \text{ext}(B_X)$, $y \in B_X$, $0 < \lambda \leq 1$ and $x = \lambda e + (1-\lambda)y$. In this case, $\lambda(x)$ is defined by

$$\lambda(x) = \sup\{\lambda: (e, y, \lambda) \text{ is amenable to } x\}.$$

X is said to have the λ-property if each $x \in B_X$ admits an amenable triple. Informally, if X has the λ-property, then the λ-function $\lambda: B_X \to [0, 1]$ gives a measure of the "largest" positive weight assigned to an extreme point e in the different representations $x = \lambda e + (1 - \lambda)y$ as x ranges over the closed unit ball B_X of X. If X has the λ-property, define $\lambda(X) = \inf\{\lambda(x): x \in B_X\}$. X is said to have the uniform λ-property in case $\lambda(X) > 0$.

These ideas were introduced and studied in [2]. In particular, spaces with the λ-property or uniform λ-property were shown to have geometric features which may be of potential use in applications. For example, if X has the λ-property, then every real-valued convex function on B_X which attains its maximum value must attain this value at a member of $\text{ext}(B_X)$. Consequently, if X is a Banach space with the λ-property, then B_X is the closed convex hull of its set of extreme points (Theorem 3.3 of [2]). A much stronger conclusion holds if X has the uniform λ-property (Theorem 3.1 of [2]). Namely, if $0 < \lambda < \lambda(X)$, then for each $x \in B_X$ there is a sequence (e_k) in $\text{ext}(B_X)$ such that for every n, we have

$$\|x - \sum_{k=1}^{n} \lambda(1 - \lambda)^{k-1} e_k\| \leq (1 - \lambda)^n.$$

Evidence is given in [2] that neither the λ-property nor uniform λ-property is rare. It is shown that many spaces of the type $\ell_1(X)$, $\ell_\infty(X)$, $c(X)$, $C_X(T)$ possess one of these properties and formulas for the λ-functions of these spaces are obtained. Also, every finite-dimensional normed space X possesses the uniform λ-property and, for each $x \in B_X$, we have $\lambda(x) \geq (1 + n)^{-1}$, where n is the dimension of X considered as a real vector space.

Since normed spaces with the λ-property or uniform λ-property satisfy strong geometric conditions, it is of interest to expand the catalog of spaces with one of these properties. That is the purpose of this note. We first consider ℓ_∞-direct sums, noting that parts of the following result were obtained independently by David Trautman.

Theorem 1. Let (X_n) be a sequence of normed spaces and let $X = (\oplus \sum X_n)_{\ell_\infty}$. Assume that each space X_n has the λ-property and let λ_n denote its λ-function.

(a) X has the λ-property if and only if there is a finite subset P_0 of the set P of positive integers such that $\inf \{\lambda_n(X_n): n \in P \setminus P_0\} > 0$. In this case, if $x = (x_n) \in B_X$, then $\lambda(x) = \inf_n \lambda_n(x_n)$.

(b) X has the uniform λ-property if and only if $\inf_n \lambda_n(X_n) > 0$.

Proof. (a) "⇐" Write $B_n = B_{X_n}$. Let $x = (x_n) \in B_X$ and note that we have $\alpha = \inf_n \lambda_n(x_n) > 0$. If $0 < \epsilon < \alpha$, then for each n, there exist $e_n \in \text{ext}(B_n)$, $y_n \in B_n$ and λ_n, where $\lambda_n(x_n) - \epsilon < \lambda_n \leqq 1$, such that $x_n = \lambda_n e_n + (1 - \lambda_n)y_n$. Since $0 < \alpha - \epsilon \leqq \lambda_n(x_n) - \epsilon$, Proposition 1.2 of [2] guarantees that we can write $x_n = (\alpha - \epsilon)e_n + (1 - \alpha - \epsilon)z_n$, where $z_n \in B_n$. Thus, $x = (\alpha - \epsilon)e + (1 - \alpha - \epsilon)z$, where $e = (e_n) \in \text{ext}(B_X)$, $z = (z_n) \in B_X$. Since $\alpha - \epsilon > 0$, X has the λ-property and $\lambda(x) \geqq \alpha - \epsilon$. Letting ϵ tend to 0 yields $\lambda(x) \geqq \alpha$. On the other hand, assume $x = \lambda e + (1 - \lambda)y$, where $e = (e_n) \in \text{ext}(B_X)$, $y = (y_n) \in B_X$ and $0 < \lambda \leqq 1$. For each n, we have $e_n \in \text{ext}(B_n)$, $y_n \in B_n$ and $x_n = \lambda e_n + (1 - \lambda)y_n$. Therefore, $\lambda \leqq \lambda_n(x_n)$ for all n. This implies $\lambda \leqq \alpha$ and, taking the supremum over all such λ, shows that $\lambda(x) \leqq \alpha$.

"⇒" Assume that $\inf \{\lambda_n(X_n): n \in P \setminus P_0\} = 0$ for each finite subset P_0 of P. Then there is an increasing sequence (n_k) in P and vectors $z_{n_k} \in B_{n_k}$ such that $\lambda_{n_k}(z_{n_k}) < k^{-1}$ for all k. If we define $(x_n) \in B_X$ by $x_{n_k} = z_{n_k}$, $k \in P$, and $x_n = 0$ otherwise, then there is no triple amenable to x. Consequently, X fails to have the λ-property.

(b) "⇐" Follows from (a), noting that $\alpha \geqq \inf_n \lambda_n(X_n)$ and hence is bounded away from 0, independent of x.

"⇒" Fix $m \in P$ and let $x \in B_m$. Define $\tilde{x} = (x_n) \in B_X$ by $x_n = x$ if $n = m$ and $x_n = 0$ otherwise. Given $0 < \epsilon < \lambda(\tilde{x})$, we can write $\tilde{x} = \lambda\tilde{e} + (1 - \lambda)\tilde{y}$, where $\tilde{e} = (e_n) \in \text{ext}(B_X)$, $\tilde{y} = (y_n) \in B_X$ and $0 < \lambda(\tilde{x}) - \epsilon \leqq \lambda \leqq 1$. Since $e_m \in \text{ext}(B_m)$, $y_m \in B_m$ and $x = \lambda e_m + (1 - \lambda)y_m$, we have $\lambda_m(x) \geqq \lambda \geqq \lambda(\tilde{x}) - \epsilon \geqq \lambda(X) - \epsilon$. Since ϵ is arbitrary, we obtain $\lambda_m(x) \geqq \lambda(X)$. Therefore, $\inf_n \lambda_n(X_n) \geqq \lambda(X) > 0$.

Remarks 2. If each space X_n is strictly convex, then $\lambda_n(x_n) = (1 + \|x_n\|)/2$ and the formula for $\lambda(x)$ given in part (a) of Theorem 1 coincides with that given in [2] (see Remark 1.14). It is also clear that Theorem 1 holds for ℓ_∞-direct sums of arbitrary families of normed spaces.

Next, we consider ℓ_1-direct sums of normed spaces. The situation is more complicated than for ℓ_∞-direct sums and we obtain upper and lower estimates for $\lambda(x)$. A precise formula for $\lambda(x)$ is not yet known.

Theorem 3. Let (X_n) be a sequence of normed spaces and let $X = (\oplus \sum X_n)_{\ell_1}$. Assume that each space X_n has the λ-property and let λ_n denote its λ-function. Then X has the λ-property and if $x = (x_n) \in B_X$, $x \neq 0$, we have $\underline{\lambda}(x) \leq \lambda(x) \leq \overline{\lambda}(x)$, where

$$\underline{\lambda}(x) = (1 - \|x\| + 2\max\{\lambda_n(x_n/\|x_n\|)\|x_n\| : x_n \neq 0\})/2$$

and

$$\overline{\lambda}(x) = \min\{\sup_n \lambda_n(x_n), (1 - \|x\| + 2\max_n\|x_n\|)/2\}.$$

Proof. Since $\|x_n\| \to 0$, the sequence $\{\lambda_n(x_n/\|x_n\|)\|x_n\| : x_n \neq 0\}$ is either finite or is a null sequence. In either case, there is a positive integer N such that

$$\lambda_N(x_N/\|x_N\|)\|x_N\| = \max\{\lambda_n(x_n/\|x_n\|)\|x_n\| : x_n \neq 0\}.$$

Let $0 < \epsilon < \lambda_N(x_N/\|x_N\|)\|x_N\|$. There is a triple (e_N, u_N, λ_N) amenable to $x_N/\|x_N\|$ such that $\lambda_N(x_N/\|x_N\|) - \epsilon < \lambda_N \leq 1$. By making λ_N slightly smaller, if necessary, we may assume that $1 - \lambda_N\|x_N\| \neq 0$. Since $x_N/\|x_N\| = \lambda_N e_N + (1 - \lambda_N)u_N$, we obtain $x_N = \lambda_N\|x_N\|e_N + (1 - \lambda_N\|x_N\|)v_N$, where $v_N = [(1 - \lambda_N)\|x_N\|/1 - \lambda_N\|x_N\|]u_N$. Define $e = (e_n) \in \mathrm{ext}(B_X)$ and $y = (y_n) \in X$ by

$$e_n = \begin{cases} 0, & n \neq N \\ e_N, & n = N \end{cases}, \qquad y_n = \begin{cases} [2/(1 + \|x\| - 2\lambda_N\|x_N\|)]x_N, & n \neq N \\ \dfrac{2(1 - \lambda_N\|x_N\|)v_N + (\|x\| - 1)e_N}{1 + \|x\| - 2\lambda_N\|x_N\|}, & n = N \end{cases}$$

and let $\lambda = (1 - \|x\| + 2\lambda_N\|x_N\|)/2$. It is clear that $x = \lambda e + (1 - \lambda)y$. Moreover, a routine computation shows

$$\begin{aligned} \|y\| &\leq \frac{2(\|x\| - \|x_N\|) + 2(1 - \lambda_N\|x_N\|)\|v_N\| + 1 - \|x\|}{1 + \|x\| - 2\lambda_N\|x_N\|} \\ &\leq \frac{2(\|x\| - \|x_N\|) + 2(1 - \lambda_N)\|x_N\| + 1 - \|x\|}{1 + \|x\| - 2\lambda_N\|x_N\|} \\ &= 1. \end{aligned}$$

Since

$$\begin{aligned} 1 \geq \lambda &= [1 - \|x\| + 2\lambda_N(x_N/\|x_N\|)\|x_N\|]/2 + [\lambda_N - \lambda_N(x_N/\|x_N\|)]\|x_N\| \\ &\geq \underline{\lambda}(x) - \epsilon\|x_N\|, \end{aligned}$$

we obtain $\lambda(x) \geq \underline{\lambda}(x) - \epsilon\|x_N\|$. Letting ε tend to zero shows $\underline{\lambda}(x) \leq \lambda(x)$, which establishes the first inequality and the fact that X has the λ-property.

That $\lambda(x) \leq (1 - \|x\| + 2\max\|x_n\|)/2$ is established in the same manner as in the proof of Theorem 1.1 of [2]. In addition, if $e = (e_n) \in \mathrm{ext}(B_X)$, $y = (y_n) \in B_X$ and $0 < \lambda \leq 1$ are such that $x = \lambda e + (1 - \lambda)y$, we have $e_m \in \mathrm{ext}(B_{X_m})$ for some m. Since $\|y_m\| \leq 1$ and $x_m = \lambda e_m + (1 - \lambda)y_m$, we have $\lambda \leq \lambda_m(x_m) \leq \sup_n \lambda_n(x_n)$. It follows that $\lambda(x) \leq \overline{\lambda}(x)$.

Remarks 4. Theorem 3 easily extends to ℓ_1-direct sums of arbitrary families of normed spaces. Also, if each X_n is strictly convex and $x_n \neq 0$, then we have $\lambda_n(x_n/\|x_n\|) = 1$ and $\lambda_n(x_n) = (1 + \|x_n\|)/2$. Hence, in this case,

$$\underline{\lambda}(x) = \overline{\lambda}(x) = \lambda(x) = (1 - \|x\| + 2\max\|x_n\|)/2,$$

which agrees with the formula given in [2] (see Remark 1.12). Observe that the space X of Theorem 3 never has the uniform λ-property. To see this, let m be a positive integer and choose $x_k \in X_k$ such that $\|x_k\| = 1$, $1 \leq k \leq m$. Then the unit vector $x = (x_1/m, \ldots, x_m/m, 0, 0, \ldots)$ of X satisfies $\lambda(x) \leq \overline{\lambda}(x) \leq 1/m$ and so $\lambda(X) = 0$. Finally, if one of the spaces X_m fails to have the λ-property, there is a unit vector $x \in X_m$ which fails to admit an amenable triple. If $\tilde{x}$ is defined as in the proof of Theorem 1, it is easy to check that $\tilde{x}$ fails to admit an amenable triple in the space X of Theorem 3. Therefore, X fails to have the λ-property.

We next consider Lorentz sequence spaces of the type d(a, 1). Before proceeding to our results, we review some notation and known facts about this class of spaces. For our purposes, let $a = (a_n) \in c_0 \backslash \ell_1$ be a positive, strictly decreasing sequence with $a_1 = 1$. The space d(a, 1) consists of all real sequences $x = (x_n)$ in c_0 such that

$$\sup \sum |x_{\pi(n)}| a_n < \infty,$$

where the supremum is taken over all permutations π of the set of positive integers. If $\|x\|$ is taken to be this supremum, then d(a, 1) is a Banach space. There is considerable literature on the Lorentz sequence spaces d(a, 1) and, more generally, on the Lorentz sequence spaces d(a, p). The interested reader should refer to references [1], [3]-[6].

If $x = (x_n) \in d(a, 1)$ and $x \neq 0$, write

$$M_1(x) = \|x\|_\infty, \qquad F_1(x) = \{n : |x_n| = M_1(x)\}.$$

If $c_{F_1(x)}$ denotes the characteristic function of $F_1(x)$, define

$$M_2(x) = \|x - xc_{F_1(x)}\|_\infty, \qquad F_2(x) = \{n : |x_n| = M_2(x)\}$$

and write $M_3(x) = \|x - xc_{F_1(x)} - xc_{F_2(x)}\|_\infty$, etc. Then $M_1(x)>0$, $M_k(x)\downarrow 0$ and if $M_k(x)>0$, then $M_k(x) > M_{k+1}(x)$. Also, if $M_j(x) > 0$, $M_k(x) > 0$ and $j \neq k$, then $F_j(x)$ and $F_k(x)$ are disjoint. We let $N(x) = \{k : M_k(x) - M_{k+1}(x) > 0\}$ and, if $k \in N(x)$, we define

$$n_k(x) = \operatorname{card}\left(\bigcup_{i=1}^{k} F_i(x)\right), \qquad s_k(x) = \sum_{n=1}^{n_k(x)} a_n.$$

If we take $n_0(x) = 0$, then we can write $\|x\|$ as

$$\|x\| = \sum_{k \in N(x)} M_k(x)[s_k(x) - s_{k-1}(x)].$$

The extreme points of $B_{d(a,\,1)}$ are precisely the vectors of the form

$$e = \left(\sum_{n=1}^{k} a_n\right)^{-1}\left(\sum_{n=1}^{k} \epsilon_n e_{i_n}\right),$$

where $i_1 < \ldots < i_k$, $\epsilon_n = \pm 1$ and (e_i) is the standard unit vector basis of $d(a, 1)$ (see [3] and [7]).

Also, observe that if $b = (b_n)$ is a sequence of $\pm$ 1's, then the mapping $T_b: d(a, 1) \to d(a, 1)$, defined by $T_b((x_n)) = (b_n x_n)$, is a linear isometry of $d(a, 1)$ onto $d(a, 1)$. In addition, any rearrangement of coordinates is a linear isometry of $d(a, 1)$ onto $d(a, 1)$. Finally, note that if $T: X \to Y$ is a linear isometry of the normed space X onto a normed space Y and $x \in B_X$, then x admits an amenable triple if and only if $T(x)$ admits an amenable triple. In this case, $\lambda(x) = \lambda(T(x))$.

Theorem 5. Let $x = (x_n) \in B_{d(a,\,1)}$, $x \neq 0$. Then

$$\lambda(x) \geqq \sup_{k \in N(x)} [M_k(x) - M_{k+1}(x)]s_k(x).$$

Consequently, the space $d(a, 1)$ has the λ-property.

Proof. For each n, choose $b_n = \pm 1$ such that $b_n x_n = |x_n|$. Then, by considering $T_b(x)$, where $b = (b_n)$, we may assume $x_n \geqq 0$ for all n.

If $k \in N(x)$, let $e \in \operatorname{ext}(B_{d(a,\,1)})$ be defined by

$$e = (s_k(x))^{-1}\left(\sum_{i=1}^{k} \sum_{n \in F_i(x)} e_n\right)$$

and let $\lambda = [M_k(x) - M_{k+1}(x)]s_k(x)$. There are two cases to consider.

First, if $\lambda < 1$, define $y = (y_n) \in d(a, 1)$ by

$$y_n = \begin{cases} x_n/(1 - \lambda), & n \notin \bigcup_{i=1}^{k} F_i(x) \\ \dfrac{M_i(x) - [M_k(x) - M_{k+1}(x)]}{1 - \lambda}, & n \in F_i(x) \text{ for some } i \in \{1, \ldots, k\} \end{cases}$$

If $1 \leqq i \leqq k - 1$, then

$$M_i(x) - [M_k(x) - M_{k+1}(x)] \geqq M_{k+1}(x)$$

and

$$M_i(x) - [M_k(x) - M_{k+1}(x)] \geqq M_{i+1}(x) - [M_k(x) - M_{k+1}(x)].$$

It follows that

$$\|y\| = \{\|x\| - [M_k(x) - M_{k+1}(x)]s_k(x)\}/(1 - \lambda) = (\|x\| - \lambda)/(1 - \lambda) \leqq 1.$$

Since $x = \lambda e + (1 - \lambda)y$, we have $\lambda(x) \geqq \lambda$, completing the proof in this case.

Next, assume $\lambda = 1$. We have

$$1 = \lambda = [M_k(x) - M_{k+1}(x)]s_k(x) \leqq M_k(x)s_k(x) \leqq \sum_{i=1}^{k} M_i(x)[s_i(x) - s_{i-1}(x)] \leqq \|x\| \leqq 1.$$

This implies $M_{k+1}(x) = 0$, all inequalities collapse to equality and all $M_i(x)$'s equal $M_k(x)$; that is, we have $k = 1$. Therefore, x is of the form

$$x = (s_1(x))^{-1}\Big(\sum_{n \in F_1(x)} e_n\Big).$$

Consequently, $x \in \operatorname{ext}(B_{d(a,\,1)})$ and $\lambda(x) = 1$, completing the proof.

Although Theorem 5 gives a lower estimate for the λ-function in $d(a, 1)$, we do not yet have a good upper estimate for this function, in general. If, however, we restrict our attention to unit vectors whose support is finite, then the exact value of $\lambda(x)$ can be determined.

Theorem 6. Let $x = (x_n) \in d(a, 1)$ with $\|x\| = 1$. If x has finite support, then

$$\lambda(x) = \max_{k \in N(x)} [M_k(x) - M_{k+1}(x)]s_k(x).$$

Consequently, $d(a, 1)$ fails to have the uniform λ-property.

Proof. As in the proof of Theorem 5, we may assume $x_n \geqq 0$ for all n. Since any rearrangements of coordinates is a linear isometry of $d(a, 1)$ onto $d(a, 1)$, we may also assume that (x_n) is nonincreasing. Therefore, there is an integer N such that $x_1 \geqq \cdots \geqq x_N > 0$ and $x_n = 0$ for $n \geqq N + 1$. If $x \in \operatorname{ext}(B_{d(a,\,1)})$, then $\lambda(x) = 1$ and, since the nonzero values of x are the constant $M_1(x)$ and $M_2(x) = 0$, we obtain

$$1 = \|x\| = M_1(x)\Big(\sum_{n=1}^{n_1(x)} a_n\Big) = [M_1(x) - M_2(x)]s_1(x).$$

Thus, we may assume $x \notin \operatorname{ext}(B_X)$.

Suppose that $x = \lambda e + (1 - \lambda)y$, where $0 < \lambda < 1$ and $\|y\| = 1$. There are positive integers $i_1 < \cdots < i_k$ such that

$$e = c_k\Big(\sum_{n=1}^{k} \epsilon_n e_{i_n}\Big),$$

where $\epsilon_n = \pm 1$ and $c_k = \Big(\sum_{n=1}^{k} a_n\Big)^{-1}$. Then

$$x_{i_n} = \lambda c_k \epsilon_n + (1 - \lambda)y_{i_n}, \quad n = 1, \ldots, k$$

$$x_n = (1 - \lambda)y_n, \quad n \notin \{i_1, \ldots, i_k\}.$$

Define the sets I, J, K, and L by

$$I = \{n : i_n \leqq N\} \qquad K = \{n : i_n > N\}$$

$$J = \{1, \ldots, N\} \setminus \{i_1, \ldots, i_k\} \qquad L = \{N+1, N+2, \ldots\} \setminus \{i_1, \ldots, i_k\}.$$

In what follows, we will denote the nth coordinate of a sequence z by either z_n or $z(n)$. Then, noting that the nth coordinate of the sequence $x = \lambda e + (1 - \lambda)y$ is zero if $n \geqq N + 1$ and $e(n) = 0$ if $n \notin \{i_1, \ldots, i_k\}$, we have

$$1 = \|x\| = \sum_{n=1}^{N} x_n a_n$$

$$= \sum_{n\in I}[\lambda c_k \epsilon_n + (1-\lambda)y_{i_n}]a_{i_n} + \sum_{n\in J}(1-\lambda)y_n a_n$$
$$+ \sum_{n\in K}[\lambda e(i_n) + (1-\lambda)y_{i_n}]a_{i_n} + \sum_{n\in L}[\lambda e(n) + (1-\lambda)y_n]a_n$$

$$= \lambda c_k(\sum_{n\in I}\epsilon_n a_{i_n}) + \lambda c_k(\sum_{n\in K}\epsilon_n a_{i_n}) + (1-\lambda)\sum_{n=1}^{\infty} y_n a_n$$

$$= \lambda c_k(\sum_{n=1}^{k}\epsilon_n a_{i_n}) + (1-\lambda)\sum_{n=1}^{\infty} y_n a_n.$$

Since $c_k(\sum_{n=1}^{k}\epsilon_n a_{i_n}) \leqq 1$ and $\sum_{n=1}^{\infty} y_n a_n \leqq \sum_{n=1}^{\infty}|y_n|a_n \leqq \|y\| \leqq 1$, it follows that

(1) $$c_k(\sum_{n=1}^{k}\epsilon_n a_{i_n}) = 1$$

and

(2) $$\sum_{n=1}^{\infty} y_n a_n = 1.$$

From (1), we obtain $\sum_{n=1}^{k}\epsilon_n a_{i_n} = \sum_{n=1}^{k} a_n$, which forces $\epsilon_n = 1$ for $1 \leqq n \leqq k$ and $i_n = n$ for $1 \leqq n \leqq k$. If $y_n < 0$ for some n, then $\sum_{n=1}^{\infty} y_n a_n < \sum_{n=1}^{\infty}|y_n|a_n$, forcing $\sum_{n=1}^{\infty} y_n a_n < 1$. Therefore, (2) implies $y_n \geqq 0$ for all n.

This means that $e = c_k(\sum_{n=1}^{k} e_n)$ and

(3) $$x_n = \lambda c_k + (1-\lambda)y_n, \quad 1 \leqq n \leqq k$$

(4) $$x_n = (1-\lambda)y_n, \quad n \geqq k+1.$$

If $N < k$, then $N + 1 \leqq k$ and (3) implies $x_{N+1} \geqq \lambda c_k > 0$, which is impossible. Therefore, we have $k \leqq N$. By (3) and (4), y is of the form

$$y = \left(\frac{x_1 - \lambda c_k}{1-\lambda}, \ldots, \frac{x_k - \lambda c_k}{1-\lambda}, \frac{x_{k+1}}{1-\lambda}, \ldots, \frac{x_N}{1-\lambda}, 0, 0, \ldots\right).$$

Then

$$1 = \|y\| \geqq [\sum_{n=1}^{k-1}(x_n - \lambda c_k)a_n + x_{k+1}a_k + (x_k - \lambda c_k)a_{k+1} + \sum_{n=k+2}^{N} x_n a_n](1-\lambda)^{-1}$$

$$= [\sum_{n=1}^{N} x_n a_n - x_k a_k - x_{k+1}a_{k+1} + x_{k+1}a_k + x_k a_{k+1}$$
$$- \lambda c_k(\sum_{n=1}^{k} a_n) + \lambda c_k a_k - \lambda c_k a_{k+1}](1-\lambda)^{-1}$$
$$= [1 - \lambda + (\lambda c_k + x_{k+1} - x_k)(a_k - a_{k+1})](1-\lambda)^{-1}.$$

Since $a_k - a_{k+1} > 0$, it must be the case that $\lambda c_k + x_{k+1} - x_k \leqq 0$. Therefore, we have $\lambda \leq (x_k - x_{k+1})c_k^{-1}$. But $\lambda > 0$ implies $x_k > x_{k+1}$. Hence, there is an integer j such that $M_j(x) = x_k$, $M_{j+1}(x) = x_{k+1}$ and $n_j(x) = k$ (i.e., $s_j(x) = c_k^{-1}$). This yields

$$\lambda(x) \leq \max_{j \in N(x)} [M_j(x) - M_{j+1}(x)] s_j(x).$$

An appeal to Theorem 5 proves the asserted equality for $\lambda(x)$.

To prove that $d(a, 1)$ fails to have the uniform λ-property, let $m = 2k$ be an even positive integer. Define the unit vector x by

$$x = [\sum_{n=1}^{m} (\frac{m - n + 1}{m}) a_n]^{-1} [\sum_{n=1}^{m} (\frac{m - n + 1}{m}) e_n].$$

From the first part of Theorem 6, we have

$$\begin{aligned} \lambda(x) &= \frac{1}{m}[\sum_{n=1}^{m} (\frac{m - n + 1}{m}) a_n]^{-1} (\sum_{n=1}^{m} a_n) \\ &= (\sum_{n=1}^{m} a_n)/[\sum_{n=1}^{m} a_n + \sum_{n=1}^{m} (m - n) a_n] \\ &\leq (\sum_{n=1}^{m} a_n)/[\sum_{n=1}^{m} a_n + k(\sum_{n=1}^{k} a_n)] \\ &\leq (\sum_{n=1}^{m} a_n)/[\sum_{n=1}^{m} a_n + (\frac{m}{4}) \sum_{n=1}^{m} a_n] \\ &= 4/(4 + m). \end{aligned}$$

As a consequence, $\lambda(d(a, 1)) = 0$, completing the proof.

REFERENCES

1. Altshuler, Z., Casazza, P. G. and Lin, B. L., On symmetric basic sequences in Lorentz sequence spaces, Israel J. Math., 15(1973), 140-155.
2. Aron, R. M. and Lohman, R. H., A geometric function determined by extreme points of the unit ball of a normed space, (preprint).
3. Calder, J. R. and Hill, J. B., A collection of sequence spaces, Trans. Amer. Math. Soc., 152(1970), 107-118.
4. Casazza, P. G. and Lin, B. L., On symmetric basic sequences in Lorentz sequence spaces. II, Israel J. Math., 17(1974), 191-218.
5. Casazza, P. G. and Lin, B. L., On Lorentz sequence spaces, Bull. Inst. Math. Acad. Sinica, 2(1974), 233-240.
6. Casazza, P. G. and Lin, B. L., Some geometric properties of Lorentz sequence spaces, Rocky Mount. J. Math., 7(1977), 683-698.
7. Davis, W. J., Positive bases in Banach spaces, Rev. Roum. Math. Pures Et Appl., 16(1971), 487-492.

Existence of Positive Eigenvectors and Fixed Points for A-Proper Type Maps in Cones

W. V. PETRYSHYN Department of Mathematics, Rutgers University, New Brunswick, New Jersey

0. Introduction

Let X be a real Banach space, K a cone in X, $\Gamma = \{X_n, P_n\}$ a projectionally complete scheme for X with $P_n(K) \subset K$ for each $n \in N^+$, and D a bounded neighborhood of 0 in X with $D_K = D \cap K$, ∂D_K the boundary and $\bar{D}_K$ the closure of D_K relative to K.

The purpose of this paper is two-fold. First, using the approach of [14, 23] and the index theory for A-proper vector fields developed by Fitzpatrick and Petryshyn in [8], in Section 1 we give a complete proof of Theorem 1, which is an improved version of Theorem 3 stated without proof in [23] and which asserts that if $T, F:\bar{D}_K \to K$ are bounded maps such that $T_\mu = I-T-\mu F$ is A-proper w.r.t. Γ for each $\mu \in [0,1/\gamma]$ and some fixed $\gamma > 0$ with $I_K(T,D_K) \neq \{0\}$, then there exist $x \in \partial D_K$ and $\lambda \in (0,d/\delta]$ such that $x-Tx = \lambda Fx$ provided
(i): $\delta = \inf\{\|Fx\| : x \in \partial D_K\} > d\gamma/\sigma$, where $\sigma = \sigma(K) \in [1/2,1]$ is the "quasinormality" constant associated with K and $d = \sup\{\|x-Tx\| : x \in \partial D_K\}$.

Note that if $T = 0$, then clearly $I_K(\hat{0},D_K) = \{1\} \neq \{0\}$, $d = b = \sup\{\|x\| : x \in \partial D_K\} > 0$, (i) reduces to $\delta > b\gamma/\sigma$, and the requirement that $I-\mu F$ be A-proper for each $\mu \in (0,1/\gamma]$ is equivalent to the assumption that $qI-F: \bar{D}_K \to X$ is A-proper for each $q \geq \gamma$, i.e., F is P_γ-compact in the sense of [17, 18]. In this case, Theorem 1 yields a new result, i.e., Corollary 1, which establishes the existence of an eigenvalue $\eta \geq \delta/b$ and the corresponding eigenvector $x \in \partial D_K$ of F such that $Fx = \eta x$. Since every k-ball-contractive map $F:\bar{D}_K \to K$ is P_γ-compact for any fixed $\gamma > k$, it follows from Corollary 1 that there exist $\eta \geq \delta/b$ and $x \in \partial D_K$ such that $Fx = \eta x$ provided $\delta = \inf\{\|F,x\| : x \in \partial D_K\} > bk/\sigma$. The latter result was proved by Massabo and Stuart in [14] for the case when $F:\bar{D}_K \to K$ is k-set-contractive and

Supported by NSF Grant MCS80-3002

K is a normal cone with σ being the constant of normality. For the extension of some eigenvalue results in [12, 9, 14] to multivalue k-set-contractive maps see [23].

Let us add that when $T_\mu \equiv I-T-\mu F$ is A-proper for each $\mu > 0$ and $\delta > 0$ (i.e., when $\gamma = 0$), then Theorem 1 and its Corollary 1 (when $T = 0$) were first proved by Fitzpatrick and Petryshyn in [8] by a method which cannot be extended to the more general class of maps studied in this paper. In particular, Corollary 1 contains Theorem 5.5 of Krasnoselskii [12] when F is completely continuous.

Using Theorem 1 and the fixed point index properties in [8], it is shown in Theorem 2 in Section 2 that if D^1 and D^2 are bounded neighborhoods of 0 in X, T, F, C: $\bar{D}_K^2 \to K$ are bounded mappings such that $I-T-\mu F$ is A-proper for each $\mu \in [0,1/\gamma]$, condition (i) holds on ∂D_K^2, then under very general boundary conditions (see (C1) and (C2)) the map T has a fixed point $x_0 \in \bar{D}_K^2 \setminus \bar{D}_K^1$. In case $T\mu = I-T-\mu F$ is A-proper for each $\mu > 0$ and $\delta > 0$, Theorem 2 was proved by the author in [24] by a method which cannot be extended to maps treated in this paper. Various special cases of Theorem 2, when T and/or F are completely continuous, k-ball-contractive, or T is P_1-compact, will be considered. In particular, it will be shown that, for suitable choices of C and F, Theorem 2 contains, on the one hand, fixed point theorems which are of "cone-expansion and cone-compression" type in the sense of Krasnoselskii [12] for various classes of maps T and, on the other hand, it contains fixed point theorems for T when T has a Frechet derivative T_0 at 0 and/or T_∞ at ∞ along K with T_0 and/or T_∞ being of type $L_1^+(X)$ or $L_1^-(X)$ in the sense of Amann [1]. The exact references to various authors will be given in Section 2.

1. Basic eigenvalue results

We first introduce some definitions and state those results which will be needed in the sequel. Let X be a real Banach space, $\{X_n\} \subset X$ a sequence of increasing finite dimensional subspaces, and for each n in N^+ let P_n be a linear projection of X onto X_n such that $P_n(x) \to x$ in X for each x in X. We use "$\to$" to denote the strong convergence in X and set $\Gamma_\alpha = \{X_n, P_n\}$ when $\alpha = \sup_n \|P_n\|$.

The following class of A-proper (and, in particular, P_γ-compact) mappings introduced and studied by the author in [18, 19] (see [20] for the survey) proved to be quite useful since, on the one hand, this class includes completely continuous, P_γ-compact and ball-condensing vector fields, strongly accretive maps and their perturbations by k-ball-contractive operators and others and, on the other hand, the notion of the A-proper mapping proved to be also useful in the constructive solvability of operator equations; furthermore, the theory of A-proper maps can be applied directly to the solvability of differential equations when the latter are formulated as operator equations with operators acting between two different spaces.

(1) Definition.

Let D be any subset of X. A map $T:D \subset X \to X$ is said to be A-proper w.r.t. $\Gamma_\alpha = \{X_n, P_n\}$ iff $T_n = P_n T|_{D_n} : D_n \equiv D \cap X_n \to X_n$ is continuous for each $n \in N^+$ and if $\{x_{n_j} | x_{n_j} \in D_{n_j}\}$ is any bounded sequence such that $T_{n_j}(x_{n_j}) \to g$ for some g in X, then there exist a subsequence $\{x_{n_k}\}$ and $x \in D$ such that $x_{n_k} \to x$ as $k \to \infty$ and $Tx = g$. For a fixed $\gamma \geq 0$, the map $F:D \to X$ is said to be P_γ-compact if $qI-F$ is A-proper for each q dominating γ (i.e., $q \geq \gamma$ if $\gamma > 0$ and $q > 0$ if $\gamma = 0$).

The notion of an A-proper mapping evolved from the concept of a P-compact map introduced by the author in [17] to obtain constructive fixed point theorems and surjectivity theorems for monotone type maps. In terms of Definition 1 , we say that $F:D \to X$ is P-compact if $qI-F$ is A-proper for each $q > 0$, (i.e., F is P_o-compact). For subsequent use, we recall that, for a bounded set Q, $\beta(Q) = \inf\{r > 0: Q \subset \bigcup_{j=1}^{n} B_j$, B_j is a ball in X with radius $r\}$ and $\alpha(Q) = \inf\{d > 0: Q \subset \bigcup_{j=1}^{n} Q_j$ with $\mathrm{diam}(Q_j) \leq d\}$ are ball-measure and set-measure if noncompactness of Q, respectively. A continuous map $F:D \subset X \to X$ is said to be k-ball-contractive (k-set-contractive) if $\beta(F(Q)) \leq k\beta(Q)$ $(\alpha(F(Q)) \leq k\alpha(Q))$ for each bounded set $Q \subset D$ and some $k \geq 0$. Clearly, if $F:D \to X$ is completely continuous, then F is 0-ball-contractive (and 0-set-contractive). If D is a open set and $F:\overline{D} \to X$ is strictly semicontractive in the sense of Browder [3], then F is k-ball-contractive for some $k \in (0,1)$ (see [21,26]). A map $F:D \to X$ is said to be ball-condensing if $\beta(F(Q)) < \beta(Q)$ for each $Q \subset D$ such that $\beta(Q) \neq 0$. For various properties of $\beta(Q)$ and $\alpha(Q)$ and examples of the above classes of mappings see [6, 16, 25].

It is known that A-properness is invariant under compact perturbations. In some cases we can say more to indicate the generality of

A-proper maps. Thus, for example, if $F:D \to X$ is k-ball-contractive and $T:X \to X$ is c-accretive and continuous, then $T_\mu \equiv T+\mu F$ is A-proper w.r.t. Γ_1 for each $\mu \in (-ck^{-1}, ck^{-1})$. If $c = \alpha = 1$, the same holds when F is ball-condensing and $\mu \in [-1,1]$. In particular, if $F:D \to X$ is k-ball-contractive, then F is P_γ-compact for any fixed $\gamma > k$. For other examples of A-proper maps see [4, 6, 15, 20, 27] and the more recent papers by the author, Browder, Fitzpatrick, Milojevič, Webb, Fitzpatrick-Petryshyn, Petryshyn-Yu, Massabo-Nistri, Xianling, Dupuis, Toland, and others.

Since the fixed point index for A-proper vector fields introduced in [8] will play an essential role in what follows, we introduce this notion here and state those of its properties which we shall use.

Let $K \subset X$ be a closed and convex set and suppose that $P_n(K) \subset K$ for each $n \in N^+$. Let D be a bounded and open set in X such that $D_K \equiv D \cap K \neq \emptyset$ and let $\bar{D}_K$ and ∂D_K the closure and the boundary of D_K relative to K. Suppose $T:\bar{D}_K \to K$ is such that $I-T$ is A-proper w.r.t. Γ_α and $x-Tx \neq 0$ for $x \in \partial D_K$. Then the A-properness of $I-T$ implies the existence of $N_o \in N^+$ such that $x-T_n(x) \neq 0$ for $x \in \partial D_{K_n}$ and $n \geq N_o$, where $K_n = K \cap X_n$, $D_n = D \cap X_n$, $D_{K_n} = D_n \cap K_n$, $T_n = P_nT|\bar{D}_{K_n}$ and ∂D_{K_n} is the boundary of D_{K_n} relative to K_n. Following in the spirit of the multivalued topological degree defined by Browder and Petryshyn in [4], the following notion was introduced in [8].

(2) Definition.

Let X, Γ_α, K, D_K and $T:\bar{D}_K \to K$ be as above and $x \neq Tx$ for $x \in \partial D_K$. Then the fixed point index of T on D with respect to K, denoted by $I_K(T,D_K)$, is a subset of $Z' = N \cup \{-N\} \cup \{\pm\infty\}$ determined as follows:

(i) An integer $m \in I_K(T,D_K)$ if there is an infinite sequence $\{n_j\} \subset N^+$ such that $i_{K_{n_j}}(T_{n_j}, D_{K_{n_j}}) = m$ for each $j \in N^+$.

(ii) $\pm\infty \in I_K(T,D_K)$ if there is an infinite sequence $\{n_j\} \subset N^+$ such that $\lim_j i_{K_{n_j}}(T_{n_j}, D_{K_{n_j}}) = \pm\infty$.

Here "i_{K_n}" is the finite dimensional fixed point index defined via the Brouwer degree. Note that the observation immediately preceding Definition 2 implies that $I_K(T,D_K) \neq \emptyset$.

In what follows we need the following properties:

(P1) If $I_K(T,D_K) \neq \{0\}$, then T has a fixed point in D_K.

(P2) If $T(x) \equiv 0$ for $x \in \bar{D}_K$ and $0 \in D_K$, then $I_K(\hat{0},D_K) = \{1\}$.

(P3) If $F:[0,1]\times\bar{D}_K \to X$ is an A-proper homotopy such that $x-F(x,t) \in K$ for $x \in \bar{D}_K$ and $t \in [0,1]$, $F(t,x) \neq 0$ for $x \in \partial D_K$ and $t \in [0,1]$, then $I_K(I-F(T,\cdot),D_K)$ is constant in $t \in [0,1]$.

(P4) If $D \supset D_1 \cup D_2$, $\bar{D} = \bar{D}_1 \cup \bar{D}_2$, $D_1 \cap D_2 = \emptyset$ and if $x \notin \partial D_K = \partial D_{1K} \cup \partial D_{2K}$, then $I_K(T,D_K) \subset I_K(T,D_{1K}) + I_K(TD_{2K})$ with equality holding if either $I_K(T,D_{1K})$ or $I_K(T,D_{2K})$ is a singleton integer (by convention: $\infty + (-\infty) = Z'$).

We say that if $Q \subseteq X$, then $F:[0,1]\times Q \to X$ is an A-<u>proper homotopy</u> (w.r.t. Γ) if F_n is continuous and if $\{t_{n_j}\} \subset [0,1]$, $\{x_{n_j} | x_{n_j} \in Q_{n_j}\}$ is bounded, and $P_{n_j}F(x_{n_j},t_{n_j}) \to g$ for some g in X, then $\{n_j\}$ has a subsequence $\{n_k\}$ such that $t_{n_k} \to t_o \in [0,1]$, $x_{n_k} \to x_o \in Q$, and $F(x_o,t_o) = g$.

We recall that $K \subset X$ is called a <u>cone</u> provided K is closed, $ax+by \in K$ for $a,b \in R^+ = [0,\infty)$ and $x,y \in K$, and $K \cap (-K) = \{0\}$. If K is a cone, we write $x \leq y$ iff $y-x \in K$; in particular, $y \geq 0$ if 0 is the zero element in X and $y \in K$. The norm $\|\cdot\|$ in X is said to be <u>semimonotone</u> w.r.t. K if for each $x,y \in K$ with $0 \leq x \leq y$ we have $\|x\| \leq \tau\|y\|$ for some constant $\tau \geq 1$ independent of x and y. In case $\tau = 1$, the norm $\|\cdot\|$ is said to be monotone. A cone K is called <u>normal</u> if the norm $\|\cdot\|$ of X is semimonotone w.r.t. K; this is equivalent to the assertion that $\|x+y\| \geq \gamma\|x\|$ for all $x,y \in K$ and some $\gamma \in (0,1]$ (in fact, $\gamma = \tau^{-1}$). K is said to be <u>total</u> if $\overline{K-K} = X$.

To extend some eigenvalue results obtained in [14] for normal cones, the author introduced in [23] the concept of a "quasinormal cone"; a cone K in X is said to be quasinormal iff there exist a vector $y \in \overset{\circ}{K} = K\setminus\{0\}$ and a constant $\sigma > 0$ such that

$$\|x+y\| \geq \sigma\|x\| \quad \text{for all } x \text{ in } K. \tag{1.1}$$

It is easy to see that if K is quasinormal, then $\sigma \leq 1$ and that $\|x+\lambda y\| \geq \sigma\|x\|$ for all $x \in K$ and $\lambda > 0$. Note also that every normal cone K in X is quasinormal with $\sigma = \gamma$ and <u>any</u> $y \in \dot{K}$. However, as was noted in [23], the cones of nonnegative functions in the Banach spaces $C^m(\bar{Q})$ for $m \geq 1$, Hölder spaces $C^\mu(\bar{Q})$ for $\mu \in (0,1)$, or Sobolev spaces $W_p^m(Q)$ for $m \geq 1$ and $p \in [1,\infty)$ are not normal but they are quasinormal with $\sigma = 1$. The same cones in $C(\bar{Q})$ and $L_p(Q)$ are known to be normal. Here $Q \subset R^n$ is a bounded domain.

It was shown by the author in [23] that the idea of quasinormality is useful in the study of the existence of positive eigenvalues and eigenvectors in $\overset{\circ}{K}$ for the problem

$$x - Tx = \lambda Fx, \tag{1.2}$$

where $T,F:\bar{D}_K \to K$ are operators in the class of maps for which a topological index is defined. In [23] we obtained the existence results for (1.2) for the case when T and F are multivalued and k-set-contractive maps under conditions which correspond to those used in [14] when $T = 0$ and F is singlevalued.

The usefulness of the idea of quasinormality of a cone K in X stimulated further study. Thus, it was shown by Lami-Dozo in [13] that if X is a Hilbert space and K is an arbitrary cone in X, then K is quasinormal and $\sigma = 1$. Furthermore, it was shown in [13] that if K is any cone in a Banach space X and $y \in \dot{K}$, then there is a number $\sigma = \sigma_y \in (0,1]$ such that (1.1) holds for all x in K. Thus, every cone is quasinormal, and the real problem is to find the optimal constant σ. This problem was studied in [5]. If one sets $\sigma(y) = \inf\{\|x+y\| \,(\|x\|^{-1}): x \in \overset{\circ}{K}\} > 0$ for each $y \in \overset{\circ}{K}$ and defines the quasinormality constant $\sigma = \sigma(K)$ by $\sigma(K) = \sup\{\sigma(y): y \in \overset{\circ}{K}\}$, then it was shown in [5] that for any cone in X one has the estimate

$$1/2 \le \sigma \le 1 \tag{1.3}$$

and in general the constant $1/2$ in (1.3) is the best possible.

In view of the above results, our first theorem in this paper is the following improvement of Theorem 3 stated in [23] without proof.

<u>Theorem 1</u>. Let K be a cone in X, $\Gamma_\alpha = \{X_n, P_n\}$ a projection scheme in X with $P_n(K) \subset K$ for each $n \in N^+$, and let $D \subset X$ be a bounded neighborhood of $0 \in X$. Suppose $T,F:\bar{D}_K \to K$ are bounded maps such that $I-T:\bar{D}_K \to X$ is A-proper w.r.t. Γ_α and the following conditions hold:

(H1) $I_K(T,D_K) \neq \{0\}$.

(H2) There is a constant $\gamma \ge 0$ such that $T_\mu = I-T-\mu F:\bar{D}_K \to X$ is A-proper w.r.t. Γ for each $\mu \in (0,1/\gamma]$ if $\gamma > 0$ and for each $\mu > 0$ if $\gamma = 0$.

(H3) $\delta = \inf\{\|Fx\| : x \in \partial D_K\} > \frac{d\gamma}{\sigma}$, where $\sigma \in [1/2,1]$ is the quasinormality constant associated with K and $d = \sup\{\|x-Tx\| : x \in \partial D_K\}$.

Then there are $x \in \partial D_K$ and $\lambda \in (0,d/\delta]$ such that $x - Tx = \lambda Fx$, i.e., (1.2) holds.

<u>Proof</u>. We shall first consider the case when $\gamma > 0$. Now, choose

$\varepsilon \in (0,\sigma)$ such that $\delta > \gamma d/(\sigma-\varepsilon)$. Then, by definition of $\sigma(y)$ and $\sigma = \sigma(K)$, there exists $y = y_\varepsilon \in \overset{\circ}{K}$ such that $\sigma(y)+\varepsilon > \sigma(K)$ and $\|x+y\| \geq \sigma(y)\|x\| > (\sigma-\varepsilon)\|x\|$ for all x in X. Hence, for all $t > 0$ and x in $\overset{\circ}{K}$, $\|x+ty\| = \|t(1/t)x+y)\| = t\|(1/t)x+y\| > t(\sigma-\varepsilon)\|(1/t)x\| = (\sigma-\varepsilon)\|x\|$.

Now, if we let $\eta = 1/\gamma$ and $T_\eta = I-T-\eta F$, then T_η is A-proper by assumption (H2). Let $m \in N^+$ be arbitrary and consider the map $H:[0,1]\times\bar{D}_K \to X$ given by $H(t,x) = T_\eta(x)-tmy$. It is obvious that H is an A-proper homotopy. We claim that $H(t,x) \neq 0$ for $x \in \partial D_K$ and $t \in [0,1]$. If not, then there exist $t_o \in [0,1]$ and $x_o \in \partial D_K$ such that $H(t_o,x_o) = 0$. Suppose first that $t_o = 0$. Then $H(0,x_o) = T_\eta(x_o) = 0$, i.e., $x_o-Tx_o = \eta Fx_o$. Hence, by the definition of d and (H3), $d \geq \|x_o-Tx_o\| = \eta\|Fx_o\| \geq \eta\delta > \eta\,\frac{\gamma d}{\sigma} = \frac{d}{\sigma}$ since $\eta = \frac{1}{\gamma}$. Thus $\sigma > 1$, in contradiction to (1.3). Suppose now that $t_o \in (0,1]$ and $x_o-Tx_o = \eta Fx_o+t_omy$. Then the preceding discussion, the choice of $\varepsilon > 0$, and condition (H3) imply that

$$d \geq \|x_o-Tx_o\| = \|\eta Fx_o+t_omy\| > (\sigma-\varepsilon)\eta\|Fx_o\| \geq (\sigma-\varepsilon)\eta\delta$$

$$> (\sigma-\varepsilon)\eta\cdot\gamma d/(\sigma-\varepsilon) = d \quad \text{since} \quad \eta = 1/\gamma,$$

which is impossible. Thus, $H(t,x) \neq 0$ for $x \in \partial D_K$ and $t \in [0,1]$. Therefore, by the homotopy property (P3), $I_K(T+\eta F,D_K) = I_K(T+\eta F+my,D_K)$. Now, the above equality holds for any $m \in N^+$ and this, we claim, implies that $I_K(T+\eta F,D_K) = \{0\}$. Indeed, if this were not the case, then for each m in N^+ we could choose $x_m \in \bar{D}_K$ such that $(I-T-\eta F)(x_m) = my$. Since $y \neq 0$ and $(I-T-\eta F)(\bar{D}_K)$ is bounded because T and F are bounded maps, we have a contradiction when m is large. Consequently, $I_K(T+\eta F,D_K) = \{0\}$.

Consider now the mapping $H_o:[0,1]\,\bar{D}_K \to X$ given by $H_o(t,x) = x-Tx-t\eta Fx$ for $t \in [0,1]$ and $x \in \bar{D}_K$. It follows from (H2) and the boundedness of F that H_o is an A-proper homotopy. By (H1) we have $I_K(T,D_K) \neq \{0\}$, while we have just proven that $I_K(T+\eta F,D_K) = \{0\}$ Hence there exist some $t \in (0,1)$ and $x \in \partial D_K$ such that $H_o(t,x) = 0$, i.e., $x = Tx+\lambda Fx$ with $x \in \partial D_K$ and $\lambda = t\eta \in (0,d/\delta]$.

In case $\gamma = 0$, we let $\eta = 1/\varepsilon$ and choose $\varepsilon \in (0,\sigma)$ such that $\delta > d\varepsilon/(\sigma-\varepsilon)$. As before, there exists $y = y_\varepsilon \in \overset{\circ}{K}$ such that $\sigma(y) > \sigma-\varepsilon$ and $\|x+y\| \geq (\sigma-\varepsilon)\|x\|$ for all x in K. The rest of the argument is the same as in the case when $\gamma > 0$.

Q.E.D.

We note that Theorem 1 includes a number of special cases. In what follows we only state here two new special cases which extend some known results.

Corollary 1. Let K, σ, Γ and D be as in Theorem 1. Let $\gamma \geq 0$ be fixed and let $F:\bar{D}_K \to K$ be a bounded and P_γ-compact map such that

$$\delta = \inf\{\|Fx\| : x \in \partial D_K\} > b\gamma/\sigma, \; b = \sup\{\|x\| : x \in \partial D_K\}. \tag{1.4}$$

Then there exist $x \in \partial D_K$ and $\lambda \in (0, b/\delta]$ such that $x = \lambda Fx$.

Proof. Corollary 1 follows from Theorem 1 when $T = 0$. Indeed, when $T = 0$, then $I_K(\hat{0}, D_K) = \{1\} \neq \{0\}$ by (P2), i.e., (H1) holds. The hypothesis (H2) reduces in this case to the condition that $F:\bar{D}_K \to K$ is P_γ-compact, while (H3) reduces to (1.4). Consequently, Corollary 1 follows from Theorem 1.

Remark 1. When $\gamma = 0$, Theorem 1 and Corollary 1 were proved in [8] by a method which cannot be extended to the more general class of maps when $\gamma > 0$. Since a completely continuous map $F:\bar{D}_K \to K$ is P_o-compact, Theorem V1.1 in [12] is a special case of Corollary 1 when the Banach space X has a projectionally complete scheme $\Gamma = \{X_n, P_n\}$. Moreover, since every k-ball-contractive map F can be easily shown to be P_γ-compact for any fixed $\gamma > k$ when $\Gamma = \Gamma_1$ by using the argument of Webb, it follows that Corollary 1 in [23] and, in particular, Theorem 1 in [14] (for K normal) follow from Corollary 1 for k-ball-contractive maps. Furthermore, since a ball-condensing map $F:\bar{D}_K \to K$ is P_1-compact w.r.t. Γ_1, Corollary 1 shows that there exist $x \in \partial D_K$ and $\lambda \in (0, b/\delta]$ such that $x = \lambda F$ provided $\delta > b/\sigma$.

Corollary 2. Let K, σ and D be as in Theorem 1 and let $\Gamma = \Gamma_1$. Suppose that $T:\bar{D}_K \to K$ is k_o-ball-contractive with $k_o \in [0,1)$ and $I_K(T,D) \neq \{0\}$. Let $F:\bar{D}_K \to K$ be a k-ball-contractive map with $k \geq 0$ and such that

$$\delta = \inf\{\|Fx\| : x \in \partial D_K\} > \frac{dk}{\sigma(1-k_o)},$$

where $d - \sup\{\|x-Tx\| : x \, c \, \partial D_K\}$. Then there exist $x \in \partial D_K$ and $\lambda \in (0, d/\delta]$ such that $x - Tx = \lambda Fx$.

Proof. Note that when we take γ such that $\gamma > \frac{k}{1-k_o}$ and $\delta > d\gamma/\sigma$, then to deduce Corollary 2 from Theorem 1, it suffices to show that $T_\mu = I-T-\mu F$ is A-proper w.r.t. Γ_1 for each $\mu \in (0, 1/\gamma]$. But, for any such μ, the map $T+\mu F$ is (k_o+k/γ)-ball-contractive with

$k_o + k/\gamma < k_o + k(1-k_o)/k = 1$. Hence, by the results of Webb [27], T_μ is A-proper. Q.E.D.

Let us note that for some applications, and from the practical point of view, the following version of Theorem 1 may prove to be useful since we don't have to compute $d = \inf\{\|x-Tx\| : x \in \partial D_K\}$ in case K is a normal cone.

Theorem 1'. If K is a normal cone in X, then the conclusion of Theorem 1 remains valid with $\lambda \in (0, \tau b/\delta]$ if condition (H3) is replaced by

(H3') $\delta = \inf\{\|Fx\| : x \in \partial D_K\} > b\gamma/\sigma$, where $b = \sup\{\|x\| : x \in \partial D_K\}$ and $\sigma = 1/\tau$ with $\tau \geq 1$ such that $\|u\| \leq \tau\|v\|$ if $u,v \in K$ and $0 \leq u \leq v$.

Proof. The proof of Theorem 1' is similar to that of Theorem 1. It suffices to show that if we let $\eta = 1/\gamma$ if $\gamma > 0$ (and $\eta = 1/\varepsilon$ with $\varepsilon > 0$ such that $\delta > b\varepsilon/\sigma = b\varepsilon\tau$ if $\gamma = 0$) and define $H:[0,1]\times\bar{D}_K \to X$ by $H(t,x) = T_\eta(x) - tmy$, then $H(t,x) \neq 0$ for $t \in [0,1]$ and $x \in \partial D_K$. If we can show that $H(t,x) \neq 0$, then $I_K(T+\eta F, D_K) = \{0\}$ and this together with (H1) yields the conclusion. To show that $H(t,x) \neq 0$ for $t \in [0,1]$ and $x \in \partial D_K$, we first consider the case when $\gamma > 0$ (the case when $\gamma = 0$ is handled similarly). Suppose that there are $x_o \in \partial D_K$ and $t_o \in [0,1]$ such that $H(t_o,x_o) = x_o - Tx_o - \eta Fx_o - t_o my = 0$. Then $x_o = Tx_o + \eta Fx_o + t_o my > \eta Fx_o$ and thus, since $\|\cdot\|$ semimonotone on X and (H3') holds, we have $b \geq \|x_o\| \geq \sigma\eta\|Fx_o\| \geq \sigma\eta\delta > \sigma\eta\cdot b\gamma/\sigma = b$, a contradiction.

Q.E.D.

2. Existence of positive fixed points

Using Theorem 1 and the properties of the fixed point index stated in Section 1, we are now in position to extend Theorem 2 for cones proved in [24] so as to obtain the following new theorem concerning the existence of positive fixed points.

Theorem 2. Let K be a cone in X, $\Gamma = \{X_n, P_n\}$ a projectional scheme as in Theorem 1, D^1 and D^2 bounded neighborhoods of 0 in X with $\bar{D}^1 \subset D^2$, and let $T:\bar{D}^2_K \to K$ be a bounded map such that $I-T:\bar{D}^2_K \to X$ is A-proper w.r.t. Γ. Assume also that the following conditions hold:

(C1) There is a bounded map $C:\bar{D}^1_K \to K$ and $N_o \in N^+$ such that the restriction C_n of P_nC to $\bar{D}^1_{K_n}$ is continuous and $C_n(x) \neq \mu x$ for $x \in \partial D^1_{K_n}$, $\mu \geq 1$ and $n \geq N_o$; furthermore, $x \neq tT_n(x)+(1-t)C_n(x)$ for $x \in \partial D^1_{K_n}$, $t \in [0,1]$ and $n \geq N_o$.

(C2) There is a bounded map $F:\bar{D}^2_K \to K$ which satisfies conditions (H2) and (H3) of Theorem 1 and $x-Tx \neq \lambda Fx$ for all $x \in \partial D^2_K$ and $\lambda \in (0,d/\delta]$.

Then there exists $x_o \in \bar{D}^2_K \setminus \bar{D}^1_K$ such that $x_o = T(x_o)$. The same assertion is true if we assume that condition (C1) holds on D^2_K while (C2) holds on D^1_K.

Proof. First, we may assume without loss of generality that T has no fixed points on ∂D^1_K and on ∂D^2_K. Second, we will only provide a detailed proof for the case where $\gamma > 0$ in (H2) since the case when $\gamma = 0$ is handled in a similar way.

Now, for each $n \geq N_o$, define a map $M_n:[0,1]\times\bar{D}^1_{K_n} \to K_n$ by $M_n(t,x) = tC_n(x)$ for $t \in [0,1]$ and $x \in D^1_{K_n}$. It follows from the first part of condition (C1) that $x-M_n(t,x) \neq 0$ for $t \in [0,1]$, $x \in \partial D^1_{K_n}$ and $n \geq N_o$. Hence, by the properties of the finite dimensional fixed point index, $i_{K_n}(C_n,D^1_{K_n}) = \{1\}$ for all $n \geq N_o$. Now, in view of the second part in condition (C1), the map $G_n:[0,1]\times\bar{D}^1_{K_n} \to K_n$ given for each $n \geq N_o$ by $G_n(t,x) = tT_n(x)+(1-t)C_n(x)$ is such that $x \neq G_n(t,x)$ for $t \in [0,1]$ and $x \in \partial D^1_{K_n}$. Hence $i_{K_n}(T_n,D^1_{K_n}) = i_{K_n}(C_n,D^1_{K_n}) = \{1\}$ for all $n \geq N_o$ and, therefore, $I_K(T,D^1_K) = \{1\}$ by Definition 2 since $I-T:\bar{D}^1_K \to X$ is A-proper and $x-Tx \neq 0$ for $x \in \partial D^1_K$. On the other hand, in virtue of condition (C2), Theorem 1 (with $D_K = D^2_K$) implies that $I_K(T,D^2_K) = \{0\}$. Indeed, if $I_K(T,D^2_K)$ were not equal to $\{0\}$, then since the first part of condition (C2) implies (H2) and (H3) of Theorem 1, it would follow from that theorem that there exist $x \in \partial D^2_K$ and $\lambda \in (0,d/\delta]$ such that $x-Tx = \lambda Fx$, in contradiction to the second assumption in (C2). Thus $I_K(T,D^2_K) = \{0\}$.

Now, let $Q = D^2\setminus\bar{D}^1$ and observe that $\partial Q_K = \partial D^1_K \cup \partial D^2_K$ and that T has no fixed points on ∂Q_K. Thus, by the additivity property (P4),

$$I_K(T,D^2_K) = I_K(T,D^1_K) + I_K(T,Q_K)$$

since $I_K(T,D_K^1)$ is the singleton. It follows from the last equality that $I_K(T,Q_K) = \{-1\}$. Hence, by property (P1), there exists $x_o \in \bar{Q}_K = \bar{D}_K^2 \setminus \bar{D}_K^1$ such that $x_o = T(x_o)$.

To prove the second part of Theorem 2, note that when condition (C1) holds on D_K^2 then, by the same arguments as above, we see that $I_K(T,D_K^2) = \{1\}$. Similarly, when (C2) holds on D_K^1 then, in view of Theorem 1 with $D_K = D_K^1$, we see that $I_K(T,D_K^1) = \{0\}$. The property (P4) implies in this case that $I_K(T,Q) = \{1\}$ and so again, as in the preceding case, there exists $x_o \in \bar{Q}_K$ such that $x_o = T(x_o)$.

Q.E.D.

For the purpose of applications we state Theorem 2 in the following practically useful form which, as we will indicate below, unifies and extends a number of results obtained earlier by this and other authors for special classes of maps by using various methods.

Theorem 2'. Let K and T be as in Theorem 2, let $r_1,r_2 \in (0,\infty)$ with $r = \max\{r_1,r_2\}$, let $B_K(0,r) = B(0,r) \cap K$ and let $T:\bar{B}_K(0,r) \to K$ be a bounded map such that $I-T:\bar{B}_K(0,r) \to X$ is A-proper. Suppose also that (C1) and (C2) of Theorem 2 hold with $D^1 = B(0,r_1)$ and $D^2 = B(0,r_2)$. Then there exist $x_o \in K$ such that $\min\{r_1,r_2\} \leq \|x_o\| \leq \max\{r_1,r_2\}$ and $x_o = T(x_o)$.

Special cases. We shall now show that, for suitable choices of the maps T, F and C, Theorems 2 and 2' contain two classes of fixed point theorems: (A) Variations of the cone-expansion and the cone-compression theorems of the type established in [12] when T is compact and Nussbaum, Potter, Reich, Hahn, Fitzpatrick-Petryshyn, Milojevic and others when T is k-ball-contractive (see [2, 22]) for exact references.

(B) The fixed point theorems when $T:K \to K$ has a Frechet derivative T_o at 0 and/or T_∞ at ∞ along K such that T_o and/or T are of type $L_1^+(X)$ or $L_1^-(X)$ in the sense of [1].

(A) Cone-expansion and cone-compression type theorems. When $C = 0$ in (C1), then condition (C1), which in this case reduces to "$x \neq tP_nTx$ for $x \in \partial D_{K_n}^1$ $\forall t \in [0,1]$ $\forall$ $n \geq N_o$", holds when $T:\bar{D}_K^2 \to K$ is P_1-compact and $x \neq tTx$ for $t \in [0,1]$ and $x \in \partial D_K^1$. Choosing $h \in \overset{\circ}{K}$ such that $\|h\| > d\gamma/\sigma$ and $x-Tx \neq \lambda h$ for $x \in \partial D_K^2$ and $\lambda \in (0, d/\|h\|]$, and then defining $F:\bar{D}_K^2 \to K$ by $Fx = h$ for $x \in \bar{D}_K^2$ we see that (H3) holds, while (H2) also holds since $I-T-\mu F$ is A-proper for each $\mu > 0$

because F is compact. Thus, in this case we deduce from Theorem 2' the earlier results in Goncharov [10] when $T:\bar{B}_K(0,r) \to K$ is P-compact and of Hamilton [11] when T is P_1-compact with $D^1 = B(0,r_1)$ and $D^2 = B(0,r_2)$, and of Milojevic [15] when D^1 and D^2 are as in Theorem 2 and T is singlevalued. Theorem 2 in [28] is also a special case of Theorem 2.

An immediate consequence of Theorem 2' is the following extension of Corollary 1 in [24] which is a new result.

Corollary 3. Let $\Gamma = \Gamma_1$ and $T:\bar{B}_K(0,r) \to K$ be k_o-ball-contractive with $k_o \in (0,1)$. Suppose further that:

(D1) There is a bounded ball-condensing map $C:\bar{B}_K(0,r_1) \to K$ such that $Cx \neq \mu x$ for $x \in \partial B_K(0,r_1)$ and $\mu \geq 1$, and $x \neq tCx+(1-t)Tx$ for $x \in \partial B_K(0,r_1)$ and $t \in [0,1]$.

(D2) There is a bounded k-ball-contractive map $F:\bar{B}_K(0,r_2) \to K$ with $k \geq 0$ such that $\delta = \inf\{\|Fx\| : x \in \partial B_K(0,r_2)\} > \frac{dk}{\sigma(1-k_o)}$ and $x-Tx \neq \lambda Fx$ for $x \in \partial B_K(0,r_2)$ and $\lambda \in (0,d/\delta]$.

Then there is $x_o \in K$ with $\min\{r_1,r_2\} \leq \|x_o\| \leq \max\{r_1,r_2\}$ such that $x_o = T(x_o)$.

Proof. To deduce Corollary 3 from Theorem 2', it suffices to show that (D1) implies (C1), and that the map $T_\mu = I-T-\mu F$ is A-proper w.r.t. Γ_1 for each $\mu \in (0,1/\gamma]$, where γ is chosen such that $\gamma > \frac{k}{1-k_o}$ and $\delta > d\gamma/\sigma$. The latter fact has been shown to be the case in our proof of Corollary 2, while the fact that (D1) implies (C1) follows in the standard way from the fact that C is ball-condensing and $Cx \neq \mu x$ for $x \in \partial B_K(0,r_1)$ and $\mu \geq 1$ and that $F(t,\cdot) \equiv tC+(1-t)T$ is also ball-condensing (in the sense that $\beta(F([0,1]\times A)) < \beta(A)$ for each $A \in \bar{B}_K(0,r)$ with $\beta(A) \neq 0$) and $x \neq Cx + (1-t)Tx$ for $t \in [0,1]$ and $x \in \partial_K B(0,r_2)$.

Q.E.D.

When F is compact (i.e., $k = 0$ in (D2)) and $x-Tx \neq \lambda Fx$ for all $\lambda > 0$ and $x \in \partial B_K(0,r_2)$, Corollary 3 was first proved by the author in [22] for a general Banach space X where it was also shown that in this case it extends and unifies the corresponding results of Krasnoselskii, Gustafson-Schmitt, Turner, Gatica-Smith, Edmunds-Potter-Stuart, Nussbaum, Potter, Fitzpatrick-Petryshyn, Amann, Milojevic, Petryshyn and others. See [22] for the exact references of the above mentioned authors.

(B) Nonzero fixed points of differentiable maps. In order to deduce from Theorem 2 the existence of nonzero fixed points for P_1-compact maps, which are Frechet differentiable either at 0 or at ∞ along K, we recall first some definitions. We omit some details except for one lemma which is necessary but not proved in [24].

A map $T:K \to X$ is said to be (Frechet) differentiable at 0 along K if there exists a map $T_o \in L(X)$ such that for each $h \in K$

$$T(h) = T(0)+T_o h+w(0,h) \quad \text{with} \quad w(0,h) = o(\|h\|) \quad \text{as} \quad \|h\| \to 0. \tag{2.1}$$

The map T is said to be asymptotically linear along K if there is $T_\infty \in L(X)$ such that for each h in K

$$T(h) = T_\infty h + w(h) \quad \text{with} \quad w(h) = o(\|h\|) \quad \text{as} \quad \|h\| \to \infty. \tag{2.2}$$

It is not hard to show that if K is a total cone, then the maps T_o and T_∞, the derivatives of T at 0 and ∞ along K respectively, are uniquely determined. It is easy to see that if $T:K \to K$, then $T_o(K) \subset K$ and $T_\infty(K) \subset K$ whenever they exist.

Before we apply Theorem 2' to obtain the existence of nonzero fixed points for P_1-compact maps $T:K \to K$ satisfying either (2.1) or (2.2), we will need the following simple fact (see [12]): If $u,v \in K$, $u \neq 0$ and $\{\lambda_n\} \subset (0,\infty)$ is a sequence such that $\lambda_n \to \infty$, then there exists $N_o \in N^+$ such that $v-\lambda_{N_o} u \notin K$. Indeed, if $v-\lambda_n u \in K$ for all n in N^+, then $1/\lambda_n v-u \in K$ for all n and this implies that $-u \in K$ since K is closed, which is impossible.

Following [1] we say that $A \in L(X)$ is of type $L_1^+(X)$ (respectively of type $L_1^-(X)$) if there is some $\lambda_o > 1$ and $h_o \in \overset{\circ}{K}$ such that $Ah_o = \lambda_o h_o$ and $Ah \neq h$ for all $h \in \overset{\circ}{K}$ (respectively $Ah \neq \lambda h$ for all $\lambda \geq 1$ and $h \in \overset{\circ}{K}$).

Let $B(0,r)$ be an open ball in X. Then $B_K(0,r) = B(0,r) \cap K$ is an open neighborhood of 0 in K, the closure $\overline{B}_K(0,r)$ of $B_K(0,r)$ in K coincides with $\overline{B}(0,r) \cap K$, and the boundary $\partial B_K(0,r)$ of $B_K(0,r)$ in K equals $\partial B(0,r) \cap K$.

In what follows we shall need the following lemma which is related to Proposition 1.10 in [20].

Lemma 1. If $G:\overline{B}_K(0,r) \to X$ is continuous and A-proper w.r.t. $\Gamma_1 = \{X_n,P_n\}$, then the restriction of G to any closed subset M of $\overline{B}_K(0,r)$ is proper. In particular, $G(M)$ is closed.

Proof. Let $\{x_k\} \subset M$ be any sequence such that $G(x_k) \to g$ as

$k \to \infty$ for some g in K. Since $P_n(K) \subset K$, $\|P_n\| = 1$, $P_n \subset P_{n+1}$ for all n, $P_n(x) \to x$ for each x in K, and G is continuous, it follows that for each k and $\delta_k = 1/k$ there exists $n(k) \in N^+$ with $n(k) > k$ such that

$$\|x_k - w_{n(k)}\| < \delta_k \quad \text{with} \quad w_{n(k)} = P_{n(k)}x_k \in X_{n(k)} \cap \bar{B}_K(0,r) \qquad (2.3)$$

for sufficiently large k and $\|Tx_k - Tw_{n(k)}\| \le \varepsilon_k \to 0$ as $k \to \infty$. This and our properties of the scheme $\Gamma_1 = \{X_n, P_n\}$ imply that $P_{n(k)}Gw_{n(k)} \to g$ as $k \to \infty$, whence on account of the A-properness of G it follows that there exist a subsequence $w_{n(j)}$ and $x \in \bar{B}_K(0,r)$ such that $w_{n(j)} \to x$ as $j \to \infty$ and $Gx = g$. This and (2.3) imply that $x_j \to x$ as $j \to \infty$ and $x \in M$ since M is closed. This proves Lemma 1.

We now are in position to deduce from Theorem 2 the following corollaries for differentiable maps.

Assuming (C1) of Theorem 2, our first corollary in this section shows how (C2) of Theorem 2 is verified when T_o or T_∞ is of type $L_1^+(X)$.

Corollary 4. Let $T:K \to K$ be a bounded map such that $I-T:K \to X$ is A-proper w.r.t. Γ_1 and assume also that:

(a1) There exist $r > 0$, $N_o \in N^+$ and a bounded map $C:\bar{B}_K(0,r) \to K$ such that (C1) of Theorem 2 holds for $D_K^1 = B_K(0,r)$.

Suppose further that either (b1) or (d1) holds, where

(b1) $T(0) = 0$, T has the derivative T_o at 0 along K such that $(I-T_o)|_K$ is A-proper w.r.t. Γ_1 and $T_o \in L_1^+(X)$

(d1) T has the derivative T_∞ at ∞ along K such that $(i-T_\infty)|_K$ is A-proper w.r.t. Γ_1 and $T_\infty \in L_1^+(X)$.

Then, in either case, T has a fixed point in $\mathring{K}$.

Proof. Since (a1) holds, to deduce Corollary 4 from Theorem 2, it suffices to show that condition (C2) of Theorem 2 is implies by either (b1) or (d1) for a suitable ball $B = D^2$. Since the arguments are almost identical in both cases, we prove these implications simultaneously for T satisfying (b1) and (d1). First note that (b1) and (d1) can be stated as:

(d_β) $T(h) = T_\beta(h) + Q_\beta(h)$ with $\|Q_\beta(h)\| = o(\|h\|) \to \beta$ $(h \in K)$, where $\beta = 0$ or $\beta = \infty$ and $T_\beta \in L(X)$ is such that $T_\beta(K) \subseteq K$, $(I-T_\beta)|_K$ is A-proper and $T_\beta \in L_1^+(X)$.

Now, since $T_\beta \in L_+^1(X)$, there are $\lambda_\beta > 1$ and $h_\beta \in \mathring{K}$ such that

$T_\beta(h_\beta) = \lambda_\beta h_\beta$. We claim that we can choose $r_\beta > 0$ with $r_o < r$ and $r_\infty > r$ and define a map $F:\overline{B}_K(0,r_\beta) \to K$ by $F_\beta(x) \equiv h_\beta$ for $x \in \overline{B}_K(0,r_\beta)$ such that (C2) of Theorem 2 holds with $\gamma = 0$ in (H2) and (H3). Indeed since $F_\beta:\overline{B}_K(0,r_\beta) \to K$ is compact, it follows that $I-T-\mu F:\overline{B}_K(0,r_\beta) \to X$ is A-proper w.r.t. Γ_1 for each $\mu > 0$, i.e., (H2) holds for $\gamma = 0$. Since $\delta = \inf\{\|Fx\| : x \in \partial B_K(0,r_\beta)\} = \|h_\beta\| > 0$ we see that (H3) holds when $\gamma = 0$. Finally, we show that $r_\beta > 0$ chosen above is such $x-Tx \neq \lambda Fx$ for $x \in \partial B_K(0,r_\beta)$ and $\lambda > 0$. Indeed, if this were not the case, then there would exist sequences $\{r_n^\beta\}$, $\{\lambda_n^\beta\} \subset (0,\infty)$ and $\{K_n^\beta\} \subset \overset{\circ}{K}$ such that $r_n^\beta \to \beta$ as $n \to \infty$ with $\beta = 0$ or $\beta = \infty$, $\lambda_n^\beta > 0$ and $\|x_n^\beta\| = r_n^\beta$ for each $n \in N^+$ and $x_n^\beta - T(x_n^\beta) = \lambda_n^\beta h_\beta$. It follows from last equality and (d_β) that

$$(I-T_\beta)(z_n^\beta) = (\lambda_n^\beta/r_n^\beta)h_\beta + Q_\beta(x_n^\beta)/\|x_n^\beta\|, \quad z_n^\beta = x_n^\beta/\|x_n^\beta\|$$

Since $\|z_n^\beta\| = 1$ and $\|Q_\beta(x_n^\beta)\|/\|x_n^\beta\| \to 0$ as $\|x_n^\beta\| \to \beta$, it follows that $\{\lambda_n^\beta/r_n^\beta\}$ is bounded. Hence we may assume that $\lambda_n^\beta/r_n^\beta \to \eta^\beta \geq 0$ as $n \to \infty$ and so $(I-T_\beta)(g_n^\beta) \to \eta^\beta h_\beta$. Since $(I-T_\beta)|_K$ is A-proper w.r.t. Γ_1, it follows from Lemma 1 that there is a subsequence $\{z_{n_j}^\beta\}$ and $z^\beta \in \partial B_K(0,1)$ such that $z_{n_j}^\beta \to z^\beta$ and $(I-T_\beta)(z^\beta) = \eta^\beta h_\beta$ with $\eta^\beta > 0$ since $z^\beta - T_\beta(z^\beta) \neq 0$ because $T_\beta \in L_1^+(X)$. Since $T_\beta h_\beta = \lambda_\beta h_\beta$, it follows from the last equality, by induction, that $z^\beta - \eta^\beta(1+\lambda_\beta+\lambda_\beta^2+\dots+\lambda_\beta^n)(h_\beta) \in K$ for each n in N^+. Since $\eta^\beta > 0$ and $\lambda_\beta > 1$, the last relation is impossible by the remark preceding Corollary 4. Thus, (C2) holds and so Corollary 4 follows from Theorem 2 provided that (a1) and either (b1) or (d1) hold.

Q.E.D.

Remark 2. Under the stronger assumption that $T:K \to K$ is P_1-compact and the restrictions of T_o and T_∞ to K are also P_1-compact, Corollary 4 was first obtained in [24]. For the proof given in [24] to be correct it seems that one should use Lemma 1 instead of Proposition 1.1C from [20].

In our next corollary, assuming (C2) of Theorem 2, we show how (C1) of that theorem is verified when $T:K \to K$ is P_1-compact, $T_o|_K$ or $T_\infty|_K$ is also P_1-compact and either T_o or T_∞ lies in $L_1^-(X)$.

Corollary 5. Suppose $T:K \to K$ is a bounded P_1-compact map such that:

(a2) There exist $r > 0$ and a bounded map $F:\bar{B}_K(0,r) \to K$ such that condition (C2) of Theorem 2 holds.

Suppose further that either (b2) or (d2) holds, where:

(b2) $T(0) = 0$ and T has the derivative T_o at 0 along K such that $T_o|_K$ is P_1-compact and $T_o \in L_+^-(X)$.

(d2) T has the derivative T_∞ at ∞ along K such that $T_{\infty}|_K$ is P_1-compact and $T_\infty \in L_+^-(X)$. Then, in either case, T has a fixed point in $\overset{o}{K}$.

Proof. To deduce Corollary 5 from Theorem 2, it suffices to show that (C1) of Theorem 2 is implied by either (b2) or (d2) for a suitable ball $B = D^1$.

As before, the proof that either (b2) or (d2) implies (C1) will be carried simultaneously. Let $\beta = 0$ or $\beta = \infty$ and let T_β be the derivative at β along K. Since $T_\beta \in L_+^-(X)$, it follows that $T_\beta(x) \neq \mu x$ for $\mu \geq 1$ and $x \in \overset{o}{K}$. Now, since $T_\beta|_K$ is P_1-compact and $x-T_\beta x \neq 0$ for $x \in \partial_K B(0,1)$, it follows from Lemma 1 that there exist $N_\beta \in N^+$ and $m_\beta > 0$ such that $\|x-P_nT_\beta(x)\| \geq m_\beta\|x\|$ for $x \in K_n$ and $n \geq N_\beta$. Choose $r_\beta > 0$ with $r_o < r$ and $r_\infty > r$ such that $\|Tx-T_\beta x\| < m_\beta/2\|x\|$ for $x \in \partial B_K(0,r_\beta)$. The last two inequalities imply that for all $t \in [0,1]$, all $x \in \partial B_{K_n}(0,r_\beta)$ and each $n \geq N_\beta$ we have

$$\|x-tP_nt(x)-(1-t)P_nT_\beta(x)\| \geq \|x-P_nTx\| - \|P_n(Tx-T_\beta x)\| \geq \tfrac{1}{2}m_\beta r_\beta .$$

Moreover, since $T_\beta|_K$ is P_1-compact and $T_\beta \in L_+^-(X)$, it follows from Lemma 1 that $P_nT_\beta(x) \neq \mu x$ for all $x \in \partial B_K(0,r_\beta)$, all $n \geq \tilde{N}_\beta$ and some $\bar{N}_\beta \in N^+$. Thus, (C1) holds with $C = T_\beta$ and so Corollary 5 follows from Theorem 2 provided that (a2) and either (b2) or (d2) hold.

Q.E.D.

Remark 3. When $\gamma = 0$ in condition (a2), Corollary 5 was proved by the author in [24].

An immediate consequence of Corollaries 4 and 5 and Remarks 2 and 3 is the following special case established in [24].

Corollary 6. Suppose $T:K \to K$ is a bounded and P_1-compact map with $T(0) = 0$ and the derivatives T_o and T_∞ at 0 and at ∞ along K are such that $T_o|_K$ and $T_\infty|_K$ are P_1-compact.

Then T has a fixed point in $\overset{o}{K}$ provided one of the following conditions hold:

(a) $T_o \in L_1^-(X)$ and $T_\infty \in L_1^+(X)$

(b) $T_o \in L_1^+(X)$ and $T_o \in L_1^-(X)$.

Corollary 6 contains the corresponding results of [12] when T is completely continuous and of [1], [7] and [22] when T is k-ball-contractive with $k < 1$.

REFERENCES

[1] Amann, H.: Fixed points of asymptotically linear maps in ordered Banach spaces, J. Functional Anal., 14 (1973), 162-171.

[2] Amann, H.: Fixed point equations and nonlinear eigenvalue problems in ordered Banach spaces, SIAM Review, 18 (1976), 620-709.

[3] Browder, F. E.: Fixed point theorems for nonlinear semicontractive mappings in Banach spaces, Arch. Rat. Mech. Anal. 21 (1966), 259-269.

[4] Browder, F. E. and Petryshyn, W. V.: Approximation methods and the generalized topological degree for nonlinear maps in Banach spaces, J. Funct. Anal. 3 (1969), 217-245.

[5] Dancer, E. N., Nussbaum, R. D. and Stuart, C.: Quasinormal cones in Banach spaces, Nonlinear Anal. 7 (1983), 539-553.

[6] Deimling, K.: Nonlinear Functional Analysis, Springer-Verlag, Berlin, Heidenberg, 1985.

[7] Edmunds, D. E., Potter, A. J. B. and Stuart, A. C.: Noncompact positive operators, Proc. R. Soc. London, A. 328 (1972), 67-81.

[8] Fitzpatrick, P. M. and Petryshyn, W. V.: On the nonlinear eigenvalue problem $T(u) = \lambda C(u)$, involving noncompact abstract and differential operators, Bollettino U.M.I. (5) 15-B (1968), 80-107.

[9] Fitzpatrick, P. M. and Petryshyn, W. V.: Positive eigenvalues for nonlinear multivalued noncompact operators with applications to differential operators, J. Diff. Equations, 22 (1976), 428-441.

[10] Goncharov, G. M.: On some existence theorems for the solution of a class of nonlinear operator equations, Mat. Zametki 7 (1970), 229-237.

[11] Hamilton, J. D.: Noncompact mappings and cones in Banach spaces, Arch. Rat. Mech. Anal. 48 (1972), 153-162.

[12] Krasnoselskii, M. A.: Positive solutions of operator equations, Noordhoff, Groningen, Netherlands, 1964.

[13] Lami-Dozo, E.: Quasinormality of cones in Hilbert spaces, Acad. Roy Belg. Bull. Cl. Sci. (5) 67 (1981), 536-541.

[14] Massabo, I. and Stuart, C. A.: Positive eigenvectors of k-set-contractions, Nonlinear Anal., 3 (1979), 35-44.

[15] Milojevic, P. S.: Multivalued mappings of A-proper and condensing type and BV Problems, Ph.D. Thesis, Rutgers Univ., New Brunswick, N.J., May 1975.

[16] Nussbaum, R. D.: The fixed point index for local condensing maps, Ann. Mat. Pura Appl., 89 (1971), 217-258.

[17] Petryshyn, W. V.: On a fixed point theorem for nonlinear P-compact operators in Banach space, Bull. A.M.S., 72 (1966), 329-334.

[18] Petryshyn, W. V.: Iterative construction of fixed points of contractive type mappings in Banach spaces, Numerical Anal. of PDE's (C.I.M.E. 2° Ciclo, Ispra, 1967), Edizione Cremoneze, Rome, 1968

[19] Petryshyn, W. V.: On the approximation-solvability of nonlinear equations in normed linear spaces, Numerical Anal. of PDE's (C.I.M.E. 2° Ciclo, Ispra, 1967), Edizione Cremoneze, Rome, 1968.

[20] Petryshyn, W. V.: On the approximation-solvability of equations involving A-proper and pseudo-A-proper mappings, Bull. AMS, 81 (1975), 223-312.

[21] Petryshyn, W. V.: Fixed point theorems for various classes of 1-set-contractive and 1-ball-contractive maps in Banach spaces, Trans. AMS, 182 (1973), 323-352.

[22] Petryshyn, W. V.: Existence of nonzero fixed points for noncompact mappings in wedges and cones, J. Functional Anal., 33 (1979), 36-46.

[23] Petryshyn, W. V.: On the solvability of $x \in Tx+\lambda Fx$ in quasi-normal cones with T and F k-set-contractive, Nonlinear Anal., 5 (1981), 585-591.

[24] Petryshyn, W. V.: Nonlinear eigenvalue problems and the existence of nonzero fixed points for A-proper mappings, in "Theory of Nonlinear Operators" (ed. R. Kluge), Academy-Verlag, Berlin, 1978, 215-227.

[25] Sadovskii, B. N.: Ultimately compact and condensing mappings, Uspehi Mat. Nauk, 27 (1972), 81-146.

[26] Webb, J. R. L.: Fixed point theorems for nonlinear semicontractive operators in Banach spaces, J. London Math. Soc. (2), (1969), 683-688.

[27] Webb, J. R. L.: Existence theorems for sums of k-ball-contractions and accretive operators via A-proper mappings, Nonlinear Anal. 8 (1981), 891-896.

[28] Williams, L. R. and Leggett, R. L., A fixed point theorem with application to an infectious disease model, J. Math. Anal. Appl., 76 (1980), 91-97.

On the Method of Successive Approximations for Nonexpansive Mappings

SIMEON REICH* Department of Mathematics, The Technion-Israel Institute of Technology, Haifa, Israel

I. SHAFRIR Department of Mathematics, The Technion-Israel Institute of Technology, Haifa, Israel

Throughout this paper $(X, |\cdot|)$ denotes a (real) Banach space, X^* its dual, D a closed subset of X, $T : D \longrightarrow X$ a nonexpansive mapping ($|Tx-Ty| \le |x-y|$ for all x and y in D), and $\{c_n : n = 0,1,2,\ldots\}$ a positive sequence such that $0 < c_n \le 1$ and $\sum_{n=0}^{\infty} c_n = \infty$. We denote the identity operator by I, the closure of a subset B of X by $cl(B)$, and its distance from a point y in X by $d(y,B)$.

Consider the sequence of mappings $\{T_n : n = 0,1,2,\ldots\}$ defined by $T_n = (1-c_n)I + c_nT$. Assuming that the range of each T_n is contained in D (this is certainly the case if, for example, T is a self-mapping of a convex D), we can define, for each $y_0 \in D$, an iteration process by

$$y_{n+1} = T_n y_n , \quad n = 0,1,2,\cdots. \tag{1}$$

Our purpose in this note is to present several new results concerning the asymptotic behavior of this process. In other words, we intend to study the behavior of the sequence $\{y_n\}$ defined inductively by

$$y_{n+1} = (1-c_n)y_n + c_nTy_n, \quad n \ge 0. \tag{2}$$

To this end, we recall that the duality map from X into the family of nonempty closed convex subsets of X^* is defined by

$$J(x) = \{x^* \in X^* : (x,x^*) = |x|^2 = |x^*|^2\}$$

for each x in X. It is single valued if and only if X is smooth. The resolvent $J_r = (I+rA)^{-1}$ of the (accretive) operator $A = I-T$ is

---

*Current affiliation: The University of Southern California, Los Angeles, California

Partially supported by the Technion VPR Fund—K. & M. Bank Mathematics Research Fund.

nonexpansive on its domain $D(J_r) = R(I+rA)$. (It should not be confused with the duality map J.) We shall use in the sequel the following range condition at infinity:

$$\liminf_{t\to\infty} d(0,R(I+t(I-T)))/t = 0. \tag{3}$$

Once again, this condition is automatically satisfied when T is a self-mapping of a convex D. In fact, in this case the stronger range condition

$$R(I+r(I-T)) \supset D \quad \text{for all} \quad r > 0 \tag{4}$$

is also satisfied.

Our first two results (Theorems 5 and 8) are valid in every Banach space. On the other hand, our next two results (Theorems 11 and 14) are valid only in special Banach spaces. As a matter of fact, in order to formulate Theorem 14 we introduce a new geometric property of (infinite-dimensional) Banach spaces. We conclude with a consequence of Theorem 8 which involves the fixed point property for nonexpansive mappings (Theorem 15).

Our first result will be preceded by several lemmas.

<u>Lemma 1</u>. If $\{y_n\}$ is defined by (2) and $x_t \in D(J_t)$, then

$$(1+1/t)^n|J_tx_t-y_n| \le \prod_{i=0}^{n-1}(1+(1-c_i)/t)|J_tx_t-y_0| + (1/t)\sum_{i=0}^{n-1}c_i(1+1/t)^i|x_t-Ty_i|\prod_{k=i+1}^{n-1}(1+(1-c_k)/t)$$

for all $n \ge 1$.

<u>Proof</u>. Since $(1+t)J_tx_t = x_t+tTJ_tx_t$, we see that for each $n \ge 0$,

$$\begin{aligned}
&(1+1/t)^{n+1}|J_tx_t - y_{n+1}| \\
&\le (1+1/t)(1-c_n)(1+1/t)^n|J_tx_t-y_n| + c_n(1+1/t)^n(1/t)(1+t)|J_tx_t-Ty_n| \\
&\le (1+1/t)(1-c_n)(1+1/t)^n|J_tx_t-y_n| \\
&\qquad + c_n(1+1/t)^n(1/t)(|x_t-Ty_n|+t|J_tx_t-y_n|) \\
&= (1+(1-c_n)/t)(1+1/t)^n|J_tx_t-y_n| + c_n(1+1/t)^n(1/t)|x_t-Ty_n|.
\end{aligned}$$

The result now follows by induction.

Although the next lemma is well known, we include a proof for completeness.

<u>Lemma 2</u>. If $\{y_n\}$ and $\{z_n\}$ are defined by (2), then

$$|y_0-y_n| \le 2|y_0-z_0| + (\sum_{i=0}^{n-1})|z_0-Tz_0|$$

for all $n \ge 1$.

Proof. We first note that $|z_{n+1}-Tz_{n+1}| \leq |z_{n+1}-Tz_n| + |Tz_n-Tz_{n+1}| \leq |(1-c_n)z_n+c_nTz_n-Tz_n| + |z_n-z_{n+1}| = |z_n-Tz_n|$, so that the sequence $\{|z_n-Tz_n|\}$ is decreasing. Hence

$$|z_0-z_{n+1}| \leq |z_0-z_n| + c_n|z_n-Tz_n| \leq |z_0-z_n| + c_n|z_0-Tz_0|.$$

Since $|y_0-y_{n+1}| \leq |y_0-z_0| + |z_0-z_{n+1}| + |z_{n+1}-y_{n+1}| \leq 2|y_0-z_0| + |z_0-z_{n+1}|$, the result now follows by induction.

The range condition (3) is satisfied if and only if there exists a sequence $\{(t_i,x_{t_i})\}$ such that

$$t_i \to \infty,\ x_{t_i} \in D(J_{t_i}) \ \text{ and } \ |x_{t_i}|/t_i \to 0. \tag{5}$$

Our next lemma is also known [8, Proposition 3], but we present a different and simpler proof. Recall that a subset A of $X \times X$ with domain $D(A)$ and range $R(A)$ is said to be accretive if $|x_1-x_2| \leq |x_1-x_2+r(y_1-y_2)|$ for all $[x_i,y_i] \in A$, $i=1,2$, and $r > 0$. It is clear that $A = I-T$ is accretive whenever T is nonexpansive. The analog of condition (3) is therefore

$$\liminf_{t\to\infty} d(0,R(I+tA))/t = 0. \tag{6}$$

Lemma 3. Let X be a Banach space, $A \subset X \times X$ an accretive operator that satisfies (6), and $\{(t_i,x_{t_i})\}$ a sequence satisfying (5). Then

$$\lim_{t\to\infty} |J_{t_i}x_{t_i}|/t_i = d(0,R(A)).$$

Proof. Let $d = d(0,R(A))$. Since $(x_t-J_tx_t)/t$ belongs to $R(A)$, it is clear that

$$\liminf_{t\to\infty} |J_{t_i}x_{t_i}|/t_i \geq d.$$

On the other hand, for $0 < s < t$ and $[x,y] \in A$ we have

$$\begin{aligned}|x-J_tx_t| &= |J_s(x+sy)-J_s((s/t)x_t+(1-(s/t))J_tx_t)| \\ &\leq |x+sy-((s/t)x_t+(1-(s/t))J_tx_t)| \\ &\leq (1-(s/t))|x-J_tx_t| + (s/t)|x-x_t| + s|y|.\end{aligned}$$

Hence

$$|x-J_tx_t|/t \leq |x-x_t|/t + |y|$$

and

$$\limsup_{i\to\infty} |J_{t_i}x_{t_i}|/t_i \leq |y|.$$

Therefore $\limsup_{i\to\infty} |J_{t_i}x_{t_i}|/t_i \leq d$ and the result follows.

Finally, we note the following simple fact.

Lemma 4. For each $n \geq 1$, $\lim_{t\to\infty} t(1 - \prod_{i=0}^{n-1} (1-c_i/(1+t))) = \sum_{i=0}^{n-1} c_i$.

We are now in a position to establish our first result.

Theorem 5. Let D be a closed subset of a Banach space X and $T : D \to X$ a nonexpansive mapping which satisfies (3). If the sequence $\{y_n\}$ is defined by (2), then

$$\lim_{n\to\infty} |y_n|/(\sum_{i=0}^{n-1} c_i) = d(0,R(I-T)).$$

Proof. Denote $\sum_{i=0}^{n-1} c_i$ by a_n and $d(0,R(I-T))$ by d. On the one hand, Lemma 2 shows that $\limsup_{n\to\infty} |y_n|/a_n \leq d$. On the other hand, Lemma 1 shows that $|y_0-y_n| \geq |y_0-J_tx_t| - |J_tx_t-y_n| \geq$
$(1 - \prod_{i=0}^{n-1} (1-c_i/(1+t)))|J_tx_t-y_0| - (1/(1+t)) \sum_{i=0}^{n-1} c_i |x_t-Ty_i| \prod_{k=i+1}^{n-1} (1-c_k/(1+t))$
for all $n \geq 1$ and $x_t \in D(J_t)$. Therefore Lemmas 3 and 4 now imply that $|y_0-y_n| \geq a_nd$ for all $n \geq 1$, and the result follows.

In order to establish our second result we need two more lemmas.

Lemma 6. If $\{c_n\}$ is bounded away from 0 and 1, and $\{y_n\}$ is defined by (2), then for all $k \geq 1$,

$$\lim_{n\to\infty} |y_{n+1}-y_n|/c_n = \lim_{n\to\infty} |y_{n+k}-y_n|/(\sum_{j=n}^{n+k-1} c_j).$$

Proof. Fix $k \geq 1$ and denote the limit of the decreasing sequence $\{|y_n-Ty_n|\}$ by L. Since

$$y_{n+k} - y_n = \sum_{j=n}^{n+k-1} (y_{j+1}-y_j) = \sum_{j=n}^{n+k-1} c_j(Ty_j-y_j),$$

it is clear that

$$\limsup_{n\to\infty} |y_{n+k}-y_n|/(\sum_{j=n}^{n+k-1} c_j) \leq L.$$

Now we note that by [4, Proposition 1]

$$\begin{aligned} |y_{n+k}-y_n| &\geq |Ty_{n+k}-Ty_n| \\ &\geq |Ty_{n+k}-y_n| - |Ty_n-y_n| \\ &\geq \prod_{j=n}^{n+k-1} (1-c_j)^{-1}(|Ty_{n+k}-y_{n+k}|-|Ty_n-y_n|)+(\sum_{j=n}^{n+k-1} c_j)|Ty_n-y_n|. \end{aligned}$$

Assuming, as we do, that the sequence $\{c_n\}$ is bounded away from 0 and 1 we conclude that

$$\liminf_{n\to\infty} |y_{n+k}-y_n|/(\sum_{j=n}^{n+k-1} c_j) \geq L.$$

Hence the result.

Lemma 7. If $\{c_n\}$ is bounded away from 0 and 1, and $\{y_n\}$ is defined by (2), then the limit of the decreasing sequence $\{|y_n - Ty_n|\}$ is independent of the initial point y_0.

Proof. Let $\{z_n\}$ be another sequence defined by (2), and denote the limits of $\{|y_n - Ty_n|\}$ and $\{|z_n - Tz_n|\}$ by $L(y_0)$ and $L(z_0)$ respectively. We first note that

$$\begin{aligned} ||y_{n+k} - y_n| - |z_{n+k} - z_n|| &\le |y_{n+k} - z_{n+k}| + |y_n - z_n| \le 2|y_n - z_n| \\ &\le 2|y_0 - z_0|. \end{aligned}$$

Since the sequence $\{c_n\}$ is bounded away from 0, we can, given a positive ε, find an integer k such that

$$\left| |y_{n+k} - y_n| / \left(\sum_{j=n}^{n+k-1} c_j\right) - |z_{n+k} - z_n| / \left(\sum_{j=n}^{n+k-1} c_j\right) \right| < \varepsilon$$

for all n. By Lemma 6 we can now find, for this k, an integer N such that

$$\left| |y_{n+k} - y_n| / \left(\sum_{j=n}^{n+k-1} c_j\right) - L(y_0) \right| < \varepsilon$$

and

$$\left| |z_{n+k} - z_n| / \left(\sum_{j=n}^{n+k-1} c_j\right) - L(z_0) \right| < \varepsilon$$

for all $n \ge N$. Thus $|L(y_0) - L(z_0)| < 3\varepsilon$, and $L(y_0) = L(z_0)$ because ε is arbitrary.

Theorem 8. Let D be a closed subset of a Banach space X and $T : D \to X$ a nonexpansive mapping which satisfies (3). If $\{c_n\}$ is bounded away from 0 and 1, and the sequence $\{y_n\}$ is defined by (2), then for all $k \ge 1$,

$$\begin{aligned} \lim_{n\to\infty} |y_{n+1} - y_n| / c_n &= \lim_{n\to\infty} |y_{n+k} - y_n| / \left(\sum_{j=n}^{n+k-1} c_j\right) \\ &= \lim_{n\to\infty} |y_n| / \left(\sum_{j=0}^{n-1} c_j\right) = d(0, R(I-T)). \end{aligned}$$

Proof. Since $\lim_{n\to\infty} |y_{n+1} - y_n| / c_n$ is independent of the initial point y_0 by Lemma 7, it must equal $d(0, R(I-T))$. The result now follows by combining Theorem 5 with Lemma 6.

We do not know if this result remains valid when condition (3) is not assumed. It is indeed true if the sequence $\{c_n\}$ is constant [1, Theorem 2.1]. In this case the limit of Theorem 5 also exists even if (3) is not satisfied [8, Lemma 1], but it is not always equal to $d(0, R(I-T))$.

We now turn our attention to special Banach spaces. We begin with another lemma.

Lemma 9. Let D be a closed subset of a Banach space X and $T : D \to X$ a nonexpansive mapping which satisfies (3). Denote $\sum_{i=0}^{n-1} c_i$ by a_n and $d(0,R(I-T))$ by d. If the sequence $\{y_n\}$ is defined by (2), then there is a functional $z \in X^*$ with $|z| = d$ such that $((y_0-y_n)/a_n, z) \geq d^2$ for all n.

Proof. For $x_t \in D(J_t)$ let z_t belong to $J((y_0-J_tx_t)/t)$, and let the sequence $\{(t_i, x_{t_i})\}$ satisfy (5). Let z be a weak-star limit of a subnet of $\{z_{t_i}\}$. Since

$$\begin{aligned}(y_0-y_n, z_t) &= (y_0-J_tx_t, z_t) + (J_tx_t-y_n, z_t)\\ &\geq |z_t|(|y_0-J_tx_t| - |J_tx_t-y_n|),\end{aligned}$$

we can use Lemmas 1, 3 and 4 to conclude that

$$(y_0-y_n, z) \geq a_nd^2 \quad \text{for all} \quad n.$$

This inequality and Theorem 5 now show that $|z| \geq d$. Since it is clear that $|z| \leq d$, the proof is complete.

Recall that the norm of a Banach space X is said to be Fréchet differentiable if for each x in its unit sphere $U = \{x \in X : |x| = 1\}$,

$$\lim_{t\to 0}(|x+ty|-|x|)/t \tag{7}$$

is attained uniformly for y in U. We shall then write that X is (F).

Our next lemma is known (cf. [2]).

Lemma 10. X^* is (F) if and only if every sequence $\{x_n\} \subset X$ for which there exists $w \in X^*$ with $|w| = 1$ such that $\lim_{n\to\infty} |x_n| = \lim_{n\to\infty} (x_n, w)$ converges.

Theorem 11. Let D be a closed subset of a Banach space X and $T : D \to X$ a nonexpansive mapping which satisfies (3). Let the sequence $\{y_n\}$ be defined by (2) and denote $\sum_{i=0}^{n-1} c_i$ by a_n. If X^* has a Fréchet differentiable norm, then $v = \text{strong} \lim_{n\to\infty} y_n/a_n$ exists, and $-v$ is the unique point of least norm in $\mathrm{cl}(R(I-T))$.

Proof. Let the sequence $\{(t_i, x_{t_i})\}$ satisfy (5), and let $z \in X^*$ be the functional obtained in Lemma 9. It is known [8, Theorem 3] that $u = \text{strong} \lim_{i\to\infty} J_{t_i}x_{t_i}/t_i$ exists, and that $-u$ is the unique point of least norm in cl(R(I-T)). Moreover, the proof of [8, Theorem 3] shows that $z \in J(-u)$. Combining Theorem 5, Lemma 9 and Lemma 10 we see that $v = \text{strong} \lim_{n\to\infty} y_n/a_n$ also exists, and that $z \in J(-v)$. Since X is certainly strictly convex, [8, Lemma 5] now shows that $u = v$. This completes the proof.

A similar argument shows that the weak $\lim_{n\to\infty} y_n/a_n$ exists when X is assumed to be only reflexive and strictly convex. These results improve upon Theorem 1 and Corollary 2 of [3] because we show that the weak range condition (3) is sufficient for the conclusions to hold. (This condition, introduced in [8], is, in fact, also necessary.) Since X is uniformly convex if and only if the norm of X^* is uniformly Fréchet differentiable, they also improve upon previous results of the first author [11, 12].

We continue with a convergence result in the setting of Theorem 8. To this end, we introduce a new geometric property of (infinite-dimensional) Banach spaces. Recall that to each functional ω in the unit sphere of X^* there corresponds a face F of the unit sphere U of X, namely all those $x \in U$ for which $(x,\omega) = 1$. We shall say that the norm of a Banach space X is locally uniformly Fréchet differentiable (LUF) if for each face F of U the limit (7) is attained uniformly for all y in U and x in F. It is clear that X is (LUF) whenever X^* is uniformly convex or X is (F) with compact faces. In analogy with Lemma 10, we also have the following characterization.

Lemma 12. X^* is (LUF) if and only if every sequence $\{x_n\} \subset X$ for which there exists $x \in X$ with $|x| = 1$ and a sequence $\{\omega_n\} \subset J(x)$ such that $\lim_{n\to\infty}|x_n| = \lim_{n\to\infty}(x_n,\omega_n)$ converges.

We also note in passing that if X is reflexive and locally uniformly convex, then X^* is (LUF).

Lemma 13. If X^* is (LUF), then every accretive operator $A \subset X \times X$ that satisfies (6) has the following property: every sequence $\{b_n\}$ in R(A) for which $\lim_{n\to\infty}|b_n| = d(0,R(A))$ converges.

Proof. Denote d(0,R(A)) by d and let -u be the unique point of least norm in cl(R(A)). The proof of [8, Theorem 4] and Lemma 3 show that to each b_n there corresponds a point z_n in J(-u) such that $(b_n,z_n) \geq d^2$. Since $\lim_{n\to\infty}|b_n| = d$, we have, in fact, $\lim_{n\to\infty}(b_n,z_n) = d^2$, and the conclusion follows by Lemma 12.

This lemma improves upon Lemma 3.2 in [11] where X was assumed to be uniformly convex and A is assumed to satisfy the stronger range condition corresponding to (4). It can be shown that the converse of Lemma 13 is also valid.

We can now establish our convergence result. Such a result has been known so far only in uniformly convex spaces [11, Theorem 3.7(c)]. It sharpens the conclusion of Theorem 11.

Theorem 14. Let D be a closed subset of a Banach space X and $T : D \to X$ a nonexpansive mapping which satisfies (3). Assume that $\{c_n\}$ is bounded away from 0 and 1, and let the sequence $\{y_n\}$ be defined by (2). If X^* is (LUF), then the strong $\lim_{n\to\infty}(y_n - Ty_n)$ exists and coincides with the unique point of least norm in $\mathrm{cl}(R(I-T))$.

Proof. Since $\{y_n - Ty_n\} \subset R(I-T)$ and $\lim_{n\to\infty}|y_n - Ty_n| = \lim_{n\to\infty}|y_{n+1} - y_n|/c_n = d(0, R(I-T))$ by Theorem 8, the result follows by Lemma 13.

It can be shown that if X^* is not (F), then the conclusion of Theorem 14 is no longer valid. We do not know, however, if this differentiability condition is sufficient for Theorem 14 to hold.

We shall say that the norm of a Banach space X with a unit sphere U is locally uniformly Gâteaux differentiable (LUG) if for each $y \in U$ and each face F of U the limit (7) is attained uniformly for all x in F. It can be shown that if, in the setting of Theorem 14, X^* is (LUG), then the weak $\lim_{n\to\infty}(y_n - Ty_n)$ exists. We expect to present a more complete discussion of the differentiability properties (LUF) and (LUG), as well as their consequences, elsewhere.

We turn now to another consequence of Theorem 8. Recall that a closed convex subset C of a Banach space has the fixed point property for nonexpansive mappings (FPP for short) if every nonexpansive $T : C \to C$ has a fixed point [6, 7, 9].

Theorem 15. Let C be a closed convex subset of a Banach space X and $T : C \to C$ a nonexpansive mapping. Assume that $\{c_n\}$ is bounded away from 0 and 1, and let the sequence $\{y_n\}$ be defined by (2). If each bounded closed convex subset of X has the FPP, then T is fixed point free if and only if $\lim_{n\to\infty}|y_n| = \infty$.

Proof. Assuming that a subsequence $\{y_{n_k}\}$ of $\{y_n\}$ is bounded, we let $R = \limsup_{k\to\infty}|y_0 - y_{n_k}|$ and $B = \{x \in C : \limsup_{k\to\infty}|x - y_{n_k}| \le R\}$. It is clear that B is a nonempty, bounded closed convex subset of C. Since $\lim_{n\to\infty}(y_{n_k} - Ty_{n_k}) = 0$ by Theorem 8, it is also invariant under T.

Hence the result.

It can also be shown that if, in the setting of Theorem 15, X is uniformly convex and (F), and T has a fixed point, then the weak $\lim_{n\to\infty} y_n$ exists and is a fixed point of T. In this case, however, a better result is already known [10, Theorem 2].

Finally, we note that Theorems 5, 8 and 15 can be shown to carry over to these self-mappings of the Hilbert ball which are nonexpansive with resepct to the hyperbolic metric [5]. These results extend several theorems in [13] and provide an affirmative answer to the question raised at the end of that paper. It is expected that a complete discussion, as well as related results on implicit iterations, will be presented elsewhere.

REFERENCES

1. J. B. Baillon, R. E. Bruck and S. Reich, On the asymptotic behavior of nonexpansive mappings and semigroups in Banach spaces, Houston J. Math. 4(1978), 1-9.
2. K. Fan and I. Glicksberg, Some geometric properties of the spheres in a normed linear space, Duke Math. J. 25(1958), 553-568.
3. T. Fujihira, Asymptotic behavior of nonexpansive mappings and some geometric properties in Banach spaces, Tokyo J. Math. 7(1984), 119-128.
4. K. Goebel and W. A. Kirk, Iteration processes for nonexpansive mappings, Contemporary Math. 21(1983), 115-123.
5. K. Goebel and S. Reich, "Uniform Convexity, Hyperbolic Geometry, and Nonexpansive Mappings," Marcel Dekker, New York and Basel, 1984.
6. W. A. Kirk, Fixed point theory for nonexpansive mappings, Lecture Notes in Math., Vol. 886, Springer, Berlin and New York, 1981, pp. 484-505.
7. W. A. Kirk, Fixed point theory for nonexpansive mappings II, Contemporary Math. 18(1983), 121-140.
8. A. T. Plant and S. Reich, The asymptotics of nonexpansive iterations, J. Functional Anal. 54(1983), 308-319.
9. S. Reich, The fixed point property for nonexpansive mappings, I, II, Amer. Math. Monthly 83(1976), 266-268; 87(1980), 292-294.
10. S. Reich, Weak convergence theorems for nonexpansive mappings in Banach spaces, J. Math. Anal. Appl. 67(1979), 274-276.
11. S. Reich, On the asymptotic behavior of nonlinear semigroups and the range of accretive operators, J. Math. Anal. Appl. 79(1981), 113-126.
12. S. Reich, On the asymptotic behavior of nonlinear semigroups and the range of accretive operators II, Mathematics Research Center Report #2198, 1981; J. Math. Anal. Appl. 87(1982), 134-146.
13. S. Reich, Averaged mappings in the Hilbert ball, J. Math. Anal. Appl. 109(1985), 199-206.

Quasilinear Ellipticity on the N-Torus

VICTOR L. SHAPIRO, Department of Mathematics and Computer Science, University of California, Riverside, Riverside, California

1. *Introduction.* In this paper, we intend to establish one new result and extend two previous results [3, Thms. 1 & 2], for second order quasilinear elliptic partial differential equations defined on the N-torus. The main point of the extension is that now the leading order coefficients $a^{ij}(x,u)$ are allowed to have also a linear growth condition with respect to u, i.e., $|a^{ij}(x,u)| \leq a(x) + \eta|u|$ where $a(x) \in L^2(\Omega)$ and η is a positive constant. Here, $\Omega = \{x : -\pi < x_j \leq \pi, \ j = 1,\ldots,N\}$ is the N-torus, $N \geq 2$.

We also adopt the following notation: $\phi \in C^\infty(\Omega)$ means that $\phi \in C^\infty(\mathbf{R}^N)$ and is periodic of period 2π in each variable. Also, $W^{1,2}(\Omega)$ will designate the familiar closure with respect to these C^∞, periodic functions (that is, with respect to $C^\infty(\Omega)$).

We shall consider second order, quasilinear elliptic operators Q, operating on $W^{1,2}(\Omega)$, of the form

$$Qu = -D_i[a^{ij}(x,u)D_j u] + b^j(x,u,Du)D_j u, \tag{1.1}$$

where Du represents the gradient of $u, D_j u = \partial u/\partial x_j$, and the summation convention is employed for $i,j = 1,\ldots,N$.

The coefficients of Q, namely, the functions $a^{ij}(x,z)$ and $b^j(x,z,p)$ are assumed to be defined for all values of $(x,z) \in (\Omega - E) \times \mathbf{R}$ and $(x,z,p) \in (\Omega - E) \times \mathbf{R} \times \mathbf{R}^N$, respectively, where $E \subset \Omega$ is a set of Lebesgue measure zero. Furthermore, we shall suppose the following throughout the paper.

(Q-1) The coefficients $a^{ij}(x,z)$ and $b^j(x,z,p)$ satisfy the usual Carath eodory conditions: For each fixed $z \in \mathbf{R}$ and $p \in \mathbf{R}^N$, the functions $a^{ij}(x,z)$ and $b^j(x,z,p)$ are measurable; for *a.e.* $x \in \Omega$ the functions $a^{ij}(x,z)$ and $b^j(x,z,p)$ are respectively continuous in $\mathbf{R}$ and $\mathbf{R} \times \mathbf{R}^N$, $i,j = 1,\ldots,N$.

(Q-2) $\exists$ a nonnegative function $a(x) \in L^2(\Omega)$ and a constant $\eta > 0$ such that $|a^{ij}(x,z)| \leq a(x) + \eta|z|$ for every $z \in \mathbf{R}$ and *a.e.* $x \in \Omega$, $i,j = 1,\ldots,N$.

(Q-3) Q is symmetric; that is, $a^{ij}(x,z) = a^{ji}(x,z)$ for every $z \in \mathbf{R}$ and *a.e.* $x \in \Omega$, $i,j = 1,\ldots,N$.

(Q-4) Q is uniformly elliptic almost everywhere in Ω; that is, there is a constant $\eta_0 > 0$ such that

$$a^{ij}(x,z)\xi_i\xi_j \geq \eta_0|\xi|^2$$

for every $z \in \mathbf{R}$, *a.e.* $x \in \Omega$, and every $\xi \in \mathbf{R}^N$ $(|\xi|^2 = \xi_1^2 + \cdots + \xi_N^2)$.

(Q-5) There is a nonnegative function $b(x) \in L^2(\Omega)$ and a positive constants η_1 and η_2 such that

$$|b^j(x,z,p)| \leq b(x) + \eta_1|z| + \eta_2|p|$$

for every $p \in \mathbf{R}^N$, $z \in \mathbf{R}$, and *a.e.* $x \in \Omega$, $j = 1, \ldots, N$.

(Q-6) For every $u \in W^{1,2}(\Omega)$, the vector $\mathbf{b}(x,u,Du) = [b^1(x,u,Du), \ldots, b^N(x,u,Du)]$ is weakly solenoidal, i.e., $\int_\Omega b^j(x,u,Du)D_jv(x)dx = 0$ for every u and $v \in W^{1,2}(\Omega)$ where the summation convention is used.

(Q-7) If $\{u^n\}_{n=1}^\infty$ is a sequence of functions in $L^2(\Omega)$ which tend strongly to $u \in L^2(\Omega)$ and $\{\mathbf{w}^n\}_{n=1}^\infty$ is a sequence of vector-valued functions which tend weakly to $\mathbf{w} \in [L^2(\Omega)]^N$, then $\{\mathbf{b}(x,u^n,\mathbf{w}^n)\}_{n=1}^\infty$ tends weakly to $\mathbf{b}(x,u,\mathbf{w}) \in [L^2(\Omega)]^N$, i.e., with $b^j(x,u,\mathbf{w}) \equiv b^j(x,u,w_1,\ldots,w_n)$ the j-th component of $\mathbf{b}(x,u,\mathbf{w})$, then $\lim\limits_{n\to\infty} \int_\Omega b^j(x,u^n,\mathbf{w}^n)vdx = \int_\Omega b^j(x,u,w)vdx$ $\forall$ $v \in L^2(\Omega)$ and $j = 1, \ldots, N$.

By strong convergence in (Q-7), we mean convergence in norm.

In this paper, we intend to establish the following new result.

THEOREM 1. *Assume* (Q-1) - (Q-7) *and let* $f \in L^2(\Omega)$. *Then a necessary and sufficient condition that the equation*

$$Qu = f \tag{1.2}$$

has a distribution solution $u \in W^{1,2}(\Omega)$ *is that* $\int_\Omega f dx = 0$.

If $Qu = -\Delta u$ where Δ is the Laplacian, the above theorem is well known. It is also well known if $b^j \equiv 0$, $j = 1, \ldots, N$ and a^{ij} does not depend on u and is smooth as a function of x. (See [2, p. 574].) However in the case considered here where the $a^{ij}(x,u)$ are unbounded in x and possibly grow linearly in u, the result obtained in Theorem 1 is new.

To be quite explicit, what we mean by $u \in W^{1,2}(\Omega)$ being a distribution solution of $Qu = f$ is the following:

$$\int_\Omega [a^{ij}(x,u)D_juD_i\phi + \phi b^j(x,u,Du)D_ju]dx = \int_\Omega f\phi dx \tag{1.3}$$

for every $\phi \in C^\infty(\Omega)$, where the summation convention is used for $i, j = 1, \ldots, N$.

2. *Proof of Theorem 1.* We first deal with the sufficiency condition of the theorem and establish this part via a Galerkin argument. In order to accomplish this, we observe that there is a sequence $\{\psi_k\}_{k=1}^\infty$ of real-valued functions in $C^\infty(\Omega)$ with the following properties:

$$\begin{aligned} &(a) \quad \psi_1 = (2\pi)^{-N/2}; \\ &(b) \quad \int_\Omega \psi_k \psi_\ell dx = \delta_{k\ell} \qquad \text{where } \delta_{k\ell} \text{ is the Kronecker-}\delta \ k, \ell = 1, 2, \ldots . \end{aligned} \tag{2.1}$$

Also, given $\psi \in C^\infty(\Omega)$ and $\varepsilon > 0$, $\exists$ constants $c_1, \ldots, c_n$ such that

$$|\psi(x) - c_q\psi_q(x)| < \varepsilon \quad \text{and} \quad |D_j\psi(x) - c_q D_j\psi_q(x)| < \varepsilon \tag{2.2}$$

uniformly for $x \in \Omega$ and $j = 1, \ldots, N$ where the summation convention is used in (2.2) with $q = 1, \ldots, n$.

Here, and in the sequel throughout the rest of the paper, we shall use the notation

$$\langle v, w\rangle_0 = \int_\Omega v(x)w(x)dx \tag{2.3}$$

for v and w in $L^2(\Omega)$. In case v and w are also in $W^{1,2}(\Omega)$, we shall set

$$\langle v, w\rangle_1 = \langle v, w\rangle_0 + \langle D_j v, D_j w\rangle \tag{2.4}$$

where the summation convention is used for $j = 1, \ldots, N$. Also we shall set

$$\|v\|_0 = \langle v, v\rangle_0^{\frac{1}{2}} \quad \text{and} \quad \|v\|_1 = \langle v, v\rangle_1^{\frac{1}{2}}. \tag{2.5}$$

$H^{\tilde{}}$ shall designate the Hilbert space spanned by $\{\psi_k\}_{k=2}^\infty$ using the $W^{1,2}(\Omega)$-norm. Thus $w \in H^{\tilde{}}$ means $\exists \{w_n\}_{n=1}^\infty$ such that $\|w - w_n\|_1 \to 0$ as $n \to \infty$ where each w_n is a finite linear combination of elements of the sequence $\psi_2, \psi_3, \ldots, \psi_k, \ldots$. As a consequence of (2.1) and (2.2), it follows that $v \in W^{1,2}(\Omega)$ can be written uniquely in the form

$$v = c_1\psi_1 + w. \tag{2.6}$$

where $w \in H^{\tilde{}}$ and c_1 is a constant. Also, it is clear that

$$\langle \psi_1, w\rangle_1 = 0 \qquad \forall \ w \in H^{\tilde{}}. \tag{2.7}$$

Furthermore, it is easy to see from (2.1)(a), (2.4), (2.5), and (2.7) and elementary Fourier analysis that

$$\|w\|_0^2 \le \langle D_j w, D_j w\rangle_0 \qquad \forall\ w \in H^{\tilde{}}, \tag{2.8}$$

where the summation convention is used for $j = 1, \ldots, N$.

The first lemma we state is the following.

LEMMA 1. *Let $f(x)$ be the function in $L^2(\Omega)$ given in the statement of Theorem 1, and assume that Q satisfies* (Q-1) - (Q-6). *Then if n is a given positive integer ≥ 2, there is a function $u^n = \gamma_2^n \psi_2 + \cdots + \gamma_n^n \psi_n$ such that*

$$\langle a^{ij}(\cdot, u^n) D_j u^n, D_i \psi_k\rangle_0 + \langle b^j(\cdot, u^n, Du^n) D_j u^n, \psi_k\rangle_0 + \langle u^n, \psi_k\rangle_0 n^{-1} = \langle \psi_k, f\rangle_0 \tag{2.9}$$

for $k = 2, \ldots, n$ where the summation convention is used for $i, j = 1, \ldots, N$ and γ_q^n are constants $q = 2, \ldots, n$.

The proof of this lemma is essentially the same as the proof given in [3, Lemma 1], and the reader should have no difficulty in filling in the details.

Next for n fixed and ≥ 2, we multiply both sides of (2.9) by γ_k^n for $k = 2, \ldots, n$ and sum on k. Observing that $D_j(u^n)^2 = 2u^n D_j u^n$ and using (Q-6), we see that

$$\langle a^{ij}(\cdot, u^n) D_j u^n, D_i u^n\rangle_0 + \langle u^n, u^n\rangle_0 n^{-1} = \langle u^n, f\rangle_0.$$

We conclude from (Q-4) and this last fact that

$$\eta_0 \langle D_j u^n, D_j u^n\rangle_0 \le \|u^n\|_0 \|f\|_0 \tag{2.10}$$

where the summation convention is used for $j = 1, \ldots, N$. Since u^n is in $H^{\tilde{}}$, it follows from (2.8) that $\|u^n\|_0 \le \langle D_j u^n, D_j u^n\rangle_0^{\frac{1}{2}}$. We therefore obtain from (2.4), (2.5), and (2.10) that

$$\|u^n\|_1 \le \sqrt{2}\eta_0^{-1} \|f\|_0 \qquad \text{for } n = 2, 3, \ldots\ . \tag{2.11}$$

From this last fact, it follows that there is a subsequence of $\{u^n\}_{n=2}^{n}$, which for ease of notation we take to be the full sequence, with the following properties:

$$\exists u \in H^{\tilde{}} \text{ such that } \lim_{n\to\infty} \|u^n - u\|_0 = 0; \tag{2.12}$$

$$\lim_{n\to\infty} u^n(x) = u(x) \qquad \text{for a.e. } x \in \Omega; \tag{2.13}$$

$$\lim_{n\to\infty} \int_\Omega w D_j u^n dx = \int_\Omega w D_j u dx \ \ \forall\ w \in L^2(\Omega), j = 1, \ldots, N. \tag{2.14}$$

Using (2.11) - (2.14), in conjunction with (Q-5) - (Q-7), it follows exactly as in the proof [3, Thm. 1] that

$$\lim_{n\to\infty} \langle b^j(\cdot, u^n, Du^n) D_j u^n, \psi_k \rangle_0 = \langle b^j(\cdot, u, Du) D_j u, \psi_k \rangle_0 \tag{2.15}$$

for $k = 2, 3, \ldots$.

Next, we fix i and j and set

$$h_n(x) = |a^{ij}(x, u^n) - a^{ij}(x, u)|^2 \tag{2.16}$$

and observe from (2.13) and (Q-1) that

$$\lim_{n\to\infty} h_n(x) = 0 \qquad \text{for a.e. } x \in \Omega. \tag{2.17}$$

With $a(x)$ and η as in (Q-2), we set

$$g_n(x) = [2a(x) + \eta|u^n| + \eta(u)]^2 \tag{2.18}$$

and observe from (2.13) that $\lim_{n\to\infty} g_n(x) = 4[a(x) + \eta u(x)]^2$ a.e. in Ω. Also we have that

$$|h_n(x)| \le g_n(x) \qquad \text{for a.e. } x \in \Omega \text{ and } n = 2, 3, \ldots . \tag{2.19}$$

Since $a(x) \in L^2(\Omega)$, it follows from (2.12) that

$$\lim_{n\to\infty} \int_\Omega g_n(x)dx = 4\int_\Omega [a(x) + \eta|u|^2 dx. \tag{2.20}$$

We invoke [2, Thm. 16, p. 89] with h_n replacing f_n and conclude from (2.16) - (2.20) that $\lim_{n\to\infty} \int_\Omega h_n(x)dx = 0$. Now i and j were fixed but arbitrary; so we record this last fact, using (2.16), as follows:

$$\lim_{n\to\infty} \|a^{ij}(x, u^n) - a^{ij}(x, u)\|_0 \tag{2.21}$$

for $i, j = 1, \ldots, N$. But then using (2.21) and proceeding exactly as in the proof of [3, Thm. 1], we see that

$$\lim_{n\to\infty} \langle D_i\psi_k, a^{ij}(\cdot, u^n) D_j u^n \rangle_0 = \langle D_i \psi_k, a^{ij}(\cdot, u) D_j u \rangle_0 \tag{2.22}$$

for $k = 2, 3, \ldots$.

Next, we see from (2.11) that $\lim_{n\to\infty} \langle u^n, \psi_k \rangle_0 n^{-1} = 0$ for $k = 2, 3, \ldots$. We consequently obtain from (2.9) on passing to the limit as $n \to \infty$ and using (2.15) and (2.22) that

$$\langle a^{ij}(\cdot, u) D_j u, D_i \psi_k \rangle_0 + \langle b^j(\cdot, u, Du) D_j u, \psi_k \rangle_0 = \langle \psi_k, f \rangle_0 \qquad (2.23)$$

for $k = 2, 3, \ldots$. Now using (Q-6) and (2.1)(a), we see that the left-hand side of (2.23) is zero when $k = 1$. Likewise from the hypothesis of the theorem, we see the right-hand side of (2.23) is zero when $k = 1$. We conclude that (2.23) also holds for $k = 1$. But then it follows from (2.23) and the uniform convergence aspect of (2.2) that

$$\langle a^{ij}(\cdot, u) D_j u, D_i \phi \rangle_0 + \langle b^j(\cdot, u, Du) D_j u, \phi \rangle_0 = \langle \phi, f \rangle_0$$

for all $\phi \in C^\infty(\Omega)$. This last fact is the same as (1.3) and the proof of the sufficiency condition part of Theorem 1 is complete.

To establish the necessary part of the theorem, we suppose that $u \in W^{1,2}(\Omega)$ and that (1.3) holds for all $\phi \in C^\infty(\Omega)$. We choose $\phi \equiv 1$ and observe from (Q-6) then that

$$\int_\Omega \phi b^j(x, u, Du) D_j u dx = 0.$$

Also, we have that under such circumstances $D_i \phi \equiv 0$, $i = 1, \ldots, N$. Consequently with $\phi \equiv 1$, we have that the left-hand side of (1.3) is equal to zero. But this implies that $\int_\Omega f dx = 0$ and the necessary condition of the theorem is established.

3. *Extensions of Previous Theorems.* As mentioned in the introduction, the two theorems of [3] can be extended along the lines of Theorem 1 above. In particular, the following extension of [3, Thm. 2] prevails.

THEOREM 2. *Assume* (Q-1) - (Q-7), *that* $g \in C(\mathbf{R}), h \in W^{1,2}(\Omega)$ *and that the limits* $\lim_{z\to\infty} g(z) = g(\infty)$ *and* $\lim_{z\to\infty} g(z) = g(-\infty)$ *exist and are finite. Suppose also that*

$$g(\infty) < g(z) < g(-\infty) \qquad \text{for } z \in \mathbf{R}. \qquad (3.1)$$

Then a necessary and sufficient condition that a distribution solution $u \in W^{1,2}(\Omega)$ *of* $Qu = g(u) - h$ *exists is that*

$$(2\pi)^N g(\infty) < h(1) < (2\pi)^N g(-\infty).$$

The statement of Theorem 2 given here is exactly the same as [3, Thm. 2], but (Q-2) of this paper differs from the (Q-2) of [3]. The part where the two proofs differ is the establishment of (4.14) in [3]. The main aspect in establishing (4.14) is to show that

$$\lim_{n\to\infty} \|a^{ij}(x,u^n) - a^{ij}(x,u)\|_0.$$

This is accomplished with the new (Q-2) exactly as (2.21) in this current paper is established. The reader should have no difficulty in filling in the details.

A similar extension of [3, Thm. 1] exists. The statement of the new theorem is exactly the same as previously, except we have a new (Q-2). Once again, the main idea is to establish (2.21) above and the procedure is the same as that given here. The rest of the proof of [3, Thm. 1] is unchanged; the reader should experience no difficulty with the details.

REFERENCES

1. J. L. Kazdan and F. W. Warner, "Remarks on some quasilinear elliptic equations," *Comm. Pure Appl. Math* 28(1975), 567-597.

2. H. L. Royden, *Real Analysis,* 2d ed., Macmillan Publ. Co., New York, 1968.

3. V. L. Shapiro, "Resonance and quasilinear ellipticity," *Trans Amer. Math. Soc.* (in press).

A Local Minimax Theorem Without Compactness

SHI SHU-ZHONG * Mathematical Research Center, University of Paris-Dauphine, Paris, France

CHANG KUNG-CHING (ZHANG GONG-QING) Department of Mathematics, Peking University, Beijing, People's Republic of China

Dedicated to Prof. Ky Fan

1. Introduction

The von Neumann Minimax Theorem plays an important role in the game of economic theory. It has been extended by many authors. A well known formulation is due to M. Sion [13] and Ky Fan [8,9]:

Theorem (von Neumann, Sion, Ky Fan). Let E, F be two compact convex sets of Hausdorff linear topological spaces, and let $f : E \times F \longrightarrow \mathbb{R}^1$ be a function satisfying

(1) $\forall\, x \in E$, $y \longmapsto f(x,y)$ is u.s.c. and quasi-concave,

(2) $\forall\, y \in F$, $x \longmapsto f(x,y)$ is l.s.c. and quasi-convex.

Then there exists a saddle point of f, $(\bar{x},\bar{y}) \in E \times F$ such that

$$f(\bar{x},\bar{y}) = \min_{x\in E} \max_{y\in F} f(x,y) = \max_{y\in F} \min_{x\in E} f(x,y).$$

A function $g(x)$ is called quasi-convex, if

$$g(\lambda x_1+(1-\lambda)x_2) \le \max\{g(x_1),g(x_2)\} \quad \forall\, \lambda \in (0,1), \quad \forall\, x_1,x_2;$$

in other words, the level sets $\{x \in E \mid g(x) \le \lambda\}$ are convex for all $\lambda \in \mathbb{R}^1$. A function g is called quasi-concave, if $-g$ is quasi-convex.

The main purpose of this paper is to extend this theorem to a form in which the compactness conditions of the sets E and F are replaced by some conditions on the function f, which are, in some sense, relevant to the so-called "Palais Smale condition" (P.S. for short).

The following generalized P.S. condition was introduced in [5], in dealing with the critical point theory on closed convex sets.

Definition 1.1. Assume that $g : F \longmapsto \mathbb{R}^1$ is a C^1-function defined on a closed convex set F of a Banach space. We say that g satis-

*Current affiliation: Nankai University, Tianjin, People's Republic of China

fies the generalized Palais Smale condition (G.P.S.), if

(G.P.S.) any sequence $\{y_n\}$, along which $g(y_n)$ is bounded and $\inf_{l\in F-y_n} \langle g'(y_n), \frac{l}{\|l\|}\rangle \longrightarrow \beta \geq 0$, possesses a convergent subsequence,

where, and hereafter, we use $\langle\cdot,\cdot\rangle$ to denote the duality pairing between a Banach space and its dual.

An analogy for double variable functions is defined for our purpose in the following.

Definition 1.2. Suppose that $f : E \times F \longrightarrow \mathbb{R}^1$ is a C^1-function defined on the product space $E \times F$ of two closed convex sets E and F of Banach spaces. We say that f satisfies the twist Palais Smale condition (T.P.S), if

(T.P.S.) any sequence $\{(x_n,y_n)\} \subset E \times F$, along which $f(x_n,y_n) = \sup_{y\in F} f(x_n,y_n)$ is bounded and $\inf_{h\in E-x_n} \langle f'_x(x_n,y_n), \frac{h}{\|h\|}\rangle \longrightarrow \alpha \geq 0$, possesses a convergent subsequence.

Remark 1.1. If F is a compact set, then any function g defined on F satisfies (G.P.S.). If E and F are compact sets, then any function f defined on $E \times F$, satisfies (T.P.S.).

Remark 1.2. If F is a Banach space, and if g satisfies P.S., then g satisfies (G.P.S.). If E and F are Banach spaces, and if f satisfies P.S. over $E \times F$, then f satisfies (T.P.S.).

The main result of this paper is the following

Theorem A. Let E, F be two closed convex sets of Banach spaces, and let $f \in C^1(E \times F, \mathbb{R}^1)$ be a function satisfying

(1) $\forall\, x \in E$, $y \longmapsto -f(x,y)$ is bounded below and quasi-convex, and satisfies (G.P.S.),

(2) $\exists\, y_0 \in F$ such that $x \longmapsto f(x,y_0)$ is bounded below,

(3) (T.P.S.) holds for f.

Then there exists $(\bar{x},\bar{y}) \in E \times F$ such that

(1) $f(\bar{x},\bar{y}) = \max_{y\in F} f(\bar{x},y) = \min_{x\in E} \max_{y\in F} f(x,y)$,

(2) $\langle f'_x(\bar{x},\bar{y}),h\rangle \geq 0 \quad \forall\, h \in E-\bar{x}$.

Moreover, if $x \longmapsto f(x,\bar{y})$ is locally pseudo-convex at $\bar{x}$, i.e., there is a neighborhood $U(\bar{x})$ of $\bar{x}$ such that

$$\langle f'_x(\bar{x},\bar{y}), x-\bar{x}\rangle \geq 0 \quad \forall\, x \in U(\bar{x}) \implies f(\bar{x},\bar{y}) \geq f(\bar{x},\bar{y}) \quad \forall\, x \in U(\bar{x})$$

then $(\bar{x},\bar{y})$ is a saddle point of f, i.e.,

$$f(\bar{x},y) \leq f(\bar{x},\bar{y}) \leq f(x,\bar{y})$$

$\forall\, x \in U(\bar{x})$, $\forall\, y \in F$.

Remark 1.3. It is well known that the crucial point in proving von Neumann-Sion Minimax Theorem, is to prove the equality:

$$\min_{x\in E}\max_{y\in F} f(x,y) = \max_{y\in F}\min_{x\in E} f(x,y)$$

which depends heavily on the convexity hypothesis of the functions $x \longmapsto f(x,y)$ $\forall\, y \in F$. However, our approach, which starts from the point of view of critical point theory, concerns merely the local behavior near the "saddle point," so that the convexity hypothesis is not necessary.

The proof of Theorem A is based upon the following two fundamental principles:

(Ekeland's Variational Principle [7]). Let E be a complete metric space, and let $G : E \longrightarrow \mathbb{R}^1_+ \cup \{+\infty\}$ be l.s.c. with $G(x) \not\equiv +\infty$. Suppose that $x_0 \in E$ and $\varepsilon > 0$ are given such that

$$G(x_0) \leq \inf_{x\in E} G(x) + \varepsilon.$$

Then $\forall\, k > 0$, there exists $x_{\varepsilon k} \in E$ such that

$$G(x_{\varepsilon k}) \leq G(x_0),$$

$$d(x_0, x_{\varepsilon k}) \leq \frac{1}{k},$$

$$G(x) > G(x_{\varepsilon k}) - k\varepsilon d(x, x_{\varepsilon k}) \quad \forall\, x \neq x_{\varepsilon k},$$

where $d : E \times E \longrightarrow \mathbb{R}^1_+$ is the metric of E.

This principle is very useful in many problems. It has been applied to give a new proof of the Mountain Pass Lemma (cf. [2] and [12]), and will be applied time to time in this paper.

(Lop-sided Minimax Theorem). Let E, F be two convex sets of Hausdorff linear topological spaces, and let $f : E \times F \longrightarrow \mathbb{R}^1$. Suppose that

(1) $\forall\, x \in E, y \longmapsto f(x,y)$ is u.s.c. and quasi-concave,

(2) $\forall\, y \in F, x \longmapsto f(x,y)$ is l.s.c. and quasi-convex,

(3) $\exists\, x_0 \in E$ and $\lambda < \inf_{x\in E}\sup_{y\in F} f(x,y)$ such that the set $\{y \in F \mid f(x_0,y) \geq \lambda\}$ is compact.

Then there exists $\bar{y} \in F$ such that

$$\inf_{x\in E} f(x,\bar{y}) = \inf_{x\in E}\sup_{y\in F} f(x,y) = \sup_{y\in F}\inf_{x\in E} f(x,y).$$

The reader is referred to [2], [3] and [4].

The paper is organized as follows: The proof of Theorem A will be given in §2. Some extensions, which need weaker smooth conditions on the function will be discussed in §3, in which the locally Lipschitzian condition for the function f are used to replace the C^1-condition. §4 deals with applications, in which a variational inequality problem, an elliptic system and an existence theorem of nonlinear programming are studied.

2. Proof of Theorem A.

The proof is divided into several steps. In order to make it easy to understand, we start with a slightly abstract version.

Theorem 2.1. Let E, F be two closed convex sets of Banach spaces, and let $f \in C^1$ $(E \times F, \mathbb{R}^1)$. Suppose that

(1) $\forall\, x \in E, y \longmapsto f(x,y)$ is bounded above and quasi-concave,

(2) $\exists\, y_0 \in F$, such that $x \longmapsto f(x,y_0)$ is bounded below,

(3) (T.P.S.) holds for f,

(4) $\forall\, x \in E$, the set

$$M(x) = \{\bar{y} \in F \mid f(x,\bar{y}) = \sup_{y\in F} f(x,y)\}$$

is compact and nonempty,

(5) $\forall\, x \in E$, $\forall\, h \in E-x$, $\forall$ positive sequence $\{t_n\}$ with $t_n \downarrow 0$, there exist $y_n \in M(x+t_n h)$, $n = 1,2,\ldots$ such that $\{y_n\}$ has an accumulate point $\bar{y} \in M(x)$.

Then there exists a pair $(\bar{x},\bar{y}) \in E \times F$ such that

(1) $f(\bar{x},\bar{y}) = \max_{y\in F} f(\bar{x},y) = \min_{x\in E}\max_{y\in F} f(x,y)$,

(2) $\langle f'_x(\bar{x},\bar{y}),h\rangle \geq 0 \quad \forall\, h \in E-\bar{x}$.

Proof of Theorem 2.1. Let

$$G(x) = \sup_{y\in F} f(x,y).$$

It is easily seen from the assumptions (1) and (2) that $G : E \longrightarrow \mathbb{R}^1$ is l.s.c. and bounded below. According to the Ekeland's Variational Principle, $\forall\, \varepsilon > 0$, $\exists\, x_\varepsilon \in E$ such that

$$G(x_\varepsilon) \leq \inf_{x\in E} G(x) + \varepsilon, \tag{2.1}$$

$$G(x_\varepsilon) < G(x) + \varepsilon\|x-x_\varepsilon\| \quad \text{for } x \neq x_\varepsilon.$$

For each $h \in E-x_\varepsilon$, we have

$$\underline{G}'_+(x_\varepsilon;h) \triangleq \underline{\lim}_{t\to+0} \frac{G(x_\varepsilon+th)-G(x_\varepsilon)}{t} \geq -\varepsilon\|h\|. \tag{2.2}$$

Since for any positive sequence $\{t_n\}$ with $t_n \downarrow 0$, we have

$$\begin{aligned} \underline{G}'_+(x_\varepsilon;h) &\leq \underline{\lim}_{n\to\infty} \frac{1}{t_n}[G(x_\varepsilon+t_n h)-G(x_\varepsilon)], \\ &\leq \underline{\lim}_{n\to\infty} \frac{1}{t_n}[f(x_\varepsilon+t_n h,y_n) - f(x_\varepsilon,y_n)] \\ &= \underline{\lim}_{n\to\infty} \langle f'_x(x_\varepsilon+t_n\theta_n h,y_n),h\rangle \end{aligned}$$

for some $\theta_n \in (0,1)$, provided by the assumption (4), with $y_n \in M(x_\varepsilon+t_n h)$. According to the assumption (5), we may so choose $\{y_n\}$ that $\{y_n\}$ has an accumulate point $\bar{y}_n \in M(x_\varepsilon)$. The C^1-continuity of f implies that

$$G'_+(x_\varepsilon;h) \leq \langle f'_x(x_\varepsilon,\bar{y}_h),h\rangle. \tag{2.3}$$

Combining (2.2) with (2.3), it follows

$$\langle f'_h(x_\varepsilon,\bar{y}_h),h\rangle + \varepsilon\|h\| \geq 0 \quad \forall\, h \in E-x_\varepsilon. \tag{2.4}$$

Let us define a function $P_\varepsilon : (E-x_\varepsilon) \times M(x_\varepsilon) \longrightarrow \mathbb{R}^1$ by

$$P_\varepsilon(h,y) = \langle f'_x(x_\varepsilon,y),h\rangle + \varepsilon\|h\|.$$

We shall verify that

(1) $\forall\, y \in M(x_\varepsilon)$, $h \longmapsto P_\varepsilon(h,y)$ is continuous and convex,

(2) $\forall\, h \in E-x_\varepsilon$, $y \longmapsto P_\varepsilon(h,y)$ is continuous and quasi-concave,

(3) $M(x_\varepsilon)$ is compact and convex.

The convexity of $M(x_\varepsilon)$ follows from the quasi-concavity of the function $y \longmapsto f(x,y)$. It remains to prove the quasi-concavity of the function $y \longmapsto P_\varepsilon(h,y)$; because all other items are easily seen.

In fact, $\forall\, y_1,y_2 \in M(x_\varepsilon)$, and $\forall\, \lambda \in [0,1]$,

$$\begin{aligned} &\langle f'_x(x_\varepsilon,(1-\lambda)y_1+\lambda y_2),h\rangle \\ &\quad = \lim_{t\downarrow 0} \tfrac{1}{t}[f(x_\varepsilon+th,(1-\lambda)y_1+\lambda y_2)-f(x_\varepsilon,(1-\lambda)y_1+\lambda y_2)] \\ &\quad \geq \lim_{t\downarrow 0} \min\{\tfrac{1}{t}[f(x_\varepsilon+th,y_1)-G(x_\varepsilon)],\ \tfrac{1}{t}[f(x_\varepsilon+th,y_2)-G(x_\varepsilon)]\} \\ &\quad = \min\{\langle f'_x(x_\varepsilon,y_1),h\rangle,\langle f'_x(x,y_2),h\rangle\}, \end{aligned}$$

i.e., the function $y \longmapsto P_\varepsilon(h,y)$ is quasi-concave.

Now P_ε satisfies all conditions of the Lop-sided Minimax Theorem, we conclude that $\exists\, y_\varepsilon \in M(x_\varepsilon)$ such that

$$\inf_{h\in E-x_\varepsilon} P_\varepsilon(h,y_\varepsilon) = \inf_{h\in E-x_\varepsilon} \sup_{y\in M(x_\varepsilon)} (\langle f'_x(x_\varepsilon,y),h\rangle + \varepsilon\|h\|) \geq 0$$

provided by (2,4), that is,

$$\langle f'_x(x_\varepsilon,y_\varepsilon),h\rangle \geq -\varepsilon\|h\|, \quad \forall\, h \in E-x_\varepsilon. \tag{2.5}$$

By definition,

$$f(x_\varepsilon,y_\varepsilon) = G(x_\varepsilon),$$

(2.1) reads as

$$\begin{aligned}\inf_{x\in E}\sup_{y\in F} f(x,y) &= \inf_{x\in E} G(x) \leq G(x_\varepsilon)\\ &= f(x_\varepsilon,y_\varepsilon) = \sup_{y\in F} f(x_\varepsilon,y)\\ &\leq \inf_{x\in E}\sup_{y\in F} f(x,y) + \varepsilon.\end{aligned} \tag{2.6}$$

Let $\{\varepsilon_n\}$ be a positive sequence with $\varepsilon_n \downarrow 0$. Then, the sequence $\{(x_{\varepsilon_n},y_{\varepsilon_n})\}$ possesses a convergent subsequence $(x_{\varepsilon_j},y_{\varepsilon_j}) \longrightarrow (\bar{x},\bar{y}) \in E \times F$, provided by (T.P.S.). Thus we have

$$f(\bar{x},\bar{y}) = \max_{y\in F} f(\bar{x},y) = \min_{x\in E}\max_{y\in F} f(x,y)$$

by (2.6); and

$$\langle f'_x(\bar{x},\bar{y}),h\rangle \geq 0 \quad \forall\, h \in E-\bar{x}$$

by (2.5). $(\bar{x},\bar{y})$ is the solution of our theorem.

Now we turn to the proof of Theorem A. It remains to prove that (G.P.S.) of the functions

$$y \longmapsto -f(x,y) \quad \forall\, x \in E$$

implies the assumptions (4) and (5) of theorem 2.1.

Lemma 2.2. Let E be a closed convex set of a Banach space, and let $g \in C^1(E,\mathbb{R}^1)$ be a function bounded below and satisfying (G.P.S.). Then the set

$$M = \{\bar{x} \in E \mid g(\bar{x}) = \inf_{x\in E} g(x)\}$$

is nonempty and compact.

If Ω_M is a neighborhood of M, with boundary $\partial\Omega_M$, then

$$\inf_{x\in E\cap\partial\Omega_M} g(x) > \inf_{x\in E} g(x). \tag{2.7}$$

Proof. The first assertion follows directly from the Ekeland's Variational Principle. In fact, $\forall\, \varepsilon > 0\ \exists\, x_\varepsilon \in E$ such that

$$g(x_\varepsilon) \leq \inf_{x\in E} g(x) + \varepsilon$$

and

$$g'_+(x_\varepsilon; h) \geq -\varepsilon\|h\| \quad \forall\, h \in E-x_\varepsilon,$$

i.e.,

$$\operatorname*{Inf}_{h\in E-x_\varepsilon} \langle g'(x_\varepsilon), \frac{h}{\|h\|}\rangle \geq -\varepsilon.$$

We obtain a convergent subsequence $\{x_{\varepsilon_i}\} \longrightarrow \bar{x}$, provided by (G.P.S.). Therefore $\bar{x} \in M$.

Again, by (G.P.S.), the set M is compact.

Denote

$$\delta = \operatorname{dist}(M, \partial\Omega_M), \tag{2.8}$$

which is positive because M is compact.

We shall prove the second assertion by contradiction. If there were $\{x_n\} \subset E \cap \partial\Omega_M$ such that

$$g(x_n) \longrightarrow \inf_{x\in E} g(x),$$

then we would have $z_n \in E$ such that

$$g(z_n) \leq g(x_n) < \inf_{x\in E} g(x) + \varepsilon, \quad \text{for all} \quad \varepsilon > g(x_n) - \inf_{x\in E} g(x) \geq 0,$$

$$\|z_n - x_n\| \leq \frac{\delta}{2},$$

and

$$g(x) > g(z_n) - \frac{2}{\delta}\varepsilon\|x - z_n\| \quad \forall\, x \neq z_n,$$

which imply

$$\langle g'(z_n), \frac{h}{\|h\|}\rangle \geq -\frac{2}{\delta}[g(x_n) - \inf_{x\in E} g(x)].$$

Again, by (G.P.S.), $\{z_n\}$ possesses a convergent subsequence $z_{nj} \longrightarrow \bar{x}$. Therefore

$$g(\bar{x}) = \inf_{x\in M} g(x),$$

i.e., $\bar{x} \in M$, but

$$\begin{aligned} \operatorname{dist}(\bar{x}, \partial\Omega_M) &\leq \|\bar{x} - z_{nj}\| + \|z_{nj} - x_{nj}\| \\ &\leq \frac{\delta}{2} + \|\bar{x} - z_{nj}\| \longrightarrow \frac{\delta}{2}, \end{aligned}$$

which contradicts with (2.8).

Remark 2.1. In lemma 2.2, the same conclusion holds true, if the condition (G.P.S.) is weakened as follows:

(G.P.S.)* any sequence $\{x_n\} \subset E$, along which $g(x_n) \longrightarrow \inf_{x\in E} g(x)$ and $\inf_{h\in E-x_n} \langle g'(x_n), \frac{h}{\|h\|}\rangle \longrightarrow \beta \geq 0$, possesses a convergent subsequence.

It is interesting to note that (G.P.S.)* is also a necessary condition for the conclusion. In fact, if $\{x_n\} \subset E$ is a sequence such that

$$g(x_n) \longrightarrow \inf_{x \in E} g(x) \quad \text{and} \quad \inf_{h \in E - x_n} \langle g'(x_n), \frac{h}{\|h\|} \rangle \longrightarrow \beta \geq 0.$$

Then we have

$$\text{dist}(x_n, M) \longrightarrow 0$$

provided by (2.7). No loss of generality, we may assume that $\forall\, n \in \mathbb{Z}_+$; we have $x'_n \in M$ such that $\|x_n - x'_n\| < \frac{1}{n}$. Since M is a compact set, $\{x'_n\}$ possesses a convergent subsequence, so does $\{x_n\}$.

Lemma 2.3. Suppose that M is an upper semi-continuous nonempty compact set-valued mapping from a metric space E to a metric space F. Then for any sequence $\{x_n\}_1^\infty \subset E$, with $x_n \longrightarrow \bar{x}$, and for any $y_n \in M(x_n)$, $n = 1, 2, \ldots$, we have an accumulation point $\bar{y}$ of $\{y_n\}$ such that $\bar{y} \in M(\bar{x})$.

Proof. For any positive integer m, let

$$U_m = \{y \in F \mid \text{dist}(y, M(\bar{x})) < \tfrac{1}{m}\}.$$

Since the set-valued mapping M is u.s.c. and $x_n \longrightarrow x$, there exists $N = N(m) > 0$ such that

$$M(x_n) \subset U_m \quad \text{for} \quad n \geq N.$$

This implies that there exist $\bar{y}_m \in M(\bar{x})$ such that

$$\text{dist}(y_{N(m)}, \bar{y}_m) < \tfrac{1}{m}, \quad m = 1, 2, \cdots.$$

However, $M(\bar{x})$ is a compact set, $\{\bar{y}_m\}$ has an accumulation point $\bar{y} \in M(\bar{x})$. Of course $\bar{y}$ is also an accumulation point of $\{y_n\}$.

Lemma 2.4. Let E, F be two closed convex sets of Banach spaces, and let $f \in C^1(E \times F, \mathbb{R}^1)$. Suppose that $\forall\, x \in E$, the function $y \longrightarrow -f(x,y)$ is bounded below and quasi-convex, and satisfies (G.P.S.). Then the set-valued mapping

$$x \longmapsto M(x) = \{\bar{y} \in F \mid f(x,\bar{y}) = \sup_{y \in F} f(x,y)\}$$

is u.s.c. and nonempty compact set-valued.

Proof. That $x \longmapsto M(x)$ is a nonempty compact set valued mapping follows directly from lemma 2.2.

As was mentioned in the proof of theorem 2.1, $M(x)$ is a convex set, provided by the quasi-convexity of the function $y \longmapsto -f(x,y)$, $\forall\, x \in E$.

It remains to prove the u.s.c. of the mapping M. We prove it by contradiction. If the mapping M were not u.s.c. at some point $x_0 \in E$, then there would be an open (relative to F) neighborhood Ω of $M(x_0)$, a sequence $\{x_n\} \subset E$, and a sequence $\{y_n\}$ with

$y_n \in M(x_n)$ such that

$$\|x_n - x_0\| < \frac{1}{n}, \quad y_n \notin \Omega.$$

Since $M(x_0)$ is compact, there exist an open (rel. to E) neighborhood $U(x_0)$ of x_0 and an open (rel. to F) neighborhood Ω' of $M(x_0)$ such that f is Lipschitzian on $U(x_0) \times \Omega'$. No loss of generality, we may assume $\Omega \subset \Omega'$. In particular, there is a constant $c > 0$ such that

$$|f(x,y) - f(x_0,y)| \le c\|x - x_0\| \quad \forall\, x \in U(x_0), \quad \forall\, y \in \partial\Omega. \tag{2.9}$$

Choosing $y_0 \in M(x_0)$ arbitrarily, and connecting y_0 with y_n by a segment, the segment must intersect with the set $\partial\Omega$, i.e., $\exists\, \lambda_n \in (0,1)$ such that

$$z_n = (1-\lambda_n)y_0 + \lambda_n y_n \in \partial\Omega, \quad n = 1,2,\cdots.$$

On one hand, by lemma 2.2, we have

$$\begin{aligned}\sup_n f(x_0,z_n) &\le \sup_{\partial\Omega} f(x_0,y) \\ &< \sup_{y\in F} f(x_0,y) = f(x_0,y_0).\end{aligned} \tag{2.10}$$

On the other hand,

$$f(x_n,z_n) \ge \min\{f(x_n,y_n), f(x_n,y_0)\} = f(x_n,y_0),$$

provided by the quasi-convexity of the function $y \longmapsto -f(x,y)$.

For large n,

$$f(x_0,z_n) + c\|x_n - x_0\| \ge f(x_n,z_n) \ge f(x_n,y_0).$$

We obtain,

$$\varliminf_{n\to\infty} f(x_0,z_n) \ge \lim_{n\to\infty} f(x_n,y_0) = f(x_0,y_0)$$

which contradicts with (2.10).

<u>Proof of Theorem A</u>. We mentioned before that we only need to verify the assumptions (4) and (5) of theorem 2.1.

Actually, assumption (4) was verified in lemma 2.4.

According to lemma 2.4, we know that $x \longmapsto M(x)$ is u.s.c. Now, $\forall\, h \in E-x$, $\forall$ positive sequence $\{t_n\}$ with $t_n \downarrow 0$, $x_n = x + t_n h$, we choose arbitrarily $y_n \in M(x_n)$. Then $\{y_n\}$ possesses an accumulation point $\bar{y} \in M(x)$, provided by lemma 2.3. Assumption (5) is verified.

Theorem A follows directly from theorem 2.1.

3. Extensions and Remarks.

It is our purpose of this section to weaken the smoothness condition of the function f and the convexity condition of the set E of Theorem A. In addition, we shall discuss how important the convexity of the function $x \longmapsto f(x,y)$ plays a role in von Neumann-Sion-Ky Fan Theorem.

(I) Locally Lip. function.

Carefully analyzing the proof of Theorem A, the C^1-condition of the function f was used in the following points:

(1) A formulation of the P.S. condition.

(2) The inequality (2.3), where a mean value property and the C^1-continuity are used.

(3) A locally Lipschitzian property in (2.9).

These inspire us to generalize our results to loc. Lip. functions.

Let us recall the Clarke directional derivative for Loc. lip. functions [6].

For each $(x_0,y_0) \in E \times F$, $\forall\ (h,l) \in (E-x_0) \times (F-y_0)$, we define

$$f_x^0(x_0,y_0;h) = \overline{\lim_{\substack{t\downarrow 0\\(x',y')\to(x_0,y_0)}}} \frac{1}{t}[f(x'+th,y')-f(x',y')]$$

$$f_y^0(x_0,y_0;l) = \overline{\lim_{\substack{t\downarrow 0\\(x',y')\to(x_0,y_0)}}} \frac{1}{t}[f(x',y'+tl)-f(x',y')]$$

to be the Clarke partial directional derivatives.

f is said to be regular in x at x_0, if

$$f'_{+x}(x_0,y_0;h) = \lim_{t\downarrow 0} \frac{1}{t}[f(x_0+th,y_0)-f(x_0,y_0)]$$

exists for each $h \in E-x_0$, and

$$f'_{+x}(x_0,y_0;h) = f_x^0(x_0,y_0;h).$$

Employing these notions, we use $-f_y^0(x,y;l)$ (or $f_x^0(x,y;h)$) to replace $\langle -f'_y(x,y),l\rangle$ (or $\langle f'_x(x,y),h\rangle$ resp.) in the definition of (G.P.S.) ((T.P.S.) resp.) of the function $y \longmapsto -f(x,y)$ $(f(x,y))$.

In the deduction of (2.3), we carry out as follows

$$\underline{G}'_+(x_\varepsilon,h) \le \underline{\lim}_{n\to\infty} \frac{1}{t_n}[f(x_\varepsilon+t_nh,y_n)-f(x_\varepsilon,y_n)]$$
$$\le f_x^0(x_\varepsilon,\bar{y}_h;h).$$

In order to obtain the quasi-concavity of the function $y \longmapsto f_x^0(x_\varepsilon, y; h)$, we shall assume that the function $f(x,y)$ is regular in x.

In summary, Theorem A holds true for loc. Lip. functions f, which is regular in x.

(II) E without convexity.

The convexity of E was used only in defining the directional derivative of the function f. Let us recall the notions of the contingent cone and the Clarke tangential cone of a nonempty subset E of a Banach space X [2]. $\forall\, x_0 \in E$, let

$$T_E(x_0) = \{h \in X \mid \exists\, t_n \downarrow +0,\ \exists\, h_n \longrightarrow h \text{ such that } x_0 + t_n h_n \in E\}$$

and

$$C_E(x_0) = \{h \in X \mid \forall\, x_n \longrightarrow x_0, x_n \in E,\ \forall\, t_n \downarrow +0,\ \exists\, h_n \longrightarrow h, \text{ such that } x_n + t_n h_n \in E\}.$$

They are called the contingent cone and the Clarke tangential cone of E at x_0 respectively.

By definition, $C_E(x_0)$ is convex, and both $T_E(X_0)$ and $C_E(x_0)$ are closed cones with $C_E(x_0) \subset T_E(x_0)$. If $C_E(x_0) = T_E(x_0)$, then we say that E is regular at x_0 in the Clarke sense. It is easily seen that

$$C_E(x_0) = T_E(x_0) = \overline{\bigcup_{\lambda>0} \frac{1}{\lambda}(E - x_0)} \quad \forall\, x_0 \in E, \text{ if } E \text{ is convex,}$$

and

$C_E(x_0) = T_E(x_0) =$ the tangent space of E at x_0, if E is C^1 at x_0.

Let $g : E \longrightarrow \mathbb{R}^1$, and let $x_0 \in E$. The contingent derivative of g at x_0 along a direction $h \in T_E(x_0)$ is defined to be

$$\underline{g}'_{+E}(x_0; h) = \varliminf_{\substack{h' \to h \\ x_0 + th' \in E \\ t \downarrow 0}} \frac{1}{t}[g(x_0 + th') - g(x_0)].$$

Similarly, the Clarke directional derivative $g^0(x_0; h)$ at x_0 for a loc. Lip. function g is defined along directions $h \in C_E(x_0)$. And g is said to be regular at x_0, if E is regular at x_0, and

$$g'_{+E}(x_0; h) = \lim_{\substack{h' \to h \\ x_0 + th' \in E \\ t \downarrow 0}} \frac{1}{t}[g(x_0 + th') - g(x_0)]$$

exists with

$$g'_{+E}(x_0;h) = g^0(x_0;h).$$

Employing these notions, we use $C_E(x)$ to replace $E-x$ and $\inf_{h \in C_E(x_n)} f'_{+x}(x_n,y_n;\frac{h}{\|h\|}) \longrightarrow \alpha \geq 0$ to replace $\inf_{h \in E-x_n} \langle f'_{+x}(x_n,y_n), \frac{h}{\|h\|}\rangle \longrightarrow \alpha \geq 0$ in the definition of (T.P.S.), Theorem A holds true also.

In summary, Theorem A holds true for any closed set E and closed convex sets F, on which $f(x,y)$ is loc. Lip. and regular for each $x \in E$ in direction x.

<u>(III) Further discussion</u>.

We shall emphasize the main difficult point of Theorem A being the lack of convexity of the function $x \longmapsto f(x,y)$.

Provided by the Lop-Sided Minimax Theorem via weak topology we have the following

<u>Theorem 3.1</u>. Suppose that E and F are two closed convex sets of reflexive Banach spaces. We assume that

(1) $\forall\, y \in F$, $x \longmapsto f(x,y)$ is l.s.c. and quasi-convex,

(2) $\exists\, y_0 \in F$ such that the function $x \longmapsto f(x,y_0)$ is bounded below and coercive,

(3) $\forall\, x \in E$, $y \longmapsto -f(x,y)$ is l.s.c. and quasi-convex,

(4) $\exists\, x_0 \in E$ such that the function $y \longmapsto -f(x_0,y)$ is bounded below and coercive.

Then there exists a saddle point $(\bar{x},\bar{y}) \in E \times F$ such that

$$f(\bar{x},y) \leq f(\bar{x},\bar{y}) \leq f(x,\bar{y}) \quad \forall\, x \in E, \quad \forall\, y \in F.$$

Noticing a result due to Li [10], that a function satisfying P.S., and being bounded from below, must be coercive, Theorem 3.1 can be applied, if the coerciveness is replaced by the P.S. condition.

Now we would like to extend Li's result to nonsmooth functions.

<u>Theorem 3.2</u>. Suppose that E is a closed set of a Banach space. Assume that $g : E \longrightarrow \mathbb{R}^1$ is l.s.c. and bounded below, and satisfies the condition:

(G.P.S.)′ Any sequence $\{x_n\} \subset E$ along which $g(x_n) \longrightarrow \alpha \in \mathbb{R}^1$ and $\inf_{h \in T_E(x_n)} \underline{g}'_{+E}(x_n;\frac{h}{\|h\|}) \longrightarrow \beta \geq 0$, possesses a convergent subsequence. (3.1)

Then g is coercive, i.e.,

$$g(x) \longrightarrow +\infty, \quad \text{as} \quad \|x\| \longrightarrow \infty, \quad x \in E.$$

Proof. According to lemma 2.2, the set

$$M = \{\bar{x} \in E \mid g(\bar{x}) = \inf_{x \in E} g(x)\}$$

is nonempty and compact. For $b \in \mathbb{R}^1$, define the level set

$$g_b = \{x \in E \mid g(x) \le b\},$$

and

$$c = \sup\{b \in \mathbb{R}^1 \mid g_b \text{ is bounded}\}.$$

We shall prove that $c = +\infty$, so that g is coercive. In fact, if not, $c < +\infty$, then $g_{c-2^{-n}}$ is a bounded set, but $g_{c+2^{-n}}$ is not. This implies $g_{c-2^{-n}} \subset B_{R_n}$, an open ball with radius $R_n \ge n$, i.e.,

$$g(x) > c-2^{-n} \quad \forall\, x \in E\backslash B_{R_n}.$$

Since $g_{c+2^{-n}}$ is unbounded, there is $x_n \in g_{c+2^{-n}}$ with $\|x_n\| > R_n+2$. Applying Ekeland's Variational Principle to the function g on the metric space $E\backslash B_{R_n}$, we obtain $x'_n \in E\backslash B_{R_n}$, such that

$$c-2^{-n} < g(x'_n) \le g(x_n) \le c+2^{-n}, \tag{3.2}$$

$$\|x'_n - x_n\| \le 1, \tag{3.3}$$

and

$$g(x) > g(x'_n) - 2^{-n+1}\|x'_n - x\|, \quad \forall\, x \ne x'_n, \quad x \in E\backslash B_{R_n}. \tag{3.4}$$

The inequalities (3.2) and (3.4) imply

$$\underline{g}'_{+E}(x'_n, \frac{h}{\|h\|}) \ge 2^{1-n}, \quad \forall\, h \in T_E(x'_n).$$

Therefore there is a convergent subsequence of $\{x'_n\}$ provided by (G.P.S.)'. However, from (3.3),

$$\|x'_n\| \ge \|x_n\| - \|x_n - x'_n\| > R_n+1 \ge n+1 \longrightarrow +\infty.$$

It is impossible. The contradiction proves the theorem.

Corollary 3.1. Theorem 3.1 holds, if the coerciveness of the functions $x \longmapsto f(x,y_0)$ and $y \longmapsto -f(x_0,y)$ is replaced by (G.P.S.)'.

We compare Theorem A with Corollary 3.1. The conditions (1) and (2) of Corollary 3.1 are weaker than the conditions (1) and (2) of Theorem A. Since now the convexity of the function $y \longmapsto -f(x_0,y)$ is not assumed, we need a more complicated condition (T.P.S.).

(IV) An improvement of Theorem A.

Employing the contingent derivative of a function, we have generalized (G.P.S.) to (G.P.S.)'. Now, we mention that (T.P.S.) in Theorem A can be weakened by (G.P.S.)' of the function $G(x) = \sup_{y \in F} f(x,y)$. One can verify it directly, the proof is even shorter.

We point out here that (G.P.S.)' plus assumption (1) of Theorem A imply (T.P.S.). In fact, if $\{(x_n, y_n)\} \subset E \times F$, along which

$$f(x_n, y_n) = G(x_n)$$

is bounded and

$$\inf_{h \in E - x_n} G'_{+E}(x_n, h)/\|h\| \longrightarrow \alpha \geq 0,$$

then we conclude that there is a convergent subsequence $x_{nj} \longrightarrow x^*$ via (G.P.S.)'. According to lemma 2.4 and lemma 2.3, $\{y_{nj}\}$ has an accumulate point, provided by $y_{nj} \in M(x_{nj})$. We arrive at (T.P.S.).

4. Applications

In this section, we present three applications of Theorem A:

(1) A variational inequality,

(2) An elliptic system BVP,

(3) Infinite dimensional nonlinear programming.

We shall give some new existence theorems, in which the solutions are saddle points rather than loc. minima of functionals.

(I) A variational inequality.

Let M, N be two bounded closed convex sets containing the origin in $\mathbb{R}^m$ and $\mathbb{R}^n$ respectively. Let $\Omega \subset \mathbb{R}^p$ be a bounded open set. Assume that

$$u = (u^1(x), u^2(x), \ldots, u^m(x)) \in H_0^1(\Omega, \mathbb{R}^m),$$

$$v = (v^1(x), v^2(x), \ldots, v^m(x)) \in H_0^1(\Omega, \mathbb{R}^n);$$

and denote

$$E = \{u \in H_0^1(\Omega, \mathbb{R}^m) \mid u(x) \in M \quad \text{a.e.}\},$$

$$F = \{v \in H_0^1(\Omega, \mathbb{R}^n) \mid v(x) \in N \quad \text{a.e.}\}.$$

Suppose that $g \in C^1(M \times N, \mathbb{R}^1)$, and that

$$\forall\, u \in M, \quad v \longmapsto g(u,v) \quad \text{is concave.} \tag{4.1}$$

Theorem 4.1. There exists a pair of solutions (u_0, v_0) of the following variational inequalities:

$$\int_\Omega [\nabla u_0 \nabla \varphi + g'_u(u_0,v_0)\varphi]dx \ge 0, \tag{4.2}$$

$$\int_\Omega [\nabla v_0 \nabla \psi - g'_v(u_0,v_0)\psi]dx \ge 0, \tag{4.3}$$

$\forall\ \varphi \in H_0^1(\Omega,\mathbb{R}^m)$ with $\varphi \in E-u_0$, and $\forall\ \psi \in H_0^1(\Omega,\mathbb{R}^n)$ with $\psi \in F-v_0$.

Proof. Define a functional

$$I(u,v) = \int_\Omega [\tfrac{1}{2}|\nabla u|^2 - \tfrac{1}{2}|\nabla v|^2 + g(u,v)]dx$$

on the closed convex set $E \times F$. We shall verify that $I(u,v)$ satisfies all assumptions of Theorem A. Actually, only the (G.P.S.) of the function $v \longmapsto -I(u,v)$ and (T.P.S.) of I are needed to prove $\forall\ u_0 \in E$; we consider the function $v \longmapsto I(u_0,v)$.

Assume that $\{v_k\} \subset F$ is a sequence, along which

$$\int_\Omega [\tfrac{1}{2}|\nabla v_k|^2 - g(u_0,v_k)]dx \tag{4.4}$$

is bounded, and

$$\int_\Omega [\nabla v_k \nabla \psi - g'_v(u_0,v_k)\psi]dx \ge 0(\|\psi\|), \quad \forall\ \psi \in F-v_k. \tag{4.5}$$

From (4.4), $\{v_k\}$ is bounded in $H_0^1(\Omega,\mathbb{R}^n)$, so that

$$v_{k_j} \rightharpoonup v^* \quad (H_0^1(\Omega,\mathbb{R}^n)),$$

$$v_{k_j} \longrightarrow v^* \quad \text{a.e.},$$

for a subsequence. Thus $v^* \in F$. Now for each $\psi_0 \in F-v^*$, we shall prove

$$\int_\Omega [\nabla v^* \nabla \psi_0 - g'_v(u_0,v^*)\psi_0]dx \ge 0. \tag{4.6}$$

In fact, let $\psi_{k_j} = v^* + \psi_0 - v_{k_j}$, then

$$\begin{aligned}\int_\Omega \nabla v^* \nabla \psi_0 dx &= \int_\Omega \nabla v_{k_j} \nabla \psi_0 dx + 0(1)\\ &= \int_\Omega \nabla v_{k_j} \nabla \psi_{k_j} dx + \int_\Omega \nabla v_{k_j} \nabla (v_{k_j} - v^*)dx + 0(1)\\ &\ge 0(1) + \int_\Omega \nabla v_{k_j} \nabla \psi_{k_j} dx,\end{aligned} \tag{4.7}$$

and

$$\begin{aligned}\int_\Omega g'_v(u_0,v^*)\psi_0 dx &= \int_\Omega g'_v(u_0,v_{k_j})\psi_0 dx + 0(1)\\ &= \int_\Omega g'_v(u_0,v_{k_j})\psi_{k_j} dx + 0(1)\end{aligned} \tag{4.8}$$

provided by Lebesgue Theorem. Combining (4.5), (4.7) with (4.8), we obtain (4.6).

Putting $\psi = v^* - v_k$ in (4.5), and $\psi_0 = v_k - v^*$ in (4.6), and adding these two inequalities, we obtain

$$\int_\Omega |\nabla(v_{k_j} - v^*|^2 dx \le 0(\|v_{k_j} - v^*\|).$$

Thus

$$v_{k_j} \longrightarrow v^* \quad (H_0^1(\Omega, \mathbb{R}^n)).$$

This verifies (G.P.S.) of $-I(u_0, v)$.

Next, we assume $\{u_k\} \subset E$, $\{v_k\} \subset F$ such that

$$\int_\Omega [\tfrac{1}{2}|\nabla u_k|^2 - \tfrac{1}{2}|\nabla v_k|^2 + g(u_k, v_k)]dx \quad \text{is bounded,} \tag{4.9}$$

$$\int_\Omega [-\nabla v_k \nabla\psi + g'_v(u_k, v_k)\psi]dx \le 0 \quad \forall\ \psi \in F - v_k, \tag{4.10}$$

and

$$\int_\Omega [\nabla u_k \nabla\psi + g'_u(u_k, v_k)\varphi]dx \ge 0(\|\varphi\|) \quad \forall\ \varphi \in E - u_k. \tag{4.11}$$

From (4.10), take $\psi = v_0 - v_k$ for any fixed $v_0 \in F$, we obtain

$$\int_\Omega |\nabla v_k|^2 \le \|v_k\|\|v_0\| + M_1(\|v_k\| + \|v_0\|)$$

for some constant M_1, so that $\{v_k\}$ is bounded in $H_0^1(\Omega, \mathbb{R}^n)$. Substituting this into (4.9), we see that $\{u_k\}$ is bounded in $H_0^1(\Omega, \mathbb{R}^m)$. Then we have subsequences

$$u_{k_j} \longrightarrow u^* \quad H_0^1(\Omega, \mathbb{R}^m), \qquad v_{k_j} \longrightarrow v^* \quad H_0^1(\Omega, \mathbb{R}^n),$$

$$u_{k_j} \longrightarrow u^* \quad \text{a.e.} \qquad v_{k_j} \longrightarrow u^* \quad \text{a.e.}$$

Similarly, we prove

$$\int_\Omega [\nabla u^* \nabla\varphi + g'_u(u^*, v^*)\varphi]dx \ge 0 \quad \forall\ \varphi \in E - u^* \tag{4.12}$$

and $u^* \in E$.

Combining (4.10) with (4.12), we have

$$\|u_{k_j} - u^*\|^2 \le 0(\|u_{k_j} - u^*\|).$$

Similarly, we have (4.6) with $u_0 = u^*$, and then $\|v_{k_j} - v^*\|^2 \le 0(\|v_{k_j} - v^*\|)$. Thus $u_{k_j} \longrightarrow u^*$, $v_{k_j} \longrightarrow u^*$. The (T.P.S.) condition is verified.

Applying Theorem A, we obtain $(u_0, v_0) \in E \times F$ such that

$$I(u_0, v_0) = \sup_{v \in F} I(u_0, v) \tag{4.13}$$

and

$$(I'_u(u_0, v_0), \varphi) \ge 0 \quad \forall\ \varphi \in E - u_0. \tag{4.14}$$

Now (4.14) is just (4.2), and (4.13) implies (4.3).

Remark 4.1. The same conclusion holds true, if g depends on $x \in \Omega$ with some dominant conditions, say,

$$|g(x,u,v)| \le g_0(x) \quad \forall (u,v) \in M \times N$$

$$|g'_u(x,u,v)|, |g'_v(x,u,v)| \le g_1(x) \quad \forall (u,v) \in M \times N$$

where $g_0 \in L^1(\Omega)$, and $g_1 \in L^2(\Omega)$.

Remark 4.2. The same conclusion holds if M, N are closed convex sets and $|g'|$ is bounded on $M \times N$.

II. An elliptic system.

Employing the same notations as before we are looking for a weak solution of the following elliptic system:

$$\begin{cases} \Delta u = g'_u(u,v) \\ \Delta v = -g'_v(u,v), \end{cases} \tag{4.15}$$

i.e.,

$$\begin{aligned} &\int_\Omega [\nabla u \nabla \varphi + g'_u(u,v)\varphi]dx = 0 \quad \forall \varphi \in H^1_0(\Omega, \mathbb{R}^m) \\ &\int_\Omega [\nabla v \nabla \psi - g'_v(u,v)\psi]dx = 0 \quad \forall \psi \in H^1_0(\Omega, \mathbb{R}^n). \end{aligned} \tag{4.16}$$

Here we assume $g \in C^1(\mathbb{R}^m \times \mathbb{R}^n, \mathbb{R}^1)$ satisfying

(a) $\forall\, u \in \mathbb{R}^m$, $v \longmapsto g(u,v)$ is concave,

(b) $\exists$ constants $C_1 > 0$, and $\alpha \in (0, \frac{p+2}{p-2})$ such that

$$|g'_u(u,v)| \le C_1(1+|u|^\alpha + |v|^\alpha),$$

and

$$|g(u,0)| \le C_1(1+|u|^\beta) \qquad 0 \le \beta < 1,$$

(c) $\exists$ constant $\gamma \in [0,1)$ such that

$$|g'_u(u,v)| \le C_1(1+|u|^\alpha + |v|^\gamma),$$

(d) $\exists$ constants $C_2 > 0$ and $\theta \in (0, \frac{1}{2})$ such that

$$\theta g'_u(u,v) \cdot u + \frac{1}{2} g'_v(u,v)v \le C_2 + g(u,v).$$

By Young's inequality, (c) implies a constant C_3 such that

$$|g(u,v)| \le C_3(1+|v|^{1+\gamma} + |u|^{\frac{\alpha}{\gamma}(1+\gamma)}) \tag{4.17}$$

Hereafter, we denote various constants by C.

Theorem 4.2. The system (4.16) has a weak solution $(u_0, v_0) \in H^1_0(\Omega, \mathbb{R}^m) \times H^1_0(\Omega, \mathbb{R}^n)$.

Proof. We introduce the functional

$$I(u,v) = \int_\Omega [\tfrac{1}{2}|\nabla u|^2 - \tfrac{1}{2}|\nabla v|^2 + g(u,v)]dx.$$

Obviously, $v \longmapsto I(u,v)$ is quasi-concave, via (a), and is bounded above via (4.17).

It remains to verify the P.S. condition and (T.P.S.). Firstly, assume that $\{v_k\} \subset H_0^1(\Omega,\mathbb{R}^n)$ along which

$$\int_\Omega [-\nabla v_k \nabla\psi + g'_v(u_0,v_k)\psi]dx = o(\|\psi\|), \quad \forall\ \psi \in H_0^1(\Omega,\mathbb{R}^n) \tag{4.18}$$

and

$$\int_\Omega [-|\nabla v_k|^2 + g(u_0,v_k)]dx \tag{4.19}$$

is bounded, for a fixed $u_0 \in H_0^1(\Omega,\mathbb{R}^m)$.

According to (4.17) and (4.19), $\{v_k\}$ is bounded in $H_0^1(\Omega,\mathbb{R}^n)$, and then by (4.18) and (c), there is a subsequence $v_{k_j} \longrightarrow v^*$ in $H_0^1(\Omega,\mathbb{R}^n)$. Therefore $v \longmapsto I(u_0,v)$ satisfies P.S. condition $\forall\ u_0 \in H_0^1(\Omega,\mathbb{R}^m)$.

Next, we turn to verify (T.P.S.). Assume that $\{u_k\} \subset H_0^1(\Omega,\mathbb{R}^m)$, $\{v_k\} \subset H_0^1(\Omega,\mathbb{R}^n)$, along which

$$\int_\Omega [\tfrac{1}{2}|\nabla u_k|^2 - \tfrac{1}{2}|\nabla v_k|^2 + g(u_k,v_k)]dx \quad \text{is bounded,} \tag{4.20}$$

$$\int_\Omega [-\nabla v_k\nabla\psi + g'_v(u_k,v_k)\psi]dx = 0 \quad \forall\ \psi \in H_0^1(\Omega,\mathbb{R}^n), \tag{4.21}$$

and

$$\int_\Omega [\nabla u_k\nabla\varphi + g'_u(u_k,v_k)\varphi]dx = o(\|\varphi\|) \quad \forall\ \varphi \in H_0^1(\Omega,\mathbb{R}^m). \tag{4.22}$$

Provided by (d) and (4.20), we have

$$(\tfrac{1}{2}-\theta)\|u_k\|^2 = \int_\Omega [\tfrac{1}{2}|\nabla u_k|^2 - \tfrac{1}{2}|\nabla v_k|^2 + \tfrac{1}{2} g'_0(u_k,v_k)v_k + \theta g'_u(u_k,v_k)u_k]dx$$
$$\le C_4.$$

$\{u_k\}$ is bounded in $H_0^1(\Omega,\mathbb{R}^m)$. Let $\psi = v_k$ in (4.21), we have

$$\|v_k\|^2 \le C_1\int_\Omega (1 + |u_k|^\alpha + |v_k|^\gamma)|v_k|$$
$$\le C_5(\|v_k\| + \|u_k\|^\alpha\|v_k\| + \|v_k\|^{1+\gamma}),$$

provided by the Hölder inequality and the Sobolev embedding theorem. This implies that $\|v_k\|$ is bounded, provided by $0 \le \gamma < 1$. Thus, we have subsequences $u_{k_j} \rightharpoonup u^*$, $v_{k_j} \rightharpoonup v^*$.

Applying the assumptions (b) and (c), (4.21) and (4.22) imply the strong convergence of u_{k_j} and v_{k_j}.

Applying Theorem A directly, we obtain a solution (u_0, v_0).

(III) An existence theorem of the nonlinear programming.

The basic problem in the nonlinear programming is to find a minimum of a function under certain equality and inequality constraints.

Let X be a Hilbert space, and let $f, g_1, \ldots, g_p, h_1, \ldots, h_q$ be C^1 functions defined on X. The problem is to find

$$\min f(x)$$

under the constraints $g_i(x) \le 0$, $i = 1, \ldots, p$; and $h_j(x) = 0$, $j = 1, \ldots, q$; where we assume that

$$K = \{x \in X \mid g_i(x) \le 0, \ i = 1, \ldots, p; \ h_j(x) = 0, \ j = 1, \ldots, q\} \ne \emptyset.$$

For each $x_0 \in K$, denote

$$I(x_0) = \{i \in [1,p] \ g_i(x_0) = 0\}.$$

We say that $g_1, \ldots, g_p; h_1, \ldots, h_q$ satisfies the Kuhn Tucker Condition (K.T. in short), if $\{g_i'(x_0); h_j'(x_0)\}$ $i \in I(x_0)$, $j = 1, \ldots, q$, is linear independent, $\forall\, x_0 \in K$.

Theorem 4.3. Assume that

(1) $\{g_1, \ldots, g_p; h_1, \ldots, h_q\}$ satisfies (K.T.),

(2) g_i', $i = 1, \ldots, p$, h_j', $j = 1, \ldots, q$, and $f' - \mathrm{id}$ are compact mappings, i.e., they map weakly convergent sequences into strongly convergent sequences,

(3) f is coercive.

Then there exist $\bar{x} \in K$, $\bar{\lambda} = (\bar{\lambda}_1, \ldots, \bar{\lambda}_p) \in \mathbb{R}_+^p$, and $\bar{\mu} = (\bar{\mu}_1, \ldots, \bar{\mu}_q) \in \mathbb{R}^q$ such that

(1) $f(\bar{x}) = \min\limits_{x \in K} f(x)$,

(2) $f'(x) + \sum\limits_{i=1}^{p} \bar{\lambda}_i g_i'(x) + \sum\limits_{j=1}^{q} \bar{\mu}_j h_j'(x) = \theta$,

(3) $\bar{\lambda}_i g_i(\bar{x}) = 0$, $i = 1, 2, \ldots, p$.

Proof. We introduce the following Lagrangian

$$L(x; \lambda, \mu) = f(x) + \lambda \cdot g(x) + \mu \cdot h(x),$$

where $\lambda \in \mathbb{R}_+^p$, $\mu \in \mathbb{R}^q$, $g(x) = (g_1(x), \ldots, g_p(x))$ and $h(x) = (h_1(x), \ldots, h_q(x))$.

Firstly, let $E = X$, $F = F_N = \{(\lambda, \mu) \in \mathbb{R}_+^p \times \mathbb{R}^q \mid 0 \le \lambda_i \le N, |\mu_j| \le N, \ i = 1, \ldots, p; \ j = 1, \ldots, q\}$ for $N \in \mathbb{Z}_+$.

Since L is linear in (λ, μ), and F_N is compact, assumption (1) of Theorem A is satisfied.

The compactness of $f'-id$ implies that f maps bounded sets into bounded sets. In addition to the coerciveness of f, we show that $L(x;\ 0,0) = f(x)$ is bounded below.

In order to apply Theorem A on the set $E \times F_N$, it remains to verify (T.P.S.). Assume that $(x^n;\ \lambda^n,\mu^n)$ is a sequence along which

$$L(x^n;\ \lambda^n,\mu^n) = f(x^n) + \sup_{(\lambda,\mu)\in F_N} [\lambda^n\cdot g(x^n) + \mu^n\cdot h(x^n)] \quad \text{is bounded,}$$

and

$$L'_x(x^n;\ \lambda^n,\mu^n) = f'(x^n) + \lambda^n\cdot g'(x^n) + \mu^n\cdot h'(x^n) \longrightarrow \theta. \tag{4.23}$$

We want to prove that $\{(x^n,\lambda^n,\mu^n)\}$ possesses convergent subsequence. Because F_N is compact, we have $(\lambda^{n_j},\mu^{n_j}) \longrightarrow (\lambda^N,\mu^N)$. In order to get a convergent subsequence x^n, we observe that $f(x^n) = L(x^n,0,0) \le L(x^n;\ \lambda^n,\mu^n)$, which is bounded, so that $\{f(x^n)\}$ is bounded above, and then $\{x^n\}$ is bounded via assumption (3). We obtain $x^{n_j} \rightharpoonup x^N$. From (4.23) and the assumption (2), it follows that $x^{n_j} \longrightarrow x^N$. (T.P.S.) is verified.

Applying Theorem A, we have $(x^N;\ \lambda^N,\mu^N)$ satisfying

$$\begin{aligned} f(x^N) + \lambda^N\cdot g(x^N) + \mu^N\cdot h(x^N) &= f(x^N) + \sup_{(\lambda,\mu)\in F_N} [\lambda\cdot g(x^N) + \mu\cdot h(x^N)] \\ &\le \inf_{x\in X}\{f(x) + \sup_{(\lambda,\mu)\in F_N} [\lambda\cdot g(x)+\mu\cdot h(x)]\}, \end{aligned} \tag{4.24}$$

and

$$f'(x^N) + \lambda^N\cdot g'(x^N) + \mu^N\cdot h'(x^N) = \theta. \tag{4.25}$$

From (4.24), it follows

$$\lambda_i^N\cdot g_i(x^N) = \sup_{\lambda_i\in 0,N} \lambda_i g_i(x^N), \quad i = 1,\dots,p,$$

therefore,

$$\lambda_i^N g_i(x^N) = \begin{cases} N\ g_i(x^N) & \text{if } g_i(x^N) > 0 \\ 0 & \text{if } g_i(x^N) \le 0, \end{cases} \tag{4.26}$$

$i = 1,2,\dots,p$. Similarly,

$$\mu_j^N h_j(x^N) = N|h_j(x^N)|, \qquad j = 1,\dots,q. \tag{4.27}$$

Next, let $N \longrightarrow +\infty$. We shall prove that $\{(x^N;\ \lambda^N,\mu^N)\}$ possesses a convergent subsequence, which converges to the solution $(\bar{x},\bar{\lambda},\bar{\mu})$.

From (4.24), (4.26) and (4.27), we obtain

$$f(x^N) \le f(x_0) + \sup_{(\lambda,\mu)\in F_N} [\lambda\cdot g(x_0) + \mu\cdot h(x_0)] = f(x_0) \tag{4.28}$$

for any $x_0 \in K$.

Thus $f(x^N)$ is bounded above, again via the coerciveness, $\{x^N\}$ is bounded. Assume $x^{N_k} \rightharpoonup x$, then we have

$$f'(x^{N_k}) - x^{N_k} \longrightarrow f'(\bar{x}) - \bar{x},$$

$$g_i'(x^{N_k}) \longrightarrow g_i'(\bar{x}), \quad i = 1,\dots,p;$$

$$h_j'(x^{N_k}) \longrightarrow h_j'(\bar{x}), \quad j = 1,\dots,q.$$

We claim $\bar{x} \in K$. In fact, $\forall$ i, if $g_i(x^{N_k}) \le 0$ for infinitely many k, then $g_i(\bar{x}) \le 0$, provided by the compactness of g_i' and then the compactness of g_i. Similarly, $\forall$ j, if $h_j(x^{N_k}) = 0$ for infinitely many k, then $h_j(\bar{x}) = 0$.

So we shall restrict ourselves to those i and j to which

$$g_i(x^{N_k}) > 0 \quad \text{and} \quad h_j(x^{N_k}) \ne 0 \quad \text{for all } k,$$

and we shall prove that $g_i(\bar{x}) = h_j(\bar{x}) = 0$ for these i and j. Similar to (4.8), we now obtain

$$f(x^N) + \sum_{g_i(x^N)>0} Ng_i(x^N) + N \sum_{h_j(x^N)=0} h_j(x^N) \le f(x_0).$$

This implies

$$N_k\Big(\sum_{g_i(x^{N_k})>0} g_i(x^{N_k}) + \sum_{h_j(x^{N_k})=0} h_j(x^{N_k})\Big) \quad \text{is bounded.}$$

Therefore

$$g_i(\bar{x}) = \lim_{k\to\infty} g_i(x^{N_k}) = 0$$

and

$$h_j(\bar{x}) = \lim_{k\to\infty} h_j(x^{N_k}) = 0.$$

Thus we proved $\bar{x} \in K$.

Now we turn to prove that (λ^N,μ^N) possesses convergent subsequence. Combining (4.25) with (4.26), we have

$$f'(x^{N_k}) + \sum_{i\in I(\bar{x})} \lambda_i^{N_k}\cdot g_i'(x^{N_k}) + \mu^{N_k}\cdot h'(x^{N_k}) = \theta. \tag{4.29}$$

According to the K.T. condition (assumption (1)), $[g'(x^{N_k}), h_j(x^{N_k})]$

$i \in I(x)$, $j = 1,\ldots,q$ are linearly independent for k large; their Gram determinants do not vanish. Therefore we can solve $\lambda_i^{N_k}$ and μ^{N_k} by (4.28). Since $\{\|f'(x^{N_k})\|\}$ is bounded, all $\lambda_i^{N_k}$ and μ^{N_k} are bounded for $i \in I(\bar{x})$. We obtain a convergent subsequence, no loss of generality, denote by $(\lambda^{N_k}, \mu^{N_k}) \longrightarrow (\bar{\lambda}, \bar{\mu})$. By the way, again via (4.28), we have $x^{N_k} \longrightarrow \bar{x}$.

Finally, we shall verify that $(\bar{x}, \bar{\lambda}, \bar{\mu})$ solves our problem. In fact, from (4.24), we have

$$\begin{aligned} f(\bar{x}) + \bar{\lambda}\cdot g(\bar{x}) + \bar{\mu}\cdot h(\bar{x}) &= f(\bar{x}) + \sup_{(\lambda,\mu)\in \mathbb{R}_+^p \times \mathbb{R}^q} [\lambda\cdot g(\bar{x}) + \mu\cdot h(\bar{x})] \\ &= f(\bar{x}) \\ &\le f(x) + \sup_{(\lambda,\mu)\in \mathbb{R}_+^p \times \mathbb{R}^q} [\lambda\cdot g(x) + \mu\cdot h(x)] \end{aligned}$$

$\forall\, x \in X$. This implies

$$f(\bar{x}) \le f(x), \quad \forall\, x \in K,$$

and

$$\bar{\lambda}\cdot g(\bar{x}) = 0, \quad \text{and then} \quad \bar{\lambda}_i g_i(\bar{x}) = 0, \quad i = 1,2,\ldots,p.$$

The conclusion (2) follows from a limiting process of (4.25). The proof is complete.

As an example, we consider the following problem.

Suppose that $\Omega \in \mathbb{R}^n$ is a bounded open domain, and that $\varphi, \psi_1, \ldots, \psi_p, \eta_1, \ldots, \eta_q$ are Caratheodory functions defined on $\Omega \times \mathbb{R}^1$, satisfying the following growth condition:

$$|g(x,t)| \le C(1+|t|^\alpha)$$

where C is a constant, and $\alpha \in (1, \frac{n+2}{n-2})$, if $n \ge 3$, and $\alpha \ge 1$ if $n = 1,2$.

Assume that $\{\psi_i(x,t), \eta_j(x,t)\}$, $i = 1,\ldots,p$; $j = 1,\ldots,q$, is linear independent for each fixed $t \in \mathbb{R}^1$. Let

$$\Phi(x,\xi) = \int_0^\xi \varphi(x,t)\,dt, \quad f(u) = \int_\Omega \tfrac{1}{2}|\nabla u|^2 + \Phi(x,u)\,dx,$$

$$\Psi_i(x,\xi) = \int_0^\xi \psi_i(x,t)\,dt, \quad g_i(u) = \int_\Omega \Psi_i(x,u)\,dx,$$

$$\Lambda_j(x,\xi) = \int_0^\xi \eta_j(x,t)\,dt, \quad h_j(u) = \int_\Omega \Lambda_j(x,u)\,dx,$$

$i = 1,\ldots,p$; $j = 1,\ldots,q$; and $u \in H_0^1(\Omega)$.

If

$$\Phi(x,\xi) \geq \alpha|\xi|^2 - C_1$$

where C_1 is a constant, and $-\alpha < \lambda_1$, the first eigenvalue of $-\Delta$ with 0-Dirichlet data in Ω, then there exists $(\bar{u},\bar{\lambda},\bar{\mu}) \in H_0^1(\Omega) \times \mathbb{R}_+^p \times \mathbb{R}^q$ such that

(1) $f(\bar{u}) = \min\{f(u) \mid u \in H_0^1(\Omega),\ g_i(u) \leq 0,\ h_j(u) = 0,\ i = 1,2,\ldots,p;\ j = 1,2,\ldots,q\}$,

(2) $-\Delta\bar{u}(x) + \varphi(x,\bar{u}(x)) + \sum_{i=1}^{p} \bar{\lambda}_i\psi_i(x,\bar{u}(x)) + \sum_{i=1}^{q} \mu_j\eta_j(x,\bar{u}(x)) = 0$,

(3) $\bar{\lambda}_i\int_\Omega \Psi_i(x,u(x))dx = 0$,

(4) $\int_\Omega \Lambda_j(x,u(x))dx = 0$,

(5) $\int_\Omega \Psi_i(x,u(x))dx \leq 0$.

Remark 4.3. If $\dim X < +\infty$, then Theorem 4.3 is obviously true. However, in case $\dim X = +\infty$, it seems worth proving.

REFERENCES

1. Aubin, J.P., Mathematical Methods of Game and Economic Theory, North-Holland, Amsterdam-New York-Oxford, Rev. ed. 1982.

2. Aubin, J.P. and Ekeland, I., Applied Nonlinear Analysis, Wiley-Interscience, New York, 1984.

3. Barbu, V. and Precupanu, Th., Convexity and Optimization in Banach Spaces, Sijthoff & Noordhoff, Bucharest, 1978.

4. Brezis, H., Nirenberg, L. and Stampacchia, G., A remark on Ky Fan's minimax principle, Boll. Un. Math. Ital. 5(1973), 293-300.

5. Chang, Kung-ching and Eells, J., Unstable minimal surface coboundaries, preprint, Univ. of Warwick, 1985. (To appear in Acta Math. Sinica.)

6. Clarke, F.H., Optimization and Nonsmooth Analysis, Wiley-Interscience, New York, 1983.

7. Ekeland, I., Nonconvex minimization problems, Bull. Amer. Math. Soc. (n.s.) 1(1979), 443-474.

8. Fan, Ky, Minimax theorems, Proc. Nat. Acad. Sci. U.S.A. 39(1953), 42-47.

9. Fan, Ky, A minimax inequality and applications, in "Inequalities III" (Shisha, O., ed.), Academic Press, New York, 1972, 151-156.

10. Li, Shujie, A multiple critical point theorem and its application in nonlinear partial differential equations (in Chinese), Acta Mathematica Scientia, 4(1984), 135-340.

11. Nirenberg, L., Variational and topological methods in nonlinear problems, Bull. Amer. Math. Soc. (n.s.), 4(1981), 267-302.

12. Shi, Shuzhong, Ekeland's variational principle and the mountain pass lemma, Cahiers de CEREMADE, Univ. de Paris-Dauphine, n°8425. (To appear in Acta Math. Sinica.)

13. Sion, M., On general minimax theorems, Pacific J. Math. 8(1958), 171-176.

Covering Theorems of Convex Sets Related to Fixed-Point Theorems

MAU-HSIANG SHIH Department of Mathematics, Chung Yuan University, Chung-Li, Taiwan, Republic of China

KOK-KEONG TAN Department of Mathematics, Statistics and Computing Science, Dalhousie University, Halifax, Nova Scotia, Canada

1. A Common Generalization of Theorems of Knaster-Kuratowski-Mazurkiewicz, Shapley and Ky Fan.

The aim of this paper is to prove the following general covering theorem of convex sets in locally convex topological vector spaces.

THEOREM 1. *Let X be a paracompact convex subset of a real locally convex Hausdorff topological vector space E, X_0 a non-empty compact convex convex subset of X and K a non-empty compact subset of X. Let $\{A_i : i \in I\}$ be a locally finite family of closed subsets of X such that $X = \bigcup_{i \in I} A_i$, and let $\{C_i : i \in I\}$ be a family (indexed also by I) of non-empty subsets of E. Let $s: X \to P(E)$ be an upper hemi-continuous set-valued map with each $s(x)$ a weakly compact convex set. If for each $x \in (K \cap \partial X) \cup (X \setminus K)$, the closed convex hull of $\cup\{C_i + s(x): x \in A_i\}$ meets the tangent cones*

$$\begin{cases} x + \overline{\bigcup_{\lambda > 0} \lambda(X - x)} & \text{if } x \in K \cap \partial X, \\ x + \overline{\bigcup_{\lambda > 0} \lambda(X_0 - x)} & \text{if } x \in X \setminus K; \end{cases}$$

then there exist a non-empty finite subset I_0 of I and a point $\hat{x} \in X$ such that $\hat{x}$ is in the closed convex hull of $\cup\{C_i + s(\hat{x}): i \in I_0\}$ as well as in $\cap\{A_i : i \in I_0\}$.

In Theorem 1, $P(E)$ stands for the collection of all non-empty subsets of E; the symbols "∂" and "$^-$" denote the boundary operation and the closure operation, respectively. Theorem 1 remains valid if the tangent cones are replaced by $x + \overline{\bigcup_{\lambda < 0} \lambda(X - x)}$ if $x \in K \cap \partial X$, and $x + \overline{\bigcup_{\lambda < 0} \lambda(X_0 - x)}$ if $x \in X \setminus K$, respectively.

The statement of Theorem 1 becomes simpler when X is compact and $X = X_0 = K$. It may be noted that each locally finite family $\{A_i : i \in I\}$ of sets in a compact space must be finite. We state this special case of Theorem 1 explicitly as:

THEOREM 2. *Let* X *be a non-empty compact convex set in a real locally convex Hausdorff topological vector space,* $\{A_1, A_2, \ldots, A_n\}$ *be a finite system of* n *closed subsets of* X *such that* $X = \bigcup_{i=1}^{n} A_i$ *, and* $\{C_1, C_2, \ldots, C_n\}$ *be a system of* n *non-empty sets in* E *. Let* $s: X \to P(E)$ *be an upper hemi-continuous set-valued map with each* $s(x)$ *a weakly compact convex set. If for each* $x \in \partial X$ *, the closed convex hull of* $\cup\{C_i + s(x): i \in \{1,2,\ldots,n\}$ *and* $x \in A_i\}$ *meets the tangent cone* $x + \overline{\bigcup_{\lambda>0} \lambda(X-x)}$ *, then there exists a nonempty subset* I_0 *of* $\{1,2,\ldots,n\}$ *and a point* $\hat{x} \in X$ *such that* $\hat{x}$ *is in the closed convex hull of* $\cup\{C_i + s(\hat{x}): i \in I_0\}$ *as well as in* $\cap\{A_i : i \in I_0\}$.

In the case when $s(x) := \{x\}$ for each $x \in X$ and $X_0 = K$, Theorem 1 reduces to a somewhat genreal form of a covering theorem of Ky Fan [8, Theorem 12]. Also Theorem 2 extends an earlier theorem of Ky Fan [7, Theorem 2], which is seen in [7] to imply the classical Knaster-Kuratowski-Mazurkiewicz theorem [10] and its generalization by Shapley [11]. Thus Ky Fan's theorem [7, Theorem 2] is a far-reaching generalization of Shapley's theorem, and is a generalization free from the facial structure of a simplex. Another Ky Fan's covering theorem [8, Theorem 2] can be derived from a special case when $s(x) := \{0\}$ for each $x \in X$, and each C_i is a one-point subset of X in Theorem 2.

The geometric meaning of Theorem 1 or Theorem 2 is clear. To illustrate Theorem 2, let X be a compact convex set in $\mathbb{R}^2$ covered by four closed sets A_1 , A_2 , A_3 and A_4 , and let C_1 , C_2 , C_3 and C_4 be four non-empty subsets of $\mathbb{R}^2$ as indicated in the Figure. Let $s: X \to X$ be the identity map. Then the conclusion of Theorem 2 holds; in fact: Each $\hat{x}$ in the boldface shown in the Figure, is in the closed convex hull of $\cup\{C_i + \hat{x}: i = 3,4\}$ as well as in the intersection $A_3 \cap A_4$.

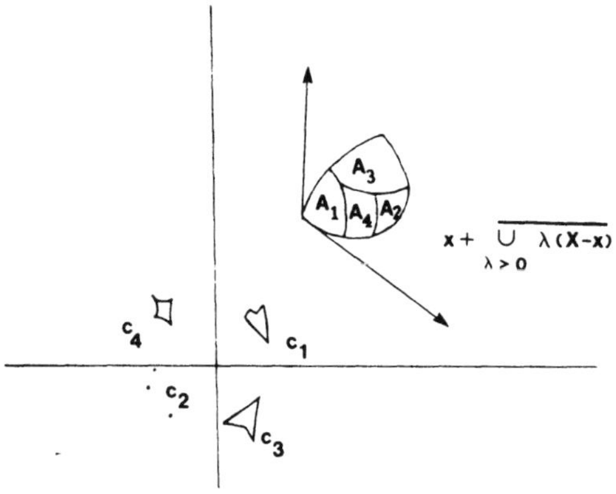

Figure

2. Upper Semi-continuity, Upper Demi-continuity and Upper Hemi-continuity.

To prove the generalized covering result Theorem 1, we shall need a generalized form of the classical Fan-Glicksberg fixed-point theorem [5], [9] involving the upper-hemi-continuity introduced by Aubin.

Let M, Y be topological spaces, and let $f: M \to P(Y)$, where $P(Y)$ is the family of all non-empty subsets of Y. As usual, we say that f is upper semi-continuous on M if for each $x \in M$ and for each open set G in Y with $f(x) \subset G$, there exists an open neighborhood N_x of x in M such that $f(x) \subset G$ for all $z \in N_x$. Let E be a real Hausdorff topological vector space and let E' be the dual space of E (i.e., the vector space of all continuous linear functionals on E). According to Ky Fan [6], we say that $f: M \to P(E)$ is upper demi-continuous on M if for every $x \in M$ and any open half-space H in E containing $f(x)$, there exists an open neighborhood N_x of x in M such that $f(u) \subset H$ for all $u \in N_x$. Recall that an open half-space H in E is a set of the form $H := \{v \in E: \phi(v) < t\}$ for some non-zero $\phi \in E'$ and some real number t. As in Aubin [2, p. 68] and Aubin-Ekeland [3, p. 122], $f: M \to P(E)$ is upper hemi-continuous on M if for each $p \in E'$ and for each $\lambda \in \mathbb{R}$, the set $\{x \in M: \sup_{u \in f(x)} \langle p,u\rangle < \lambda\}$ is open in M.

It is obvious that every upper semi-continuous map is upper demi-continuous. The following result shows that every upper demi-continuous map is upper hemi-continuous.

PROPOSITION 1. *Let M be a topological space, E a real Hausdorff topological vector space, and let $f: M \to P(E)$. If f is upper demi-continuous, then f is upper hemi-continuous.*

Proof. Let $p \in E'$ and $\lambda \in \mathbb{R}$ be given. Let $x_0 \in M$ be such that $\sup_{u \in f(x_0)} \langle p,u\rangle < \lambda$. Clearly we may assume without loss of generality that $p \neq 0$. Choose $\beta > 0$ so that

$$\sup_{u \in f(x_0)} \langle p,u\rangle + \beta < \lambda .$$

Define

$$H := \{y \in E : \langle p,y\rangle < \sup_{u \in f(x_0)} \langle p,u\rangle + \beta\};$$

then H is an open half-space in E which contains $f(x_0)$. By upper demi-continuity of f, there exists an open neighborhood N of x_0 in M such that $f(x) \subset H$ for all $x \in N$. But then for each $x \in N$,

$$\sup_{u \in f(x)} \langle p,u\rangle \leqq \sup_{u \in f(x_0)} \langle p,u\rangle + \beta < \lambda .$$

Therefore the set $\{x \in M: \sup_{u \in f(x)} \langle p,u\rangle < \lambda\}$ is open in M. Hence f is upper hemi-continuous on M. □

We observe that if f is a compact-valued map, then the concepts of upper hemi-continuity and upper demi-continuity coincide. In what follows, we shall show that under certain conditions, the concepts of upper semi-continuity, upper demi-continuity and upper hemi-continuity are the same.

PROPOSITION 2. *Let* M *be a topological space, let* Z *be a non-empty compact subset of a real locally convex Hausdorff topological vector space* E *, and let* $f: M \to P(Z)$ *be such that each* $f(x)$ *is convex. Then the following statements are equivalent:*

(1) f *is upper semi-continuous.*

(2) f *is upper demi-continuous.*

(3) f *is upper hemi-continuous.*

Proof. In view of Proposition 1 and the remark preceding it, we need only show that (3) => (1). We argue by contradiction, so assume that f is not upper semi-continuous. Then there exist a point $x_0 \in M$ and an *open absolutely convex neighborhood* U of 0 in E such that for each neighborhood N of x_0, there is a point $x_N \in N$ for which $f(x_N)$ is not contained in $f(x_0) + U$, so that we can choose an $y_N \in f(x_N)$ with $y_N \notin f(x_0) + U$. We now observe that $\{x_N : N$ is a neighborhood of $x_0\}$ is a net in M which converges to x_0 and $\{y_N : N$ is a neighborhood of $x_0\}$ is a net in Z. By compactness, we may (by taking a subnet, if necessary) assume that $y_N \to y$ for some $y \in Z$. It follows that $y \notin \overline{f(x_0) + \frac{1}{2} U}$, and hence $y \notin \overline{f(x_0)}$. As $\overline{f(x_0)}$ is closed and convex, the Hahn-Banach theorem ensures that there exist $p \in E'$ and $t \in \mathbb{R}$ satisfying

$$\sup_{u \in f(x_0)} \langle p,u \rangle = \sup_{u \in \overline{f(x_0)}} \langle p,u \rangle < t < \langle p,y \rangle .$$

By continuity of p at y, there exists an open neighborhood W of y such that $\langle p,w \rangle > t$ for all $w \in W$. As $y_N \to y$, there exists a neighborhood N_1 of x_0 such that $y_N \in W$ for all neighborhood N of x_0 with $N \subset N_1$. Hence for all neighborhood N of x_0 with $N \subset N_1$, we have

$$\sup_{u \in f(x_0)} \langle p,u \rangle < t < \langle p,y_N \rangle . \tag{1}$$

On the other hand, since f is upper hemi-continuous and $\sup_{u \in f(x_0)} \langle p,u \rangle < t$, there exists a neighborhood N_2 of x_0 for which $\sup_{u \in f(x)} \langle p,u \rangle < t$ for all $x \in N_2$. Take $N_0 := N_1 \cap N_2$; then

$$\langle p,y_{N_0} \rangle \leqq \sup_{u \in f(x_{N_0})} \langle p,u \rangle < t ,$$

which contradicts (1). This completes the proof. □

It is of interest to observe a result of Castaing (see [3, Theorem 10, p. 128]) which asserts that if M is a topological space, E is a real locally convex

Hausdorff topological vector space equipped with the weak topology and $f: M \to P(E)$ is upper hemi-continuous such that each $f(x)$ is a weakly compact convex set in E, then f is upper semi-continuous. We remark that Castaing's result and our Proposition 2 are not comparable.

Let $f,g: M \to P(E)$ be upper hemi-continuous. As

$$\sup_{y\in[f(x)+g(x)]} \langle p,y\rangle = \sup_{y\in f(x)} \langle p,y\rangle + \sup_{y\in g(x)} \langle p,y\rangle ,$$

$f+g$ is also upper hemi-continuous. It may be emphasized here that the sum of two upper semi-continuous maps, however, need not even be upper demi-continuous as the following example shows.

EXAMPLE 1. Let $E := \mathbb{R}^2$, $X := \{t \in \mathbb{R}: 0 \le t \le 1\}$. Define $f,g: X \to P(\mathbb{R}^2)$ by

$$f(t) := \{(u,v) \in \mathbb{R}^2: (u-1)(v-1) \geqq 1 \text{ and } u > 1\} ,$$

$$g(t) := \{(-z,z) \in \mathbb{R}^2: 0 \le z \le t/\sqrt{2}\}$$

for all $t \in X$. (Note that f is a constant set-valued map). Then it is not hard to verify that f and g are both upper semi-continuous but $f+g$ is not upper demi-continuous.

Incidently, the above example also shows that an upper hemi-continuous map need not be upper demi-continuous, since $f+g$ in Example 1 is necessarily upper hemi-continuous. We also observe that an upper demi-continuous map need not be upper semi-continuous as the following example shows.

EXAMPLE 2. Define $f: \mathbb{R}^2 \to \mathbb{R}^2$ by

$$f(x,y) := \{(u,v) \in \mathbb{R}^2: u \geqq x \text{ and } v \geqq y\}$$

for all $(x,y) \in \mathbb{R}^2$. Then f is upper demi-continuous, but not upper semi-continuous.

3. Fixed-point Theorems

To prove the fixed-point theorems, we shall need the following minimax inequality [8, Theorem 6] which is an improvement of a result in [1], generalizing the well-known Ky Fan minimax principle [6].

THEOREM 3. *Let X be a non-empty convex set in a Hausdorff topological vector space. Let ψ be a real-valued function defined on $X \times X$ such that*

(a) *For each fixed $x \in X$, $\psi(x,y)$ is a lower semi-continuous function of y on X.*

(b) *For each fixed $y \in X$, $\psi(x,y)$ is a quasi-concave function of x on X.*

(c) *$\psi(x,x) \leqq 0$ for all $x \in X$.*

(d) *X has a non-empty compact convex subset X_0 such that the set $\{y \in X: \psi(x,y) \leqq 0 \text{ for all } x \in X_0\}$ is compact.*

Then there exists a point $\hat{y} \in X$ such that $\psi(x,\hat{y}) \leqq 0$ for all $x \in X$.

We recall that a real-valued function ψ defined on a convex set X is said to be quasi-concave if for every real number t the set $\{x \in X: \psi(x) > t\}$ is convex. We remark that in Theorem 3, the coercive condition (d) is a unification of the two coercive conditions given in Allen [1, Theorem 2, condition (d)] and in Brezis-Nirenberg-Stampacchia [4, Theorem 1, condition (5)]. Notice also that in Theorem 3, conditions (b) and (c) imply that the set $\{y \in X: \psi(x,y) \leq 0$ for all $x \in X_0\}$ is non-empty.

THEOREM 4. *Let X be a paracompact convex subset of a real locally convex Hausdorff topological vector space E, X_0 a non-empty compact convex subset of X, and K a non-empty compact subset of X. Let $f: X \to P(E)$ be upper hemi-continuous with each $f(x)$ a closed convex subset of E such that*

(a) *For each* $x \in K \cap \partial X$, $f(x)$ meets $x + \overline{\bigcup_{\lambda>0} \lambda(X-x)}$.

(b) *For each* $x \in X\backslash K$, $f(x)$ *meets* $x + \overline{\bigcup_{\lambda>0} \lambda(X_0-x)}$.

Then there exists $\hat{x} \in X$ *such that* $\hat{x} \in f(\hat{x})$.

Proof. We first observe that if x is an interior point of X, then $x + \overline{\bigcup_{\lambda>0} \lambda(X-x)} = E$; thus condition (a) is equivalent to the following condition:

(a') For each $x \in K$, $f(x)$ meets $x + \overline{\bigcup_{\lambda>0} \lambda(X-x)}$.

Suppose that the assertion of the theorem is false. Let $x \in X$; then $x \notin f(x)$, so by the Hahn-Banach theorem, there exists $\zeta_x \in E'$ and $t_x \in \mathbb{R}$ such that

$$\sup\{\langle\zeta_x,v\rangle : v \in f(x)\} < t_x < \langle\zeta_x,x\rangle .$$

As f is upper hemi-continuous and ζ_x is continuous, there is an open neighborhood U_x of x satisfying

$$\sup_{y\in f(u)} \langle\zeta_x,y\rangle < t_x < \langle\zeta_x,u\rangle \quad \text{for all} \quad u \in U_x . \tag{2}$$

As $\{U_x: x \in X\}$ is an open covering of the paracompact space X, there is a continuous partition of unity $\{\phi_x: x \in X\}$ subordinate to this open covering. Thus (i) for each $x \in X$, ϕ_x is a non-negative real-valued continuous function on X such that its support $\operatorname{supp} \phi_x \subset U_x$, (ii) the family $\{\operatorname{supp} \phi_x: x \in X\}$ is a locally finite covering of X and (iii) $\sum_{x\in X} \phi_x(y) = 1$ for all $y \in X$. Define $p: X \to E'$ by setting

$$p(y) := -\sum_{z\in X} \phi_z(y)\zeta_z \quad \text{for all} \quad y \in X .$$

Let $y \in X$; note that whenever $\phi_z(y) \neq 0$, we have $y \in \operatorname{supp} \phi_z \subset U_z$ so by (2), we must have

$$\sup\{\langle\zeta_z,v\rangle : v \in f(y)\} < t_z < \langle\zeta_z,y\rangle ;$$

thus

$$\langle\zeta_z,v\rangle < t_z < \langle\zeta_z,y\rangle \quad \text{for all} \quad v \in f(y) . \tag{3}$$

As $\phi_z(y) \neq 0$ for at least one $z \in X$, it follows from (3) that

$$\langle p(y),y\rangle = -\sum_{z\in X} \phi_z(y)\langle\zeta_z,y\rangle$$

$$< -\sum_{z\in X} \phi_x(y)\langle\zeta_z,v\rangle \quad \text{for all} \quad v \in f(y) .$$

Thus we have shown that for each $y \in X$,

$$\langle p(y),y\rangle < \langle p(y),v\rangle \quad \text{for all} \quad v \in f(y) . \tag{4}$$

Now we define $\psi\colon X \times X \to \mathbb{R}$ by

$$\psi(x,y) := \langle p(y),x-y\rangle \quad \text{for all} \quad (x,y) \in X \times X .$$

We observe that

(I) For each fixed $x \in X$, $\psi(x,y)$ is a lower semi-continuous function of y on X. Indeed, let $t \in \mathbb{R}$ and let $(y_\alpha)_{\alpha\in I}$ be a net in X for which $y_\alpha \to y \in X$ and $\psi(x,y_\alpha) \leqq t$ for all $\alpha \in I$. Because $\{\text{supp}\ \phi_z\colon z \in X\}$ is locally finite, there is an open neighborhood N_y of y such that $N_y \cap \text{supp}\ \phi_z \neq \emptyset$ for at most finitely many $z \in X$, say, $\{z \in X\colon N_y \cap \text{supp}\ \phi_z \neq \emptyset\} = \{z_1,z_2,\ldots,z_n\}$. Choose an $\alpha_0 \in I$ such that $y_\alpha \in N_y$ for all $\alpha \geqq \alpha_0$, we have

$$t \geqq \psi(x,y_\alpha) = \langle p(y_\alpha)\ ,\ x - y_\alpha\rangle$$

$$= -\sum_{z\in X} \phi_z(y_\alpha)\langle\zeta_z,x-y_\alpha\rangle$$

$$= -\sum_{i=1}^{n} \phi_{z_i}(y_\alpha) \langle \zeta_{z_i}\ ,\ x - y_\alpha\rangle \quad (\text{if} \quad \alpha \geqq \alpha_0)$$

$$\to -\sum_{i=1}^{n} \phi_{z_i}(y)\langle\zeta_{z_i},x-y\rangle$$

$$= -\sum_{z\in X} \phi_z(y)\langle\zeta_z,x-y\rangle$$

$$= \psi(x,y) .$$

Hence $\psi(x,y)$ is a lower semi-continuous function of y on X.

(II) For each fixed $y \in X$, $\psi(x,y)$ is an affine function and hence a quasi-concave function of x on X.

(III) Clearly $\psi(x,x) = 0$ for all $x \in X$.

(IV) The set $C := \{y \in X\colon \psi(x,y) \leqq 0 \ \text{for all}\ x \in X_0\}$ is compact. Indeed, by (I), it is sufficient to show that $C \subset K$. Suppose $y \in X\backslash K$; then by hypothesis (b), there exists $u \in f(y) \cap [y + \bigcup_{\lambda>0} \lambda(X_0-y)]$. As $u \in f(y)$, it follows from (4) that $\langle p(y),y\rangle < \langle p(y),u\rangle$. By continuity of $p(y)$ at u, there is an open neighborhood N_u of u such that

$$\langle p(y),y\rangle < \langle p(y),u'\rangle \quad \text{for all} \quad u' \in N_u . \tag{5}$$

As $u \in \overline{y + \bigcup_{\lambda>0} \lambda(X_0 - y)}$, we may choose some $u' \in N_u \cap [y + \bigcup_{\lambda>0} \lambda(X_0-y)]$; thus there exist $x \in X_0$ and $r > 0$ for which $u' = y + r(x-y)$. Therefore

$$\begin{aligned} \psi(x,y) &= \langle p(y), x-y \rangle \\ &= -\sum_{z \in X} \phi_z(y) \langle \zeta_z, x-y \rangle \\ &= -\frac{1}{r} \sum_{z \in X} \phi_z(y) \langle \zeta_z, u'-y \rangle \\ &= \frac{1}{r} \langle p(y), u'-y \rangle > 0 \quad \text{[by (5)].} \end{aligned}$$

This shows that $y \in X \backslash C$. Hence $C \subset K$.

Thus all hypotheses in Theorem 3 are satisfied, it follows that there exists $\hat{y} \in X$ such that $\psi(x,\hat{y}) \leqq 0$ for all $x \in X$; that is,

$$\langle p(\hat{y}), x \rangle \leqq \langle p(\hat{y}), \hat{y} \rangle \quad \text{for all } x \in X . \tag{6}$$

Note that $\hat{y} \in K$; thus by (a'), there exists $w \in f(\hat{y}) \cap \overline{[y + \bigcup_{\lambda>0} \lambda(X-\hat{y})]}$. We shall verify that $\langle p(\hat{y}), w \rangle \leqq \langle p(\hat{y}), \hat{y} \rangle$: Indeed, let $\varepsilon > 0$ be given; because $p(\hat{y})$ is continuous at w, there is an open neighborhood W of w such that

$$|\langle p(\hat{y}), w \rangle - \langle p(\hat{y}), u \rangle| < \varepsilon \quad \text{for all } u \in W .$$

As $w \in \overline{\hat{y} + \bigcup_{\lambda>0} \lambda(X-\hat{y})}$, we can find $u_0 \in W \cap [\hat{y} + \bigcup_{\lambda>0} \lambda(X-\hat{y})]$, it follows that $\langle p(\hat{y}), w \rangle < \langle p(\hat{y}), u_0 \rangle + \varepsilon$ and $u_0 = \hat{y} + r(\hat{u} - \hat{y})$ for some $\hat{u} \in X$ and $r > 0$. By (6), we have $\langle p(\hat{y}), u \rangle \leqq \langle p(\hat{y}), \hat{y} \rangle$, and hence

$$\langle p(\hat{y}), u_0 - \hat{y} \rangle = r \langle p(\hat{y}), \hat{u} - \hat{y} \rangle \leqq 0 .$$

Therefore

$$\begin{aligned} \langle p(\hat{y}), w \rangle &< \langle p(\hat{y}), u_0 \rangle + \varepsilon \\ &\leqq \langle p(\hat{y}), \hat{y} \rangle + \varepsilon . \end{aligned}$$

As $\varepsilon > 0$ is arbitrary, we must have $\langle p(\hat{y}), w \rangle \leqq \langle p(\hat{y}), \hat{y} \rangle$ which contradicts (4) as $w \in f(\hat{y})$. This completes the proof. □

We remark that our proof of Theorem 4 follows the method given in Ky Fan [8].

THEOREM 5. *Let X be a paracompact convex subset of a real locally convex Hausdorff topological vector space E, X_0 a non-empty compact convex subset of X, and K a non-empty compact subset of X. Let $f: X \to P(E)$ be upper hemi-continuous with each $f(x)$ a closed convex subset of E such that*

(a) For each $x \in K \cap \partial X$, $f(x)$ meets $\overline{x + \bigcup_{\lambda<0} \lambda(X-x)}$.

(b) For each $x \in X \backslash K$, $f(x)$ meets $x + \bigcup_{\lambda<0} \lambda(X_0-x)$.

Then there exists $\hat{x} \in X$ such that $\hat{x} \in f(\hat{x})$.

Proof. Consider a set-valued map $g: X \to P(E)$ given by $g(x) := 2x - f(x)$ for each $x \in X$. As $x - g(x) = -(x - f(x))$ for each $x \in X$, f and g have the same fixed points. For each $x \in K \cap \partial X$, if $f(x)$ meets $x + \overline{\bigcup_{\lambda<0} \lambda(X-x)}$, then $g(x)$ meets $x + \overline{\bigcup_{\lambda>0} \lambda(X-x)}$. Similarly, for each $x \in X \setminus K$, if $f(x)$ meets $x + \overline{\bigcup_{\lambda<0} \lambda(X_0-x)}$, then $g(x)$ meets $x + \overline{\bigcup_{\lambda>0} \lambda(X_0-x)}$. Thus by Theorem 4, g has a fixed point and therefore f has a fixed point. □

4. Proof of the Generalized Theorem

We proceed now to prove Theorem 1. For each $x \in X$, let $I(x) := \{i \in I\colon x \in A_i\}$. Because $\bigcup_{i \in I} A_i = X$ and $\{A_i\colon i \in I\}$ is locally finite, each $I(x)$ is non-empty and finite. For each $x \in X$, let $f(x)$ be the closed convex hull of $\cup\{C_i + s(x)\colon i \in I(x)\}$. Then f is a set-valued map defined on X such that for each $x \in X$, $f(x)$ is a non-empty closed convex subset of E. For each $x \in X$, let $g(x)$ be the closed convex hull of $\cup\{C_i\colon i \in I(x)\}$. Since for each $x \in X$,

$$\cup\{C_i + s(x)\colon i \in I(x)\} = \cup\{C_i\colon i \in I(x)\} + s(x),$$

and each $s(x)$ is weakly compact convex, we see that $f = g + s$. For each fixed $x \in X$, since $\{A_i\colon i \in I\}$ is locally finite, $\bigcup_{i \notin I(x)} A_i$ is closed in X and hence $U(x) := X \setminus \bigcup_{i \notin I(x)} A_i$ is an open neighborhood of x. If $y \in U(x)$, then $I(y) \subset I(x)$ and therefore $g(y) \subset g(x)$. Consequently, g is upper hemi-continuous on X. Thus $f = g + s$ is upper hemi-continuous. By hypothesis, for each $x \in K \cap \partial X$, $f(x)$ meets $x + \overline{\bigcup_{\lambda>0} \lambda(X-x)}$ and for each $x \in X \setminus K$, $f(x)$ meets $x + \overline{\bigcup_{\lambda>0} \lambda(X_0-x)}$. Applying Theorem 4, there exists a point $\hat{x} \in X$ such that $\hat{x} \in f(\hat{x})$. If we take $I_0 := I(\hat{x})$, then the proof of Theorem 1 is complete. □

REFERENCES

1. G. Allen, Variational inequalities, complementarity problems, and duality theorems, J. Math. Anal. Appl. 58(1977), 1-10.

2. J.-P. Aubin, Mathematical Methods of Game and Economic Theory, North-Holland, Amsterdam, Revised Edition, 1982.

3. J.-P. Aubin and I. Ekeland, Applied Nonlinear Analysis, John Wiley and Sons, New York, 1984.

4. H. Brezis, L. Nirenberg and G. Stampacchia, A remark on Ky Fan's minimax principle, Boll. Un. Mat. Ital. 6(1972), 293-300.

5. K. Fan, Fixed-point and minimax theorems in locally convex topological linear spaces, Proc. Nat. Acad. Sci. USA 38(1952), 121-126.

6. K. Fan, A minimax inequality and applications, in "Inequalities III", Proceedings Third Symposium on Inequalities (O. Shisha, Ed.), pp. 103-113, Academic Press, New York, 1972.

7. K. Fan, A further generalization of Shapley's generalization of the Knaster-Kuratowski-Mazurkiewicz theorem, in "Game Theory and Related Topics", (O. Moeschlin and D. Pallaschke, Eds.), pp. 275-279, North-Holland, Amsterdam, 1981.

8. K. Fan, Some properties of convex sets related to fixed point theorems, Math. Ann. 266(1984), 519-537.

9. I.L. Glicksberg, A further generalization of the Kakutani fixed point theorem, with application to Nash equilibrium points, Proc. Amer. Math. Soc. 3(1952), 170-174.

10. B. Knaster, C. Kuratowski and S. Mazurkiewicz, Ein Beweis des Fixpunktsatzes fur n-dimensionale Simplexe, Fund. Math. 14(1929), 132-137.

11. L.S. Shapley, On balanced games without side payments, in "Mathematical Programming", (T.C. Hu and S.M. Robinson, Eds.), pp. 261-290, Academic Press, New York, 1973.

Shapley Selections and Covering Theorems of Simplexes

MAU-HSIANG SHIH Department of Mathematics, Chung Yuan University, Chung-Li, Taiwan, Republic of China

KOK-KEONG TAN Department of Mathematics, Statistics and Computing Science, Dalhousie University, Halifax, Nova Scotia, Canada

§1. KKMS Theorem

Throughout this paper, σ will denote a simplex in a euclidean space. The family of all faces of σ (of all dimensions) is denoted by F. For each $\tau \in F$, let $c(\tau)$ denote the barycenter of τ. A set $\mathcal{D}$ of faces of σ is said to be balanced if the convex hull of the set $\{c(\tau): \tau \in \mathcal{D}\}$ contains the barycenter $c(\sigma)$ of σ. The collection of all subsets of σ is denoted by 2^σ.

To formulate the results, we need the following

DEFINITION. A set-valued map $B : F \longrightarrow 2^\sigma$ is called a Shapley-map if for each $\tau \in F$,

$$\tau \subset \bigcup_{\tau \supset \rho \in F} B(\rho) .$$

The following remarkable generalization, which plays an important role in game theory, of the classical Knaster-Kuratowski-Mazurkiewicz Theorem [5] is due to Shapley [6].

KKMS Theorem. *If $A : F \longrightarrow 2^\sigma$ is a Shapley-map with each $A(\rho)$ a closed subset of σ, then there exists a balanced set $\mathcal{D}$ of faces of σ such that*

$$\bigcap_{\tau \in \mathcal{D}} A(\tau) \neq \emptyset .$$

In case $A(\rho) \neq \emptyset$ only for 0-dimensional faces of σ, KKMS theorem becomes the classical Knaster-Kuratowski-Mazurkiewicz theorem [5]. Shapley's proof of the KKMS theorem is based on a generalization [6] of the Sperner combinatorial lemma [7]. A proof using a Ky Fan's coincidence theorem [1] has been recently given by Ichiishi [3]. In the present paper, we first prove a selection theorem for Shapley-maps, and then apply it together with the KKMS theorem to give two new covering theorems of simplexes with facial structures. As a consequence, we obtain a general and more direct method for dealing with a recent basic covering theorem of Ky Fan [2, Theorem 2].

§2. A Selection Theorem.

Let $A, B : F \longrightarrow 2^{\sigma}$ be set-valued maps. If $A(\rho) \subset B(\rho)$ for each $\rho \in F$, then A is called a selection for B. If A is both a Shapley-map and a selection for B, then A is called a Shapley-selection for B. The closure of a set U in σ is denoted by $\overline{U}$.

We now establish

THEOREM 1. If $B : F \longrightarrow 2^{\sigma}$ is a Shapley-map with each $B(\rho)$ an open subset of σ, then B admits a Shapley-selection $A : F \longrightarrow 2^{\sigma}$ such that each $A(\rho)$ is a a closed subset of σ.

Proof. Since B is a Shapley-map, $\sigma = \bigcup_{\rho \in F} B(\rho)$. For each $y \in \sigma$, let

$$H_y := \cap\{B(\rho) : y \in B(\rho)\}.$$

Then H_y is an open set in σ containing y, and therefore there exists an open neighbourhood U_y of y in σ such that

$$y \in U_y \subset \overline{U}_y \subset H_y .$$

For each $\tau \in F$, we have

$$\cup\{B(\rho) : \rho \subset \tau\} = \cup\{U_y : y \in B(\rho) \text{ for some } \rho \subset \tau\}.$$

As B is a Shapley-map, the compactness of τ ensures that there exists a finite set B_{τ} of $\cup\{B(\rho) : \rho \subset \tau\}$ such that

$$\tau \subset \cup\{U_y : y \in B_{\tau}\}.$$

We now let

$$K := \cup\{B_{\tau} : \tau \in F\};$$

then K is a finite set. Define $A : F \longrightarrow 2^{\sigma}$ by

$$A(\rho) := \cup\{\overline{U}_y : y \in K \text{ and } U_y \subset B(\rho)\}$$

for each $\rho \in F$. Then for each $\rho \in F$, $A(\rho)$ is a closed subset of σ such that $A(\rho) \subset B(\rho)$. It remains to verify that A is a Shapley-map. Let $\tau \in F$ be arbitrarily fixed. For each $z \in \tau$, there exist $\rho \in F$ with $\rho \subset \tau$ and $y \in B(\rho)$ such that $z \in U_y$; it follows that $y \in K$ and $U_y \subset \overline{U}_y \subset H_y \subset B(\rho)$, and hence $z \in A(\rho)$ for some $\rho \subset \tau$. Thus

$$\tau \subset \bigcup_{\tau \supset \rho \in F} A(\rho).$$

This concludes the proof of our theorem. □

§3. Some Covering Theorems of Simplexes

KKMS and Theorem 1 imply the following

THEOREM 2. *If* $B : F \to 2^{\sigma}$ *is a Shapley-map with each* $B(\rho)$ *an open subset of* σ, *then there exists a balanced set* $\mathcal{D}$ *of faces of* σ *such that*

$$\bigcap_{\tau \in \mathcal{D}} B(\tau) \neq \emptyset .$$

In what follows we shall show that Theorem 2 also implies the KKMS theorem. Indeed, let $A : F \to 2^{\sigma}$ be a Shapley-map with each $A(\rho)$ a closed subset of σ. For each $k = 1,2,\ldots,$ we define $B^{(k)} : F \to 2^{\sigma}$ by

$$B^{(k)}(\rho) := \left\{x \in \sigma : \operatorname{dist}(x,A(\rho)) < 1/k\right\}$$

for each $\rho \in F$; then $B^{(k)}$ is a Shapley-map with each $B^{(k)}(\rho)$ an open subset of σ so that by Theorem 2, there exists a balanced set $\mathcal{D}_k$ of faces of σ such that

$$\bigcap_{\tau \in \mathcal{D}_k} B^{(k)}(\tau) \neq \emptyset .$$

Let $\mathcal{D}$ be a balanced set of faces of σ such that $\mathcal{D} = \mathcal{D}_k$ for infinitely many k's. It follows that $\bigcap_{\tau \in \mathcal{D}} B^{(k)}(\tau) \neq \emptyset$ for all $k = 1,2,\ldots$. Let

$$x_k \in \bigcap_{\tau \in \mathcal{D}} B^{(k)}(\tau), \quad k = 1,2,\ldots .$$

By compactness of σ, there is a subsequence $\{x_{k_i}\}$ of $\{x_k\}$ which converges to $\hat{x} \in \sigma$. Let $\rho \in \mathcal{D}$. Since

$$\begin{aligned}\operatorname{dist}(\hat{x},A(\rho)) &\leqq d(\hat{x},x_{k_i}) + \operatorname{dist}(x_{k_i},A(\rho)) \\ &< \operatorname{dist}(\hat{x},x_{k_i}) + 1/k_i \to 0 \quad \text{as} \quad i \to +\infty ,\end{aligned}$$

we see that $\hat{x} \in A(\rho)$ as $A(\rho)$ is closed. Hence $\hat{x} \in \bigcap_{\rho \in \mathcal{D}} A(\rho)$.

COROLLARY 1. *Let* $\sigma := a_1a_2\ldots a_n$ *be an* $(n-1)$-*simplex, and let* $\{B_1,B_2,\ldots,B_n\}$ *be a family of* n *open subsets of* σ. *If*

$$a_{i_1}a_{i_2}\ldots a_{i_k} \subset \bigcup_{j=1}^{k} B_{i_j}$$

holds for every face $a_{i_1}a_{i_2}\ldots a_{i_k}$ *of* σ, *then* $\bigcap_{i=1}^{n} B_i \neq \emptyset$.

Proof. This is the case of $B(\rho) \neq \emptyset$ only for 0-dimensional faces ρ of σ in Theorem 2. □

Corollary 2 (Ky Fan). *Let* $\sigma := a_1a_2\ldots a_n$ *be an* $(n-1)$ - *simplex*, and let $\{A_1,A_2,\ldots,A_n\}$ *be a family of* n *closed subsets of* σ *such that* $\bigcup_{i=1}^{n} A_i = \sigma$. *Then there exist* k *indices* $i_1 < i_2 < \ldots < i_k$ *between* 1 *and* n *such that the* $(k-1)$-*face* $a_{i_1}a_{i_2}\ldots a_{i_k}$ *contains a point of the intersection* $\bigcap_{j=1}^{k} A_{i_j}$.

Proof. This is a restatement of Corollary 1 in its contraposition form and in terms of the complement A_i of B_i. □

Ky Fan has given two basically different proofs of Corollary 2 using Kakutani's fixed point theorem [4] or a generalization of the Knaster-Kuratowski-Mazurkiewicz theorem and a skillful combinatorial lemma (see [2, pp. 520-521 and 523-524]). We observe that Corollary 2 yields another proof of Brouwer's fixed-point theorem. To see this, let f be a continuous map from an $(n-1)$-simplex $\sigma := a_1a_2\ldots a_n$ into itself. For a point $x \in \sigma$, let $\lambda_i(x)$ $(i = 1,2,\ldots,n)$ denote the barycentric coordinates of x, i.e.,

$$x = \sum_{i=1}^{n} \lambda_i(x)a_i \quad \text{with} \quad \lambda_i(x) \geqq 0 \quad \text{and} \quad \sum_{i=1}^{n} \lambda_i(x) = 1 .$$

For each $i = 1,2,\ldots,n$, let

$$A_i := \{x \in \sigma : \lambda_i(x) \leqq \lambda_i(f(x))\} .$$

As λ_i and f are continuous, each A_i is closed in σ. Clearly $\bigcup_{i=1}^{n} A_i = \sigma$. By Corollary 2, there exists $\{i_1,i_2,\ldots,i_k\} \subset \{1,2,\ldots,n\}$ with $i_1 < i_2 < \ldots < i_k$ such that the $(k-1)$-face $a_{i_1}a_{i_2}\ldots a_{i_k}$ contains a point of the intersection $\bigcap_{j=1}^{k} A_{i_j}$. Let

$$\hat{x} \in a_{i_1}a_{i_2}\ldots a_{i_k} \cap [\bigcap_{j=1}^{k} A_{i_j}] .$$

Then $\sum_{j=1}^{k} \lambda_{i_j}(\hat{x}) = 1$ and $\lambda_{i_j}(\hat{x}) \leqq \lambda_{i_j}(f(\hat{x}))$ for each $j = 1,2,\ldots,k$. Thus $\lambda_i(\hat{x}) = \lambda_i(f(\hat{x}))$ for each $i = 1,2,\ldots,n$ and hence $\hat{x}$ is a fixed-point of f.

To compare Ky Fan's theorem (i.e., Corollary 2 stated above) with KKMS theorem, Ky Fan's theorem may be further stated as follows.

COROLLARY 2'. Let $A : F \to 2^{\sigma}$ be a set-valued map with each $A(\rho)$ a closed subset of σ and $\bigcup_{\rho\in F} A(\rho) = \sigma$ (A is not necessarily a Shapley-map). Then there exists a non-empty subset G of F such that the convex hull of the barycenters $\{c(\tau) : \tau \in G\}$ contains a point of the intersection $\bigcap_{\tau\in G} A(\tau)$.

Proof. Let σ be a $(m-1)$-simplex. Because F contains 2^m-1 elements, let $F = \{\tau_1,\tau_2,\ldots,\tau_{2^m-1}\}$ and $c(\tau_i)$ be the corresponding barycenter of τ_i. Let $\Delta^{2^m-2} := a_1a_2\ldots a_{2^m-1}$ be a (2^m-2)-simplex. Define $\phi : \Delta^{2^m-2} \to \sigma$ by

$$\phi(\sum_{i=1}^{2^m-1} \lambda_i a_i) := \sum_{i=1}^{2^m-1} \lambda_i c(\tau_i) .$$

Let $B_i := \phi^{-1}(A(\tau_i))$ for each $i = 1,2,\ldots,2^m-1$. Applying Corollary 2 to the closed covering $\{B_1,B_2,\ldots,B_{2^m-1}\}$ of Δ^{2^m-2}. Then there exist k indices

$i_1 < i_2 < \ldots < i_k$ between 1 and 2^m-1 such that the convex hull of $\{a_{i_1}, a_{i_2}, \ldots, a_{i_k}\}$ contains a point $\hat{a}$ in the intersection $\bigcap_{j=1}^{k} B_{i_j}$. Then $\phi(\hat{a})$ is in the convex hull of $\{c(\tau_{i_1}), c(\tau_{i_2}), \ldots, c(\tau_{i_k})\}$ as well as in the intersection $\bigcap_{j=1}^{k} A(\tau_{i_j})$. Take $G := \{\tau_{i_1}, \tau_{i_2}, \ldots, \tau_{i_k}\}$, the result follows. □

Theorem 2 may be restated in its contraposition form and in terms of the complement $A(\rho)$ of $B(\rho)$ in σ as follows .

THEOREM 3. Let $A : F \to 2^{\sigma}$ be a set-valued map with each $A(\rho)$ a closed subset of σ. Suppose that for each balanced set $\mathcal{D} \subset F$,

$$\bigcup_{\tau \in \mathcal{D}} A(\tau) = \sigma .$$

Then there exists $\tau \in F$ such that

$$\tau \cap [\bigcap_{\tau \supset \rho \in F} A(\rho)] \neq \emptyset .$$

To illustrate Theorem 3, we consider a 2-simplex $\sigma := a_1a_2a_3$ with $A(\sigma) := \sigma$ and all other $A(\rho)$ being closed subsets of σ as indicated in the Figure. It is easily checked that $\sigma = \bigcup_{\rho \in \mathcal{D}} A(\rho)$ for each balanced set $\mathcal{D}$ of σ. We find that the face a_2a_3 is such that $a_2a_3 \cap [A(a_2) \cap A(a_3) \cap A(a_2a_3)] \neq \emptyset$.

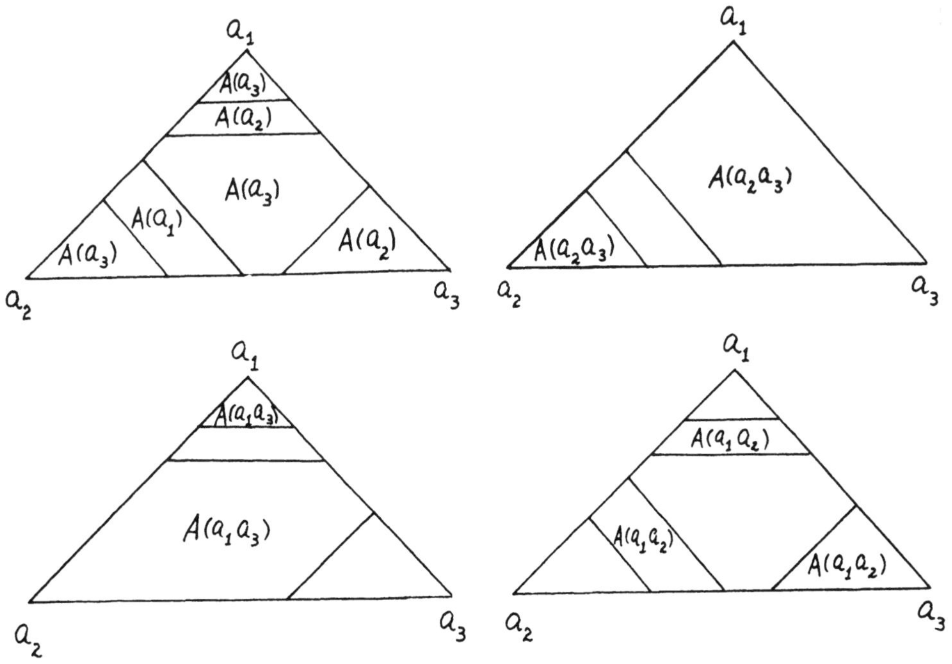

Figure

To compare the KKMS theorem with Theorem 3, we reformulate the KKMS theorem as in the following form.

THEOREM 4. *Let* $B : F \to 2^{\sigma}$ *be a set-valued map with each* $B(\rho)$ *an open subset of* σ. *Suppose that for each balanced set* $\mathcal{D} \subset F$

$$\bigcup_{\tau \in \mathcal{D}} B(\tau) = \sigma \quad .$$

Then there exists $\tau \in F$ *such that*

$$\tau \cap [\bigcap_{\tau \supset \rho \in F} B(\rho)] \neq \emptyset \ .$$

A further generalization of Theorem 2 can be obtained by applying Ky Fan's generalization [2] of KKMS theorem and Theorem 1.

THEOREM 5. *Let* $A, B : F \to 2^{\sigma}$ *be set-valued maps such that all* $A(\rho), B(\rho)$ *are open subsets of* σ. *Suppose that* $\bigcup_{\tau \in F} A(\tau) = \sigma$ (*A is not necessarily a Shapley-map*) *and* B *is a Shapley-map*. *Then there exist two non-empty subfamilies* G *and* H *of* F *such that*

(a) $[\bigcap_{\tau \in G} A(\tau)] \cap [\bigcap_{\rho \in H} B(\tau)] \neq \emptyset$

and

(b) *the convex hull of the barycenters* $\{c(\tau) : \tau \in G\}$ *meets the convex hull of the barycenters* $\{c(\rho) : \rho \in H\}$.

Proof. The open covering $\bigcup_{\tau \in F} A(\tau) = \sigma$ can be shrunken to a closed covering, i.e., we can find closed sets $D(\tau) \subset A(\tau)$ for each $\tau \in F$ such that $\bigcup_{\tau \in F} D(\rho) = \sigma$. On the other hand, by Theorem 1, there exists a Shapley-selection $E : F \to 2^{\sigma}$ for B with each $E(\rho)$ a closed subset of σ. So by Ky Fan's generalization [2, Theorem 13] of KKMS theorem we can find two non-empty subfamilies G and H of F such that

$$[\bigcap_{\tau \in G} D(\tau)] \cap [\bigcap_{\rho \in H} E(\rho)] \neq \emptyset$$

and the convex hull of the barycenters $\{c(\tau) : \tau \in G\}$ meets the convex hull of the barycenters $\{c(\rho) : \rho \in H\}$, and therefore the conclusion of Theorem 5 follows. □

In the case when $A(\tau) = \emptyset$ for $\tau \neq \sigma$, $A(\sigma) = \sigma$, Theorem 5 becomes Theorem 2. If, in addition, $B(\tau) \neq \emptyset$ only for 0-dimensional faces τ of σ, the result becomes Corollary 1 of Theorem 2.

Theorem 5 may be restated in its contraposition form and in terms of the complement $D(\rho)$ of $A(\rho)$, the complement $E(\rho)$ of $B(\rho)$ as follows.

THEOREM 6. Let $D,E : F \to 2^{\sigma}$ be set-valued maps such that all $D(\rho)$, $E(\rho)$ are closed subsets of σ . Suppose that for each pair of non-empty subfamilies G and H of F , either

$$[\bigcup_{\tau\in G} D(\tau)] \cup [\bigcup_{\rho\in H} E(\rho)] = \sigma$$

or the intersection of the convex hull of the barycenters $\{c(\tau) : \tau \in G\}$ with the convex hull of the barycenters $\{c(\rho) : \rho \in H\}$ is empty .

Then : either $\bigcap_{\tau\in F} D(\tau) \neq \emptyset$ or there exists a face $\tau \in F$ such that

$$\tau \cap [\bigcap_{\tau\supset\rho\in F} E(\rho)] \neq \emptyset .$$

References

1. K. Fan, Extensions of two fixed point theorems of F.E. Browder, Math. Z. 112 (1969), 234-240.

2. K. Fan, Some properties of convex sets related to fixed point theorems, Math. Ann. 266(1984), 519-537.

3. T. Ichiishi, On the Knaster-Kuratowski-Mazurkiewicz-Shapley theorem, J. Math. Anal. Appl. 81(1981), 297-299.

4. S. Kakutani, A generalization of Brouwer's fixed-point theorem, Duke Math. J. 8(1941), 457-459.

5. B. Knaster, C. Kuratowski and S. Mazurkiewicz, Ein Beweis des Fixpunktsatzes für n-dimensionale Simplexe, Fund. Math. 14(1929), 132-137.

6. L.S. Shapley, On balanced games without side payments, In: Mathematical Programming, pp. 261-290. (Eds. T.C. Hu and S.M. Robinson.) New York, Academic Press, 1973.

7. E. Sperner, Neuer Beweis für Invarianz der Dimensionszahl und des Gebietes, Abh. Math. Sem. Univ. Hamburg, 6(1928), 265-272.

Generalizations of Convex Supremization Duality

IVAN SINGER Department of Mathematics, National Institute for Scientific and Technical Creation, Bucharest, Romania

Abstract. We give some extensions of our duality theorems of [13]-[15], to the optimization problems (P) $\alpha=\sup h(G)$ and (P) $\alpha=\sup(h-f)(F)$, where G is a subset of a set F, $h:F\to\overline{R}=[-\infty,+\infty]$ is a W-quasi-convex ([5], [21]) or a W-convex [4] functional, W being a subset of $\overline{R}^F$, and $f:F\to\overline{R}$ is arbitrary.

§1. Introduction

Given a set F, a subset G of F (assumed to be non-empty, throughout the sequel) and a functional $h:F\to\overline{R}=[-\infty,+\infty]$, we shall consider the following (global, scalar) primal supremization problem:

$$(P)=(P_{G,h}) \qquad \alpha=\alpha_{G,h}=\sup h(G). \tag{1.1}$$

In the paper [13] (see also [6]) we have proved some theorems of "unperturbational surrogate duality" type (in a sense similar to [20], [22]) for the particular case of problem (1.1), in which F is a locally convex space, G is a bounded subset of F and h is convex and lower semi-continuous, with values in $R\cup\{-\infty\}$ (where $R=(-\infty,+\infty)$). Furthermore, in [15] we have proved some theorems of "unperturbational Lagrangian duality" type (in a sense similar to [22]), for (1.1) with F a locally convex space, G a bounded subset of F and $h:F\to\overline{R}$ a proper lower semi-continuous convex functional. In [14] we have extended the main duality theorem of [15] to a duality theorem for the problem of supremization of the difference h_1-h_2 on a locally convex space F, where $h_1:F\to\overline{R}$ is a proper lower semi-continuous convex functional on F and $h_2:F\to\overline{R}$ is arbitrary, with the convention $+\infty-(+\infty)=-\infty$ (thus, taking h_2=the indicator functional of a bounded subset G of F, i.e., $h_2(y)=0$ for $y\in G$ and $h_2(y)=+\infty$ for $y\in F\setminus G$, we obtain the case of [15]). The result of [14] has been also obtained, independently, in an equivalent form (namely, as a duality theorem for the infimization problem $\inf(h_1-h_2)(F)$, where $h_1:F\to\overline{R}$ is arbitrary and $h_2:F\to\overline{R}$ is proper lower semi-continuous, with $+\infty-(+\infty)=+\infty$), and from a different starting point (namely, some non-linear problems in the calculus of variations, which arise in mechanics, such as the analysis of a steadily rotating heavy chain) by Toland [25]; moreover, Toland has developed, in [24], a theory of "perturbational Lagrangian duality" type (in a sense similar to [22]), which contains, as a special case, the duality theory of [25]. The importance of problem (1.1) with $F=R^n$, G a closed convex (possibly unbounded) subset of R^n, and $h:F\longrightarrow R$ a finite convex functional, has been stressed by Tuy [26], who has shown that it includes a wide class of mathematical programming problems (such as linear and convex programming, 0-1 integer programming, bilinear programming, linear and convex complementarity problems, and "convex-difference" programming).

Motivated by the above mentioned results of [13], [15] (see e.g. corollary 3.2 and remark 5.3 below), we introduce here the following concept of "dual problem" to (P) of (1.1) (without any assumptions on F, G, h):

Definition 1.1. By a dual problem to (P) we shall mean any supremization problem of the form

$$(Q)=(Q^{G,h}) \qquad \beta=\beta^{G,h}=\sup \lambda(W), \tag{1.2}$$

where $W=W^{G,h}$ is a set (assumed non-empty, without loss of generality) and $\lambda=\lambda^{G,h}:W\longrightarrow\overline{R}$ is a functional.

Remark 1.1. a) We assume no relation between α and β.

b) There is a marked difference between the above dual problems (1.2) and the "usual" dual problems [22] to (P) (extending the usual dual problems for concave supremization, i.e., for (P) of (1.1) with F a linear space, h concave and G convex), in which $\beta=\inf \lambda(W)$, or, equivalently (see e.g. [8]), $\beta=-\sup \lambda(W)$. Therefore, as in [23], we shall call the dual problems (1.2) "unusual" dual problems to (P).

We shall first consider "unperturbational surrogate dual problems" to (P), in a sense similar to [22] (see also [20]), namely, the case when λ of (1.2) is of the form

$$\lambda(w)=\lambda^{G,h}_{W\Delta}(w)=\inf h(\Delta_{G,w}) \qquad (w\in W), \tag{1.3}$$

where $\Delta_{G,w}\subseteq F$ $(w\in W)$ is a given family of ("surrogate constraint") sets; thus, by (1.2) and (1.3), we have

$$\beta=\sup_{w\in W} \inf h(\Delta_{G,w}). \tag{1.4}$$

Remark 1.2. If we interchange everywhere sup and inf, then (P) and (Q) become infimization problems and β of (1.4) will be replaced by

$$\beta'=\inf_{w\in W}\sup h(\Delta_{G,w}). \tag{1.5}$$

There is a marked difference between (1.5), the values $\beta'=\sup_{w\in W}\inf h(\Delta_{G,w})$ of the "usual" surrogate dual problems to

$$(P')\qquad \alpha'=\inf h(G), \tag{1.6}$$

and the values $\beta'=\inf_{w\in W}\inf h(\Delta_{G,w})$ of the "unusual" surrogate dual problems to (P') of (1.6), studied in [23] (see also [16], [7]); in [23], the latter ones and the dual problems with β' of (1.5) have been called "unusual surrogate dual problems of the first type" and "unusual surrogate dual problems of the second type", respectively.

In §2 we shall give some necessary and sufficient conditions for $\alpha\geqslant\beta$, for $\alpha\leqslant\beta$ and for $\alpha=\beta$, with $\alpha\in\overline{R}$ arbitrary and β of (1.4), and some simultaneous characterizations of "solutions" of (P) (of (1.1)) and of "weak duality" for (P),(Q) (i.e., conditions in order to have $\alpha=\beta$, with α,β of (1.1), (1.4)), involving the level sets

$$A_c(h)=\{y\in F\,|\,h(y)<c\}\qquad (c\in R), \tag{1.7}$$

$$S_c(h)=\{y\in F\,|\,h(y)\leqslant c\}\qquad (c\in R) \tag{1.8}$$

of h; we recall that, by definition, the "solutions" of (P) (of (1.1)) are the elements of the (possibly empty) set

$$\mathcal{M}_G(h)=\{g_o\in G\,|\,h(g_o)=\sup h(G)\}. \tag{1.9}$$

In §3 we shall apply the results of §2 to $\alpha=\sup h(G)$ and to certain families of surrogate constraint sets $\Delta^i_{G,w}\subseteq F$ ($w\in W$, $1=1,\ldots,6$), where $W\subseteq\overline{R}^F$ (we recall that $\overline{R}^F$ denotes the family of all functionals $w:F\longrightarrow\overline{R}$); in the particular case when F is a locally convex space and $W\subseteq F^*$ (where F^* denotes the conjugate space of F), these sets $\Delta^i_{G,w}$ admit convenient geometric interpretations.

In §4 we shall show how the results of surrogate duality of §3 can be applied to problem (P) of (1.1) with $G=u^{-1}(\Omega)$, where u is a mapping of F into a "parameter set" X and Ω is a subset of X, with $u(F)\cap\Omega\neq\emptyset$ (where $\emptyset$ denotes the empty set); in this case we shall take $W\subseteq R^X$ (rather than $W\subseteq\overline{R}^F$) and we shall define surrogate constraint sets $\Delta^i_{u^{-1}(\Omega),w}\subseteq F$ ($w\in W$) corresponding to those of §3.

Finally, in §5, considering the "Lagrangian dual problem" to (P) of (1.1), i.e., problem (1.2), with $W\subseteq\overline{R}^F$ and λ of the form

$$\lambda(w)=\lambda_W^{G,h}(w)=\sup w(G)\dotplus\inf_{y\in F}\{h(y)\dotplus-w(y)\}\qquad (w\in W) \tag{1.10}$$

(for $\dot{+}$, $\dot{\dot{+}}$, see (5.2) and (5.3)), we shall show that the main result of [15] can be extended to W-convex [4] functionals h on a set F, where $W\subseteq\overline{R}^F$, and to arbitrary subsets G of F. The usefulness of such an extension consists in the possibility of applying it to various choices of W's, which permits a unified treatment of "augmented Lagrangians" (for the corresponding theory for problem (P') of (1.6), see e.g. [4]). Also, we shall extend the main result of [14] to the W-convex case, where $W\subseteq\overline{R}^F$.

Throughout the paper, we adopt the usual conventions

$$\inf \emptyset=+\infty , \quad \sup \emptyset=-\infty . \tag{1.11}$$

Also, as in [17]-[19], we make the convention that if $A_c(h)=\emptyset$ or $S_c(h)=\emptyset$ for some $c\in R$, then the conditions involving these $A_c(h)$, $S_c(h)$ (see e.g. (2.1), (2.2), etc.) will be considered satisfied (vacuously). By "linear space" (with or without a topology) we shall mean: real linear space.

§2. Surrogate duality results in the general case

Let us first recall

Lemma 2.1 ([19], proposition 1.1 and corollary 1.1). *Let F be a set, $\Delta\subseteq F$, $h:F\longrightarrow\overline{R}$ and $c\in R$.*

a) *We have* $\inf h(\Delta)\geqslant c$ *if and only if* $\Delta\cap A_c(h)=\emptyset$.

b) *If* $\inf h(\Delta)>c$, *then* $\Delta\cap S_c(h)=\emptyset$.

Proof [19]. If $y_o\in\Delta\cap A_c(h)$, then $\inf h(\Delta)\leqslant h(y_o)<c$. The proof of b) is similar. Finally, if $\inf h(\Delta)<c$, then there exists $y_o\in\Delta$ such that $h(y_o)<c$, so $y_o\in\Delta\cap A_c(h)$.

Proposition 2.1. *Let F, G, h, W and $\Delta_{g,w}\subseteq F$ $(w\in W)$ be as in §1, and let $\alpha\in\overline{R}$ be arbitrary. The following statements are equivalent:*

1°. *We have*

$$\Delta_{G,w}\cap A_c(h)\neq\emptyset \qquad (w\in W,\ c\in R,\ c>\alpha). \tag{2.1}$$

2°. *We have*

$$\Delta_{G,w}\cap S_c(h)\neq\emptyset \qquad (w\in W,\ c\in R,\ c>\alpha). \tag{2.2}$$

3°. *We have*

$$\alpha\geqslant\beta=\sup_{w\in W}\inf h(\Delta_{G,w}). \tag{2.3}$$

Proof. The implication $1^{\circ} \Rightarrow 2^{\circ}$ is obvious.

$2^{\circ} \Rightarrow 3^{\circ}$. If 2° holds, say $y_{w,c} \in \Delta_{G,w} \cap S_c(h)$, then

$$\lambda(w) = \inf h(\Delta_{G,w}) \leq h(y_{w,c}) \leq c \qquad (w \in W,\ c \in R,\ c > \alpha),$$

whence $\beta = \sup \lambda(W) \leq \inf\limits_{c>\alpha} c = \alpha$.

$3^{\circ} \Rightarrow 1^{\circ}$. If 3° holds, then for each $c \in R$, $c > \alpha$, we have $c > \beta = \sup\limits_{w \in W} \inf h(\Delta_{G,w})$, whence, by lemma 2.1 a), we obtain (2.1).

Remark 2.1. In particular, if $\alpha = \inf h(G)$ and $G \subseteq \Delta_{G,w}$ $(w \in W)$, then $\emptyset \neq G \cap A_c(h) \subseteq \Delta_{G,w} \cap A_c(h)$ $(w \in W,\ c \in R,\ c > \alpha)$, so we have 1°-3°. Hence, proposition 2.1 permits an improvement of the results of [19].

Proposition 2.2. Let $\alpha \in \bar{R}$ *be arbitrary. The following statements are equivalent:*

1°. *For each* $c \in R$, $c < \alpha$, *there exists* $w_c \in W$ *such that*

$$\Delta_{G,w_c} \cap A_c(h) = \emptyset. \tag{2.4}$$

2°. *For each* $c \in R$, $c < \alpha$, *there exists* $w_c \in W$ *such that*

$$\Delta_{G,w_c} \cap S_c(h) = \emptyset\ . \tag{2.5}$$

3°. *We have*

$$\alpha \leq \beta = \sup_{w \in W} \inf h(\Delta_{G,w}). \tag{2.6}$$

Proof. $1^{\circ} \Rightarrow 3^{\circ}$. If c and w_c are as in 1°, then, by lemma 2.1 a), we have

$$\lambda(w_c) = \inf h(\Delta_{G,w_c}) \geq c,$$

whence $\beta = \sup \lambda(W) \geq \sup\limits_{c<\alpha} \lambda(w_c) \geq \sup\limits_{c<\alpha} c = \alpha$.

$3^{\circ} \Rightarrow 2^{\circ}$. If 3° holds and $c \in R$, $c < \alpha$, then $c < \beta$, and hence, by (1.4), there exists $w_c \in W$ such that $c < \inf h(\Delta_{G,w_c})$. Then, by lemma 2.1 b), we have (2.5).

Finally, the implication $2^{\circ} \Rightarrow 1^{\circ}$ is obvious.

Remark 2.2. For the particular case when $\alpha = \inf h(G)$ and $G \subseteq \Delta_{G,w}$ $(w \in W)$, whence $\alpha \geq \beta$, the above argument has been given, essentially, in [19], proof of theorem 1.1.

Combining propositions 2.1 and 2.2, we obtain

Theorem 2.1. Let $\alpha \in \bar{R}$ *be arbitrary. The following statements are equivalent:*

1°. <u>We have</u> (2.1), <u>and for each</u> $c \in R$, $c < \alpha$, <u>there exists</u> $w_c \in W$ <u>satisfying</u> (2.4).

2°. <u>We have</u> (2.2), <u>and for each</u> $c \in R$, $c < \alpha$, <u>there exists</u> $w_c \in W$ <u>satisfying</u> (2.5).

3°. <u>We have</u>

$$\alpha = \sup_{w \in W} \inf h(\Delta_{G,w}) . \tag{2.7}$$

Concerning simultaneous characterizations of solutions of (P) and of weak duality for $\{(P),(Q)\}$ of (1.1), (1.4), let us prove

<u>Theorem 2.2</u>. <u>For an element</u> $g_o \in G$, <u>and for</u> $\alpha = \sup h(G)$, <u>the following statements are equivalent</u>:

1°. <u>We have</u>

$$\Delta_{G,w} \cap A_c(h) \neq \emptyset \qquad (w \in W,\ c \in R,\ c > h(g_o)), \tag{2.8}$$

<u>and for each</u> $c \in R$, $c < \alpha$, <u>there exists</u> $w_c \in W$ <u>satisfying</u> (2.4).

2°. <u>We have</u>

$$\Delta_{G,w} \cap S_c(h) \neq \emptyset \qquad (w \in W,\ c \in R,\ c > h(g_o)), \tag{2.9}$$

<u>and for each</u> $c \in R$, $c < \alpha$, <u>there exists</u> $w_c \in W$ <u>satisfying</u> (2.5).

3°. <u>We have</u> $g_o \in \mathcal{M}_G(h)$ <u>and</u> (2.7).

<u>Proof</u>. $1^{\circ} \Longrightarrow 3^{\circ}$. Assume 1°. Then, by (2.8) and proposition 2.1 (with $\alpha = h(g_o)$), we have $h(g_o) \geq \beta$. Furthermore, by the second condition of 1° and by proposition 2.2, we have (2.6). Hence, by $g_o \in G$, we obtain

$$\beta \geq \alpha = \sup h(G) \geq h(g_o) \geq \beta . \tag{2.10}$$

$3^{\circ} \Longrightarrow 1^{\circ}$. If 3° holds, then $h(g_o) = \sup h(G) = \alpha$, and hence, by theorem 2.1, we have 1°.

Finally, the proof of the equivalence $2^{\circ} \Longleftrightarrow 3^{\circ}$ is similar.

<u>Remark 2.3</u>. Similarly, one can prove the following result for infimization, which extends [19], theorem 1.4 (and hence also the particular cases of [19], theorem 1.4, given in [18]): <u>For an element</u> $g_o \in G$ <u>and for</u> $\alpha = \inf h(G)$, <u>the following statements are equivalent</u>:

1°. <u>We have</u> (2.1), <u>and for each</u> $c \in R$, $c < h(g_o)$, <u>there exists</u> $w_c \in W$ <u>satisfying</u> (2.4).

2°. <u>We have</u> (2.2), <u>and for each</u> $c \in R$, $c < h(g_o)$, <u>there exists</u> $w_c \in W$ <u>satisfying</u> (2.5).

3°. <u>We have</u>

$$h(g_o)=\inf h(G)=\sup_{w\in W} \inf h(\Delta_{G,w}). \qquad (2.11)$$

Indeed, in the proof, the inequalities (2.10) are now replaced by

$$\beta\leq\alpha=\inf h(G)\leq h(g_o)\leq\beta . \qquad (2.12)$$

§3. Applications to surrogate duality for supremization

In this section we shall assume that F is a set and $W\subseteq\overline{R}^F$. Also, as before, let $G\subset F$ and $h:F\longrightarrow\overline{R}$.

1) Let us define a family of sets $\Delta^1_{G,w}\subseteq F$ ($w\in W$) by

$$\Delta^1_{G,w}=\{y\in F|\, w(y)\geq\sup w(G)\} \qquad (w\in W). \qquad (3.1)$$

Remark 3.1. a) If $0\in W$ (where 0 denotes the zero functional on F), then $\Delta^1_{G,0}=F$, whence, by (1.3), $\lambda(0)=\inf h(\Delta^1_{G,0})=\inf h(F)$. Hence,

$$\beta=\sup_{0\neq w\in W} \inf h(\Delta^1_{G,w}). \qquad (3.2)$$

b) If F is a locally convex space, then for $0\neq w\in F^*$ such that $\sup w(G)=+\infty$, we have $\Delta^1_{G,w}=\emptyset$, while for $0\neq w\in F^*$ such that $\sup w(G)<+\infty$, $\Delta^1_{G,w}$ is a closed half-space in F, supporting the set G (i.e., $G\cap\mathrm{Int}\,\Delta^1_{G,w}=\emptyset$ and the boundary of $\Delta^1_{G,w}$ is a support hyperplane of G; for the definition of support hyperplanes, see e.g. [17], I.0), and we have

$$\beta=\sup_{w\in G^s} \inf h(\Delta^1_{G,w}) , \qquad (3.3)$$

where

$$G^s=\{w\in W|\, w\neq 0,\ \sup w(G)<+\infty\} . \qquad (3.4)$$

Thus, if $W=F^*$ or $W=F^*\setminus\{0\}$, formula (3.3) means that

$$\beta = \sup_{D\in\mathfrak{D}_G} \inf h(D), \qquad (3.5)$$

where $\mathfrak{D}_G$ denotes the collection of all closed half-spaces in F, which support the set G. We shall omit the corresponding geometric interpretations of the β's occurring in the sequel, and, for simplicity, we shall work only with β's written similarly to (3.2).

For a set F and functionals $h,w:F\longrightarrow\overline{R}$, $w\neq 0$, let

$$\varphi(c)=\varphi_w(c)=\inf_{\substack{y\in F\\ w(y)\geq c}} h(y) \qquad (c\in R); \qquad (3.6)$$

in the particular case $F=R^n$, $0\neq w\in (R^n)^*$, the non-decreasing functions $\varphi_w: R \to \overline{R}$ have been studied in [3], [2], [9]. Extending [3], p.214, we shall say that h is <u>regular</u> (or, extending [9], p.66, one might use the term "semi-regular") <u>with respect to</u> w, if

$$\varphi_w(c)=\sup_{\substack{c'\in R\\ c'<c}} \varphi_w(c') \qquad (c\in R). \qquad (3.7)$$

<u>Remark 3.2</u>. If $h:R^n \to \overline{R}$ is convex, then, by [2], theorem 11 i), φ_w is convex, for all $w\in (R^n)^*\setminus\{0\}$, and hence, if h is convex and $h(R^n)\subseteq R$, then, by [1], p.48 and [9], p.66, h is regular with respect to all $w\in (R^n)^*\setminus\{0\}$ (alternatively, one can prove these statements similarly to [13], lemma 2.1). Let us also mention that, conversely, if $h:R^n \to \overline{R}$ is quasi-convex and lower semi-continuous and if all φ_w $(w\in (R^n)^*\setminus\{0\})$ are convex, then, by [2], theorem 11 ii), h is convex.

<u>Proposition 3.1</u>. <u>Let</u> F <u>be a set</u>, $W\subseteq\overline{R}^F$, G <u>a subset of</u> F, <u>and</u> $h: F \to \overline{R}$ <u>a functional, which is regular with respect to all</u> $w\in W\setminus\{0\}$. <u>Then, for</u> $\alpha=\sup h(G)$ <u>and</u> β <u>of</u> (3.2), <u>we have</u> (2.3).

<u>Proof</u>. For any $w\in W\setminus\{0\}$ we have

$$\varphi_w(w(g))=\inf_{\substack{y\in F\\ w(y)\geq w(g)}} h(y)\leq h(g)\leq\alpha \qquad (g\in G), \qquad (3.8)$$

whence, by (3.7) (with $c=\sup w(G)$) and since φ_w is non-decreasing,

$$\lambda(w)=\inf_{\substack{y\in F\\ w(y)\geq \sup w(G)}} h(y)=\varphi_w(\sup w(G))=\sup_{\substack{c'\in R\\ c'<\sup w(G)}} w(c')\leq$$

$$\leq\sup_{g\in G}\varphi_w(w(g))\leq\alpha \qquad (w\in W\setminus\{0\});$$

hence, by (3.2), we obtain $\beta\leq\alpha$.

We recall that, following Ky Fan [5], a subset M of a set F is said to be <u>W-convex</u>, where $W\subseteq\overline{R}^F$, if for each $y\notin M$ there exists $w\in W$, $w\neq 0$, such that $\sup w(M)<w(y)$. In particular, for a locally convex space F and $W=F^*$ or $W=F^*\setminus\{0\}$, from the strict separation theorem it follows that a set $M\subseteq F$ is W-convex if and only if it is <u>closed and convex</u> (see e.g. [4]).

<u>Proposition 3.2</u>. <u>Let</u> F <u>be a set</u>, $W\subseteq\overline{R}^F$, G <u>a subset of</u> F, <u>and</u> $h:F \to \overline{R}$ <u>a functional, such that for each</u> $c<\alpha=\sup h(G)$, <u>the level set</u>

$S_c(h)$ <u>is</u> W-<u>convex</u>. <u>Then, for</u> β <u>of</u> (3.2), <u>we have</u> (2.6).

<u>Proof</u>. For each $c<\alpha=\sup h(G)$ there exists $g_c \in G$ such that $h(g_c)>c$, that is, $g_c \notin S_c(h)$. Hence, since $S_c(h)$ is W-convex, there exists $w_c \in W$, $w_c \neq 0$, such that

$$w_c(g_c) > \sup w_c(S_c(h)). \tag{3.9}$$

Then, by $g_c \in G$ and (3.9), we obtain

$$\sup w_c(G) \geq w_c(g_c) > w_c(y) \qquad (y \in S_c(h)), \tag{3.10}$$

and thus w_c satisfies (2.5) (with $\Delta=\Delta^1$). Hence, by proposition 2.2, we have (2.6).

<u>Remark 3.3</u>. The assumption of proposition 3.2 is satisfied for each $h:F \to \overline{R}$ which is "W-quasi-convex" in the sense of [21], i.e., for which all level sets $S_c(h)$ $(c \in R)$ are W-convex. In particular, if F is a locally convex space and $W=F^*$ or $F^* \setminus \{0\}$, then h is W-quasi-convex if and only if it is <u>quasi-convex</u> (in the usual sense) <u>and lower semi-continuous</u> (see [21]).

Combining propositions 3.1 and 3.2, we obtain

<u>Theorem 3.1</u>. <u>Let</u> F <u>be a set</u>, $W \subseteq \overline{R}^F$, G <u>a subset of</u> F, <u>and</u> $h:F \to \overline{R}$ <u>a functional, which is regular with respect to all</u> $w \in W \setminus \{0\}$, <u>and such that, for each</u> $c < \sup h(G)$, <u>the level set</u> $S_c(h)$ <u>is</u> W-<u>convex</u>. <u>Then</u>

$$\sup h(G) = \sup_{0 \neq w \in W} \inf_{\substack{y \in F \\ w(y) \geq \sup w(G)}} h(y). \tag{3.11}$$

From theorem 3.1 and remarks 3.2, 3.3, there follows

<u>Corollary 3.1</u>. <u>Let</u> F <u>be a locally convex space</u>, G <u>a subset of</u> F <u>and</u> $h:F \to R$ <u>a finite lower semi-continuous convex functional. Then we have</u> (3.11) <u>with</u> $W=F^*$.

2) If $W=-W$, then the family of sets

$$\Delta^2_{G,w} = \{y \in F \mid w(y) \leq \inf w(G)\} \qquad (w \in W) \tag{3.12}$$

coincides with (3.1), since

$$\Delta^1_{G,w} = \Delta^2_{G,-w} \qquad (w \in W). \tag{3.13}$$

Hence, <u>if</u> $W=-W$, <u>then formula</u> (3.11) <u>is equivalent to</u>

$$\sup h(G) = \sup_{0 \neq w \in W} \inf_{\substack{y \in F \\ w(y) \leq \inf w(G)}} h(y). \tag{3.14}$$

3) Let us define a family of sets $\Delta^3_{G,w} \subseteq F$ $(w \in W)$ by

$$\Delta^3_{G,w} = \{y \in F \mid w(y) = \sup w(G)\} \qquad (w \in W). \tag{3.15}$$

<u>Remark 3.4</u>. a) If $0 \in W$, then $\Delta^3_{G,0} = F$. Hence,

$$\beta = \sup_{0 \neq w \in W} \inf h(\Delta^3_{G,w}). \tag{3.16}$$

b) If F is a locally convex space, then for $0 \neq w \in F^*$ such that $\sup w(G) = +\infty$ we have $\Delta^3_{G,w} = \emptyset$, while for $0 \neq w \in F^*$ such that $\sup w(G) < +\infty$, $\Delta^3_{G,w}$ is a support hyperplane of G.

For a set F and functionals $h, w: F \longrightarrow \overline{R}$, $w \neq 0$, let

$$\gamma(c) = \gamma_w(c) = \inf_{\substack{y \in F \\ w(y) = c}} h(y) \qquad (c \in R); \tag{3.17}$$

in the particular case when F is a linear space and $0 \neq w \in F^{\#}$ (the algebraic conjugate space of F), the functions $\gamma_w: R \longrightarrow \overline{R}$ have been studied in [13]. In contrast with φ_w (of (3.6)), the functions $\gamma_w: R \longrightarrow \overline{R}$ need not be non-decreasing. Nevertheless, we have

<u>Proposition 3.3</u> ([13], lemma 2.1 and remark 2.2 a)). <u>Let</u> F <u>be a linear space</u>, $h: F \longrightarrow R$ <u>a finite convex functional and</u> $w \neq 0$ <u>a linear functional on</u> F. <u>Then the function</u> γ_w <u>of</u> (3.17) <u>is finite, convex and continuous on</u> R. <u>Hence, for</u> $\alpha = \sup h(G)$ <u>and</u> β <u>of</u> (3.16), <u>we have</u> (2.3).

From the above, we obtain

<u>Corollary 3.2</u> ([13], theorem 2.1). <u>Under the assumptions of corollary</u> 3.1, <u>we have</u>

$$\sup h(G) = \sup_{0 \neq w \in F^*} \inf_{\substack{y \in F \\ w(y) = \sup w(G)}} h(y). \tag{3.18}$$

<u>Proof</u>. By proposition 3.3 and corollary 3.1, we have

$$\sup h(G) \geq \sup_{0 \neq w \in F^*} \inf_{\substack{y \in F \\ w(y) = \sup w(G)}} h(y) \geq$$

$$\geq \sup_{0 \neq w \in F^*} \inf_{\substack{y \in F \\ w(y) \geq \sup w(G)}} h(y) = \sup h(G).$$

4) Considering the family of sets

$$\Delta^4_{G,w} = \{y \in F \mid w(y) = \inf w(G)\} \qquad (w \in W), \tag{3.19}$$

we see that (since $F^* = -F^*$) <u>formula</u> (3.18) <u>is equivalent to</u>

$$\sup h(G)=\sup_{0\neq w\in F^*}\ \inf_{\substack{y\in F\\ w(y)=\inf w(G)}} h(y). \tag{3.20}$$

5) Let us define a family of sets $\Delta^5_{G,w}\subseteq F$ $(w\in W)$ by

$$\Delta^5_{G,w}=\{y\in F \mid w(y)>\sup w(G)\} \qquad (w\in W). \tag{3.21}$$

Remark 3.5. a) If $0\in W$, then $\Delta^5_{G,0}=\emptyset$, whence, by (1.3), $\lambda(w)=\inf\emptyset=+\infty$, and hence $\beta=\sup\lambda(W)=+\infty$; thus, in general, for β of (1.4), with $\Delta=\Delta^5$, we need not have the equality corresponding to (3.2) and (3.16). One can avoid this problem by working directly with $W\setminus\{0\}$, instead of W.

b) If F is a locally convex space, then for $0\neq w\in F^*$ such that $\sup w(G)=+\infty$ we have $\Delta^5_{G,w}=\emptyset$, while for $0\neq w\in F^*$ such that $\sup w(G)<+\infty$, $\Delta^5_{G,w}$ is an open half-space in F, supporting the set G (i.e., $G\cap\Delta^5_{G,w}=\emptyset$ and the boundary of $\Delta^5_{G,w}$ is a support hyperplane of G).

From theorem 3.1, there follows

Corollary 3.3. *Under the assumptions of theorem* 3.1, *if* F *is a locally convex space*, $W\subseteq F^*$ *and* $h:F\to\overline{R}$ *is upper semi-continuous, then*

$$\sup h(G)=\sup_{0\neq w\in W}\ \inf_{\substack{y\in F\\ w(y)>\sup w(G)}} h(y). \tag{3.22}$$

Proof. Since h is upper semi-continuous on F, for every subset M of F we have $\inf h(\overline{M})=\inf h(M)$ (where $\overline{M}$ is the closure of M). Thus, observing that for each $0\neq w\in F^*$ we have

$$\overline{\{y\in F \mid w(y)>\sup w(G)\}}=\{y\in F \mid w(y)\geq\sup w(G)\}, \tag{3.23}$$

and applying theorem 3.1, we obtain (3.22).

Similarly, from corollary 3.1 there follows

Corollary 3.4. *Under the assumptions of corollary* 3.1, *if* h *is also continuous on* F, *then we have* (3.22) *with* $W=F^*$.

6) Considering the family of sets

$$\Delta^6_{G,w}=\{y\in F \mid w(y)<\inf w(G)\} \qquad (w\in W), \tag{3.24}$$

we see that *if* $W=-W$, *then formula* (3.22) *is equivalent to*

$$\sup h(G)=\sup_{0\neq w\in W}\ \inf_{\substack{y\in F\\ w(y)<\inf w(G)}} h(y). \tag{3.25}$$

§4. Applications to surrogate dual problems for systems

By a "system" we shall mean a triple $(F \xrightarrow{u} X)$, consisting of two sets F, X and a mapping u of F into X. For a system $(F \xrightarrow{u} X)$, we shall consider now the primal supremization problem

$$(P)=(P_{u^{-1}(\Omega),h}) \qquad \alpha=\alpha_{u^{-1}(\Omega),h} = \sup_{\substack{y\in F\\ u(y)\in\Omega}} h(y), \tag{4.1}$$

i.e., (1.1) with $G=u^{-1}(\Omega)$, where $h:F\to\overline{R}$ and $\Omega\subset X$, $u(F)\cap\Omega\neq\emptyset$ (Ω is called a "target set"). Furthermore, we shall assume that $W\subseteq\overline{R}^X$.

1) Let us define a family of sets $\Delta^1_{u^{-1}(\Omega),w}\subseteq F$ $(w\in W)$ by

$$\Delta^1_{u^{-1}(\Omega),w}=\{y\in F\mid w(u(y))\geqslant \sup w(u(F)\cap\Omega)\} \qquad (w\in W). \tag{4.2}$$

The main tool in studying surrogate duality for (4.2), is the following observation:

Remark 4.1. Surrogate duality for (4.2) is equivalent to surrogate duality for a family of type (3.1). Indeed, clearly, (3.1) is the particular case $X=F$, $u=I_F$ (the identity operator) and $\Omega=G$, of (4.2). Conversely, given (4.2) as above, let

$$V=V_W=\{v_w\mid w\in W\}\subseteq\overline{R}^F, \tag{4.3}$$

where

$$v_w=wu \qquad (w\in W). \tag{4.4}$$

Then, for $G=u^{-1}(\Omega)\subseteq F$, we have

$$\sup w(u(F)\cap\Omega)= \sup_{\substack{y\in F\\ u(y)\in\Omega}} wu(y)=\sup v_w(G) \qquad (w\in W), \tag{4.5}$$

whence, by (4.2) and (3.1),

$$\Delta^1_{u^{-1}(\Omega),w}=\Delta^1_{G,v_w} \qquad (w\in W). \tag{4.6}$$

Thus, from each result of §3, on (3.1), one can obtain a corresponding result for (4.2), replacing G, W and w by $u^{-1}(\Omega)$, V and $v_w=wu$, respectively. Note that the condition $w\neq 0$ of §3 will now be replaced by $wu\neq 0$; also, the assumption occurring in some results of §3, that F is a locally convex space, will now be replaced by the assumption that F and X are locally convex spaces and $u:F\to X$ is a continuous linear mapping (which will ensure that $v_w=wu\in F^*$ for all $w\in X^*$). As an example, let us mention that formula (3.11) will be replaced by

$$\sup_{\substack{y\in F\\ u(y)\in\Omega}} h(G)=\sup_{\substack{w\in W\\ wu\neq 0}} \inf_{\substack{y\in F\\ w(u(y))\geq \sup w(u(F)\cap\Omega)}} h(y). \tag{4.7}$$

2)-6) One can define, similarly to (4.2), families of surrogate constraint sets $\Delta^2_{u^{-1}(\Omega),w},\ldots,\Delta^6_{u^{-1}(\Omega),w} \subseteq F$ $(w\in W)$, where $W\subseteq\overline{R}^X$, corresponding to 2)-6) of §3. For these sets, again, there hold similar remarks to remark 4.1 (mutatis mutandis). We omit the details.

§5. Lagrangian duality for supremization

Motivated by the results of [15] (see e.g. remark 5.3 below), we define here the "Lagrangian dual problem" to (P) of (1.1) (without any assumptions on F, G, h), as the dual problem (1.2), with λ of the form

$$\lambda(w)=\sup w(G)\dot{+}\inf_{y\in F}\{h(y)\dot{+}-w(y)\} \qquad (w\in W); \tag{5.1}$$

we recall that $\dot{+}$ and $\underset{\cdot}{+}$ denote the "upper addition" and the "lower addition" on $\overline{R}$, defined (see [10], [11]) by

$$a\dot{+}b=a\underset{\cdot}{+}b=a+b \quad \text{if} \quad R\cap\{a,b\}\neq\emptyset \quad \text{or} \quad a=b=\pm\infty, \tag{5.2}$$

$$a\dot{+}b=+\infty\ ,\ a\underset{\cdot}{+}b=-\infty \quad \text{if } a=-b=\pm\infty. \tag{5.3}$$

Remark 5.1. In [15] we have used, for problem (1.2), (5.1) above, the term "quasi-Lagrangian dual problem", since it corresponds to the Lagrangian dual problem to (P′) of (1.6), defined (see [12], [21], [22]) as problem (1.2), with λ of the form

$$\lambda'(w)=\inf w(G)\underset{\cdot}{+}\inf_{y\in F}\{h(y)\dot{+}-w(y)\} \qquad (w\in W); \tag{5.4}$$

however, in subsequent papers we have used the term "quasi-Lagrangian" in a different sense, and therefore, we call here problem (1.2), (5.1) above, simply, the "Lagrangian dual problem" to (P) of (i.1). Note that, by remark 1.1 b), this is an "unusual" dual problem to (P).

Now we shall show that the main results of [15] and [14] on Lagrangian type duality for supremization, involving proper lower semi--continuous convex functionals and bounded subsets in locally convex spaces F, can be extended to W-convex functionals on a set F, where $W\subseteq\overline{R}^F$, and to arbitrary subsets of F. We recall that the "W-convex hull" of $h:F\rightarrow\overline{R}$ is the functional $h_{\mathcal{H}(W)}:F\rightarrow\overline{R}$ defined [4] by

$$h_{\mathcal{H}(W)} = \sup_{\substack{w \in W \\ w \leq h}} w, \tag{5.5}$$

and that h is said to be "W-convex" [4], if $h_{\mathcal{H}(W)} = h$. The "W-conjugate" of $h: F \to \overline{R}$ and the "second W-conjugate" of h are (see e.g. [11], [4]) the functionals $h^W: W \to \overline{R}$ and $h^{WW}: F \to \overline{R}$ defined by

$$h^W(w) = \sup_{y \in F} \{w(y) \dot{+} -h(y)\} \qquad (w \in W), \tag{5.6}$$

$$h^{WW}(y) = \sup_{w \in W} \{w(y) \dot{+} -h^W(w)\} \qquad (y \in F). \tag{5.7}$$

<u>Lemma 5.1</u>. *Let* F *be a set*, $W \subseteq \overline{R}^F$, $y_0 \in F$, *and* $h: F \to \overline{R}$ *a* W-*convex functional. Then*

$$h(y_0) = h^{WW}(y_0) = \sup_{w \in W} \{w(y_0) \dot{+} \inf_{y \in F} \{h(y) \dot{+} -w(y)\}\}. \tag{5.8}$$

<u>Proof</u>. By [21], theorem 4.1, for any $h: F \to \overline{R}$ we have $h^{WW} = h_{\mathcal{H}(W+R)} \leq h$, where $R = (-\infty, +\infty)$ is identified with the family of all real-valued constant functionals on F; furthermore, by (5.5), $h_{\mathcal{H}(W)} \leq h_{\mathcal{H}(W+R)}$. Hence, if h is W-convex, then

$$h = h_{\mathcal{H}(W)} \leq h_{\mathcal{H}(W+R)} = h^{WW} \leq h, \tag{5.9}$$

whence $h = h^{WW}$. Finally, by [21], formula (4.26), for any $h: F \to \overline{R}$ and $y_0 \in F$ we have the second equality in (5.8).

<u>Remark 5.2</u>. a) In general, $h^{WW} \neq h_{\mathcal{H}(W)}$ (see [21]); the problem of the existence of a concept of "conjugation" for which the "second conjugate" of h coincides with $h_{\mathcal{H}(W)}$, raised in [22], has been solved, in the affirmative, in [27].

b) When F is a locally convex space, a functional $h: F \to \overline{R}$ is (F^*+R)-convex if and only if <u>either</u> $h \equiv -\infty$, <u>or</u> $h \equiv +\infty$, <u>or</u> $h(F) \subseteq R \cup \{+\infty\}$ <u>and</u> h <u>is lower semi-continuous convex</u> (see [4], p.279). Hence, observing that, by (5.8) we have

$$h^{F^*+R,\, F^*+R} = h^{F^*,\, F^*}, \tag{5.10}$$

it follows that for a locally convex space F and for $W = F^* + R$ (the family of all continuous affine functionals on F), lemma 5.1 yields again [15], lemma 2.1.

We recall that, by [11], formula (3.2), we have

$$a \dot{+} (b \dot{+} c) \geq (a \dot{+} b) \dot{+} c \qquad (a, b, c \in \overline{R}). \tag{5.11}$$

Theorem 5.1. Let F be a set, $W\subseteq\overline{R}^F$, G a subset of F, and $h:F\to\overline{R}$ a W-convex functional. Then

$$\sup h(G)=\sup_{w\in W}\{\sup w(G)\underset{\cdot}{+}\inf_{y\in F}\{h(y)\dot{+}-w(y)\}\}. \tag{5.12}$$

Proof. Let $w\in W$ and $c<\sup w(G)$. Then there exists $g'=g'_{w,c}\in G$ such that $w(g')\geq c$, whence, by [11], formula (2.1) and p.120, corollary, we have $0\geq -w(g')\underset{\cdot}{+}c$. Consequently, by (5.11),

$$\sup h(G)\geq h(g')\geq h(g')\dot{+}(-w(g')\underset{\cdot}{+}c)\geq$$

$$\geq(h(g')\dot{+}-w(g'))\underset{\cdot}{+}c\geq c\underset{\cdot}{+}\inf_{y\in F}\{h(y)\dot{+}-w(y)\}, \tag{5.13}$$

whence, since $c<\sup w(G)$ and $w\in W$ were arbitrary, we obtain

$$\sup h(G)\geq\sup_{w\in W}\{\sup w(G)\underset{\cdot}{+}\inf_{y\in F}\{h(y)\dot{+}-w(y)\}\}; \tag{5.14}$$

note that this is valid for any functional $h:F\to\overline{R}$.

On the other hand, if h is W-convex, then, by lemma 5.1, we have

$$h(g)=\sup_{w\in W}\{w(g)\underset{\cdot}{+}\inf_{y\in F}\{h(y)\dot{+}-w(y)\}\}\leq$$

$$\leq\sup_{w\in W}\{\sup w(G)\underset{\cdot}{+}\inf_{y\in F}\{h(y)\dot{+}-w(y)\}\}\qquad(g\in G). \tag{5.15}$$

Hence, by (5.14) and (5.15), we obtain (5.12).

Remark 5.3. In the particular case when F is a locally convex space and $W=F^*+R$, by remark 5.1 b) we see that theorem 5.1 yields an improvement of [15], theorem 2.1 (namely, the assumption of boundedness of G, made in [15], is omitted).

We recall that, by [11], formulae (4.8) and (2.1), for any set E and any $k:E\to\overline{R}$ and $a,b,c\in\overline{R}$ we have

$$\sup_{x\in E} k(x)\underset{\cdot}{+}c=\sup_{x\in E}\{k(x)\underset{\cdot}{+}c\}, \tag{5.16}$$

$$-(a\underset{\cdot}{+}b)=-a\dot{+}-b. \tag{5.17}$$

Theorem 5.2. Let F be a set, $W\subseteq\overline{R}^F$, $h:F\to\overline{R}$ a W-convex functional, and $f:F\to\overline{R}$ an arbitrary functional. Then

$$\sup_{y\in F}\{h(y)\underset{\cdot}{+}-f(y)\}=\sup_{w\in W}\{f^W(w)\underset{\cdot}{+}-h^W(w)\}. \tag{5.18}$$

Proof. Since h is W-convex, by lemma 5.1, (5.16), (5.6) and (5.17), we obtain

$$\sup_{y\in F}\{h(y)\dot{+}-f(y)\}=$$

$$=\sup_{y\in F}\left\{\sup_{w\in W}[w(y)\dot{+}\inf_{y'\in F}\{h(y')\dot{+}-w(y')\}]\dot{+}-f(y)\right\}=$$

$$=\sup_{y\in F}\left\{\sup_{w\in W}[w(y)\dot{+}-f(y)\dot{+}\inf_{y'\in F}\{h(y')\dot{+}-w(y')\}]\right\}=$$

$$=\sup_{w\in W}\left\{\sup_{y\in F}[w(y)\dot{+}-f(y)\dot{+}\inf_{y'\in F}\{h(y')\dot{+}-w(y')\}]\right\}=$$

$$=\sup_{w\in W}\left\{\sup_{y\in F}\{w(y)\dot{+}-f(y)\}\dot{+}\inf_{y'\in F}\{h(y')\dot{+}-w(y')\}\right\}=$$

$$=\sup_{w\in W}\{f^W(w)\dot{+}-h^W(w)\}.$$

<u>Remark 5.4</u>. a) In the particular case when $f=\chi_G$, the indicator functional of a subset G of F (i.e., $\chi_G(y)=0$ for $y\in G$ and $\chi_G(y)=+\infty$ for $y\in F\setminus G$), we have

$$f^W(w)=\sup_{y\in F}\{w(y)\dot{+}-\chi_G(y)\}=\sup w(G) \qquad (w\in W), \qquad (5.19)$$

and hence theorem 5.2 yields again theorem 5.1.

b) In the particular case when F is a locally convex space and $W=F^*+R$, by remark 5.2 b) we see that theorem 5.2 yields the main result of [14].

REFERENCES

[1] J.-P-Crouzeix, Contributions à l'étude des fonctions quasiconvexes. Thèse. Université de Clermont, 1977.

[2] J.-P.Crouzeix, Continuity and differentiability properties of quasiconvex functions on R^n. In: Generalized concavity in optimization and economics (S.Schaible and W.T.Ziemba, eds.), Acad. Press, New York, 1981, pp.109-130.

[3] J.-P.Crouzeix, A duality framework in quasiconvex programming. In: Generalized concavity in optimization and economics (S.Schaible and W.T.Ziemba, eds.), Acad. Press, New York, 1981, pp.207-225.

[4] S.Dolecki and S.Kurcyusz, On Φ-convexity in extremal problems. SIAM J.Control Optim. 16(1978), 277-300.

[5] K.Fan, On the Krein-Milman theorem. In: Convexity (V.Klee, ed.). Proc. Symposia Pure Math. 7. Amer.Math.Soc., Providence, 1963, pp.211-220.

[6] C.Franchetti and I.Singer, Deviation and farthest points in normed linear spaces. Rev.Roum.Math.Pures Appl. 24(1979), 373-381.

[7] C.Franchetti and I.Singer, Best approximation by elements of caverns in normed linear spaces. Boll.Un.Mat.Ital. (5) 17-B(1980), 33-43.

[8] P.-J.Laurent, Approximation et optimisation. Hermann, Paris, 1972.

[9] J.-E.Martínez-Legaz, Un concepto generalizado de conjugación. Applicación a las funciones quasiconvexas. Thesis, Barcelona, 1981.

[10] J.-J.Moreau, Fonctionnelles convexes. Sémin.Eq.Dériv.Part.Collège de France, Paris, 1966-1967, no.2.

[11] J.-J.Moreau, Inf-convolution, sous-additivité, convexité des fonctions numériques. J.math.pures et appl. 49 (1950), 109-154.
[12] I.Singer, Some new applications of the Fenchel-Rockafellar duality theorem: Lagrange multipier theorems and hyperplane theorems for convex optimization and best approximation. Nonlinear Anal.Theory, Methods, Appl. 3(1979), 239-248.
[13] I.Singer, Maximization of lower semi-continuous convex functionals on bounded subsets of locally convex spaces. I: Hyperplane theorems. Appl.Math.Optim. 5(1979), 349-362.
[14] I.Singer, A Fenchel-Rockafellar type duality theorem for maximization. Bull.Austral.Math.Soc. 20(1979), 193-198.
[15] I.Singer, Maximization of lower semi-continuous convex functionals on bounded subsets of locally convex spaces. II: Quasi-Lagrangian duality theorems. Result. Math. 3(1980), 235-248.
[16] I.Singer, Minimization of continuous convex functionals on complements of convex subsets of locally convex spaces. Math.Operationsforsch. Stat. Ser. Optim. 11 (1980), 221-234.
[17] I.Singer, Optimization by level set methods. I :Duality formulae. In: Optimization: Theory and algorithms (Proceedings.Internat. Confer. in Confolant, March 1981; J.-B.Hiriart-Urruty, W.Oettli and J.Stoer, eds.), Lecture Notes in Pure and Appl. Math. 86, Marcel Dekker, New York, 1983, pp.13-43.
[18] I.Singer, Optimization by level set methods. III: Characterizations of solutions in the presence of duality. Numer.Funct.Anal. Optim. 4 (1981-1982), 151-170.
[19] I.Singer, Optimization by level set methods. IV: Generalizations and complements. Numer. Funct. Anal. Optim. 4 (1981-1982), 279--310.
[20] I.Singer, A general theory of surrogate dual and perturbational extended surrogate dual optimization problems. J.Math.Anal. Appl. 104(1984), 351-389.
[21] I.Singer, Generalized convexity, functional hulls and applications to conjugate duality in optimization. In: Selected topics in operations research and mathematical economics (G.Hammer and D.Pallaschke, eds.). Lecture Notes in Econ. and Math. Systems 226, Springer-Verlag, Berlin-Heidelberg-New York-Tokyo, 1984, pp.49-79.
[22] I.Singer, A general theory of dual optimization problems. J.Math. Anal.Appl. (to appear). Preprint INCREST (Bucharest) 67/1984.
[23] I.Singer, Optimization by level set methods. VI: Generalizations of surrogate type reverse convex duality (to appear). Preprint INCREST (Bucharest) 74/1985.
[24] J.F.Toland, Duality in nonconvex optimization. J.Math.Anal.Appl. 66(1978), 399-415.
[25] J.F.Toland, A duality principle for non-convex optimisation and the calculus of variations. Arch.Rat.Mech.Anal. 71(1979), 41-61.
[26] H.Tuy, Global maximization of a convex function, over a closed, convex, not necessarily bounded set. Preprint CEREMADE (Paris) 8223/1982.
[27] J.-E.Martínez-Legaz and I.Singer, Dualities between complete lattices (in preparation).

Addendum. 1) For problem (1.1), a theory of perturbational Lagrangian duality type, different from that of Toland [24], has been given in P. Kanniappan, "Fenchel-Rockafellar type duality for a non-convex, non-differential optimization problem," J. Math. Anal. Appl. 97(1983), 266-276. Both theories can be extended to our setting.

2) The formula for $(h-f)^*$, given recently in J.-B. Hiriart-

Urruty, "A general formula on the conjugate of the difference of functions," Sémin. Anal. Numér. 1984 (to appear in Canad. Math. Bull.), can be also extended, with slight changes in the proof, to a formula for $(h \dot{+} -f)^W$, where $h : F \to \bar{R}$ is W-convex and $f : F \to \bar{R}$ is arbitrary, which, in turn, implies again theorem 5.2.

On the Asymptotic Behavior of Almost-Orbits of Commutative Semigroups in Banach Spaces

WATARU TAKAHASHI Department of Information Science, Tokyo Institute of Technology, Tokyo, Japan

JONG YEOUL PARK Department of Mathematics, Busan National University, Busan, Republic of Korea

ABSTRACT: Let C be a closed convex subset of a real Banach space E, G a commutative semigroup with identity, and $S = \{S(t): t \in G\}$ a family of nonexpansive mappings of C into itself such that $S(t+s)x = S(t)S(s)x$ for all $t, s \in G$ and $x \in C$. Then we deal with the asymptotic behavior of almost-orbits $\{u(t): t \in G\}$ of the semigroup $S = \{S(t): t \in G\}$. That is, we first prove that if $\{u(t): t \in G\}$ is an almost-orbit of $S = \{S(t): t \in G\}$, then the closed convex set

$$\bigcap_{s \in G} \overline{co}\{u(t): t \geqq s\} \cap F(S)$$

consists of at most one point, where $\overline{co}\{u(t): t \geqq s\}$ is the closed convex hull of $\{u(t): t \geqq s\}$ and $F(S)$ is the set of common fixed points of $S(t)$, $t \in G$. Furthermore this result is applied to study the problem of weak convergence of the net $\{u(t): t \in G\}$.

1. Introduction

Let C be a nonempty closed convex subset of a real Banach space E. Then a mapping T: C → C is called nonexpansive on C if

$$|Tx - Ty| \leqq |x - y| \quad \text{for all } x, y \in C.$$

Let $S = \{S(t) : t \geqq 0\}$ be a family of nonexpansive mappings of C into itself such that $S(0) = I$, $S(t+s) = S(t)S(s)$ for all $t, s \in [0, \infty)$ and $S(t)x$ is continuous in $t \in [0, \infty)$ for each $x \in C$. Then S is said to be a nonexpansive semigroup on C. Recently, Miyadera and Kobayashi [15] introduced the notion of an almost-orbit of a nonexpansive semigroup on C and established the weak and strong almost convergence of such an almost-orbit in a Banach space. See also Bruck [7] for an almost-orbit of a nonexpansive mapping. The notion of an almost-orbit is useful in applications. For example, Consider the initial value problem:

$$\frac{du(t)}{dt} + Au(t) \ni f(t), \qquad t > 0 \qquad \ldots\ldots (1)$$
$$u(0) = x,$$

where A is an m-accretive operator in E, $f \in L^1(0, \infty; E)$ and $x \in \overline{D(A)}$. Then it is well known that (1) has a unique integral solution [3]. And also the integral solution $u(t)$ of (1) is an almost-orbit of a nonexpansive semigroup on $\overline{D(A)}$ generated by $-A$; see Section 2 or [15].

In this paper, we introduce the notion of an almost-orbit of a family $S = \{S(t) : t \in G\}$ of nonexpansive mappings of C into itself such that G is a commutative semigroup with identity and $S(t+s)x = S(t)S(s)x$ for all $t, s \in G$ and $x \in C$, and then we study the asymptotic behavior of such an almost-orbit in a Banach space. First, we prove that if $\{u(t): t \in G\}$ is an almost-orbit of S and G is directed by an order relation "$\geqq$" defined by $t \leqq s$ if and only if there is $u \in G$ with $t + u = s$, then the closed convex set

$$\bigcap_{s \in G} \overline{co}\{u(t) : t \geqq s\} \cap F(S)$$

consists of at most one point, where $\overline{co}\{u(t) : t \geqq s\}$ is the closed convex hull of $\{u(t) : t \geqq s\}$ and $F(S)$ is the set of common fixed points of $S(t)$, $t \in G$. This result is applied to study the problem of weak convergence of the net $\{u(t) : t \in G\}$. We also prove that if P is the metric projection of E onto $F(S)$, then the strong $\lim_t Pu(t)$ exists. Our proofs employ the methods of Hirano-Takahashi [11], Hulbert-Reich [12], Kido-Takahashi [13], Lau-Takahashi [14], Miyadera-Kobayashi [15] and Takahashi [18].

2. Preliminaries

Let E be a real Banach space and let E^* be its dual, that is, the space of all continuous linear functionals x^* on E. The value of $x^* \in E^*$ at $x \in E$ will be denote by $\langle x, x^* \rangle$. With each $x \in E$, we associate the set

$$J(x) = \{x^* \in E^* : \langle x, x^* \rangle = \|x\|^2 = \|x^*\|^2\}.$$

Using the Hahn-Banach theorem, it is immediately clear that $J(x) \neq \phi$ for any $x \in E$. Then multi-valued operator $J : E \to E^*$ is called the _duality mapping_ of E. Let $U = \{x \in E : \|x\| = 1\}$ be the unit sphere of E. Then a Banach space E is said to be _smooth_ provided

$$\lim_{t \to 0} \frac{\|x + ty\| - \|x\|}{t}$$

exists for each $x, y \in U$. When this is the case, the norm of E is said to be _Gâteaux differentiable_. It is said to be _Fréchet differentiable_ if for each x in U, this limit is attained uniformly for y in U. The space E is said to have _uniformly Gâteaux differentiable_ norm if for each $y \in U$, the limit is attained uniformly for $x \in U$. It is well known that if E is smooth, then the duality mapping J is single value. It is also known that if E has a Fréchet differentiable norm, then J is norm to norm continuous; see [5] or [9] for more details. Let D be a subset of E. Then we denote by d(D) the diameter of D. A point $x \in D$ is a _diametral point_ of D provided

$$\sup\{\|x - y\| : y \in D\} = d(D).$$

A closed convex subset C of a Banach space E is said to have _normal structure_ if for each closed bounded convex subset K of C, which contains at least two points, there exists an element of K which is not a diametral point of K. It is well known that a closed convex subset of a uniformly convex Banach space has normal structure and a compact convex subset of a Banach space has normal structure. When $\{x_\alpha\}$ is a net in E, then $x_\alpha \to x$ (resp. $x_\alpha \rightharpoonup x$)

will denote norm (resp. weak) convergence of the net $\{x_\alpha\}$ to x. For x and y in E, let $\langle y, x\rangle_s = \max\{\langle y, j\rangle : j \in J(x)\}$. Let C be a closed convex set in E and let $S = \{S(t): t \geqq 0\}$ be a nonexpansive semigroup on C. Then, according to Miyadera-Kobayashi [15], a continuous function u: $[0, \infty) \to C$ is called an almost-orbit of $S = \{S(t): t \geqq 0\}$ if

$$\lim_{s \to \infty} \left(\sup_{t \geqq 0} | u(t+s) - S(t)u(s) | \right) = 0 .$$

An operator $A \subset E \times E$ with domain D(A) and range R(A) is said to be accretive if $\langle y_1 - y_2, x_1 - x_2\rangle_s \geqq 0$ for all $x_i \in D(A)$ and $y_i \in Ax_i$, $i = 1, 2$. An accretive operator $A \subset E \times E$ is m-accretive if $R(I+rA) = E$ for all $r > 0$, where I is the identity operator. Let $A \subset E \times E$ be an m-accretive operator and let u(t) be an integral solution of (1). Then we know that

$$|u(t) - u|^2 - |u(s) - u|^2 \leqq 2\int_s^t \langle f(\tau) - v, u(\tau) - u\rangle_s d\tau$$

holds for each $(u, v) \in A$ and $0 \leqq s \leqq t < +\infty$. We also know that if v(t) is another integral solution of (1) corresponding to $g \in L^1(0, \infty; E)$ and $y \in \overline{D(A)}$, then

$$|u(t) - v(t)| - |u(s) - v(s)| \leqq \int_s^t |f(\tau) - g(\tau)| d\tau, \quad \ldots (2)$$

whenever $0 \leqq s \leqq t < +\infty$. If $A^{-1}0 = \{x : x \in D(A),\ Ax \ni 0\}$ is nonempty, we can take $z \in A^{-1}0$ and $g(t) \equiv 0$ in (2) to obtain

$$|u(t) - z| - |u(s) - z| \leqq \int_s^t |f(\tau)|\, d\tau. \quad \ldots (3)$$

Consequently, $u(t)$ is bounded on $[0, \infty)$ and the function $t \mapsto |u(t) - z| - \int_0^t |f(\tau)|\, d\tau$ is nonincreasing on $[0, \infty)$. Then, since $f \in L^1(0, \infty; E)$, we deduce that

$$\lim_{t\to\infty} |u(t) - z| = \rho(z)$$

exists for each $z \in A^{-1}0$. Finally, let $\{S(t) : t \geqq 0\}$ be a nonexpansive semigroup generated by $-A$ and let $u(t)$ be the integral solution of (1). Then we obtain that

$$|S(t)u(s) - u(t+s)| \leqq \int_s^{t+s} |f(\tau)|\, d\tau, \quad \ldots (4)$$

whenever $0 \leqq s,\ t < +\infty$. In fact, the function $t \mapsto v(t) = u(t + s)$ is an integral solution of (1) corresponding to $g(t) = f(t + s)$ and $u(s) \in \overline{D(A)}$. From (4), we obtain that

$$\lim_{s\to\infty} \sup_{t \geqq 0} |S(t)u(s) - u(t+s)| = 0. \quad \ldots (5)$$

This implies that the integral solution of (1) is an almost-orbit of the nonexpansive semigroup $S = \{S(t): t \geqq 0\}$ generated by $-A$.

3. Lemmas

Unless other specified, G denotes a commutative semigroup with identity and $S = \{S(t): t \in G\}$ a family of nonexpansive mappings of a closed convex subset C of a real Banach space E into C such that $S(t+s)x = S(t)S(s)x$ for all $t, s \in G$ and $x \in C$. If G is directed as in Introduction, then a function $u: G \to C$ is said to be an almost-orbit of $S = \{S(t): t \in G\}$ if

$$\lim_s \left(\sup_t \| u(t+s) - S(t)u(s) \| \right) = 0 .$$

Lemma 1. Let $\{u(t): t \in G\}$ and $\{v(t): t \in G\}$ be almost-orbits of $S = \{S(t): t \in G\}$. Then, the limit of $\| u(t) - v(t) \|$ exists. In particular, for every $z \in F(S)$, the limit of $\| u(t) - z \|$ exists.

Proof. Put $\phi(s) = \sup_t \| u(t+s) - S(t)u(s) \|$ and $\Psi(s) = \sup_t \| v(t+s) - S(t)v(s) \|$ for $s \in G$. Then $\lim_s \phi(s) = 0$ and $\lim_s \Psi(s) = 0$. Since

$$\begin{aligned} \| u(t+s) - v(t+s) \| &\leqq \| u(t+s) - S(t)u(s) \| \\ &\quad + \| S(t)u(s) - S(t)v(s) \| + \| S(t)v(s) - v(t+s) \| \\ &\leqq \phi(s) + \Psi(s) + \| u(s) - v(s) \| , \end{aligned}$$

we have

$$\inf_t \sup_{t \leqq w} |u(w) - v(w)| \leqq \phi(s) + \Psi(s) + |u(s) - v(s)|$$

for every $s \in G$ and then

$$\inf_t \sup_{t \leqq w} |u(w) - v(w)| \leqq \sup_t \inf_{t \leqq s} |u(s) - v(s)|.$$

Thus, the limit of $|u(t) - v(t)|$ exists. Putting $v(t) = z$ for every $t \in G$, it is obvious that the limit of $|u(t) - z|$ exists.

<u>Lemma</u> 2. Let C be a closed convex subset of a real reflexive Banach space which has normal structure and let $\{u(t): t \in G\}$ be an almost-orbit of $S = \{S(t): t \in G\}$. Then $F(S) \neq \phi$ if and only if there exists $t_0 \in G$ such that $\{u(t): t \geqq t_0\}$ is bounded.

<u>Proof</u>. Suppose that $\{u(t): t \geqq t_0\}$ is bounded. Then, since $\{u(t): t \in G\}$ is an almost-orbit of $S = \{S(t): t \in G\}$, there exists $t_1 \in G$ with $t_1 \geqq t_0$ such that $\{S(t)u(t_1): t \in G\}$ is bounded. Therefore by [19], $F(S)$ is nonempty. Conversely, if $F(S) \neq \phi$, then since the limit of $|u(t) - z|$ exists for each $z \in F(S)$, it follows that there exists $t_0 \in G$ such that $\{u(t): t \geqq t_0\}$ is bounded.

<u>Lemma</u> 3. Let E be a uniformly convex Banach space, let $\{u(t): t \in G\}$ be an almost-orbit of $S = \{S(t): t \in G\}$ and $F(S) \neq \phi$. Let $y \in F(S)$ and $0 < \alpha \leqq \beta < 1$. Then, for any $\varepsilon > 0$, there is $t_0 \in G$ such that

$$\| S(t)(\lambda u(s) + (1-\lambda)y) - (\lambda S(t)u(s) + (1-\lambda)y) \| < \varepsilon$$

for all $s \in G$ with $s \geqq t_0$, $t \in G$ and $\lambda \in R$ with $\alpha \leqq \lambda \leqq \beta$.

Proof. Let $r = \lim_t \| u(t)-y \|$ and $r > 0$. Then we can choose $d > 0$ so small that

$$(r + d)(1 - c\,\delta(\frac{\varepsilon}{r+d})) = r_0 < r,$$

where δ is the modulus of convexity of the norm and

$$c = \min\{2\lambda(1-\lambda) : \alpha \leqq \lambda \leqq \beta\}.$$

Let $a > 0$ with $r_0 + 2a < r$. Then we can choose $t_0 \in G$ such that $\| u(s) - y \| \geqq r - a$ for all $s \geqq t_0$ and $\| S(t)u(s) - u(t+s) \| < a$ for all $t \in G$ and $s \geqq t_0$.
Suppose that

$$\| S(t)(\lambda u(s) + (1-\lambda)y) - (\lambda S(t)u(s) + (1-\lambda)y) \| \geqq \varepsilon$$

for some $s \in G$ with $s \geqq t_0$, $t \in G$ and $\lambda \in R$ with $\alpha \leqq \lambda \leqq \beta$.
Put $u = (1-\lambda)(S(t)z - y)$ and $v = \lambda(S(t)u(s) - S(t)z)$, where $z = \lambda u(s) + (1-\lambda)y$. Then $\| u \| \leqq (1-\lambda)\lambda \| u(s) - y \|$ and $\| v \| \leqq \lambda \| u(s) - z \| = \lambda(1-\lambda) \| y - u(s) \|$. We also have that $\| u-v \| = \| S(t)z - (\lambda S(t)u(s) + (1-\lambda)y \| \geqq \varepsilon$ and $\lambda u + (1-\lambda)v = \lambda(1-\lambda)(S(t)u(s)-y)$. So by using the Lemma in [10], we have

$$\lambda(1-\lambda)|S(t)u(s)-y| = |\lambda u+(1-\lambda)v|$$

$$\leqq \lambda(1-\lambda)|u(s)-y|(1-2\lambda(1-\lambda)\delta(\frac{\varepsilon}{|u(s)-y|}))$$

$$\leqq \lambda(1-\lambda)(r+d)(1-c\delta(\frac{\varepsilon}{r+d}))$$

$$= \lambda(1-\lambda)r_0$$

and hence $|S(t)u(s)-y| \leqq r_0$. This implies

$$|u(t+s)-y| < |S(t)u(s)-y|+a \leqq r_0+a < r-a.$$

On the other hand, since $|u(t+s)-y| \geqq r-a$, this is a contradiction. In the case when $r = 0$, for any $t, s \in G$, $y \in F(S)$ and $\lambda \in R$ with $0 \leqq \lambda \leqq 1$,

$$|S(t)(\lambda u(s)+(1-\lambda)y)-(\lambda S(t)u(s)+(1-\lambda)y)|$$

$$\leqq \lambda|S(t)(\lambda u(s)+(1-\lambda)y)-S(t)u(s)|$$
$$+(1-\lambda)|S(t)(\lambda u(s)+(1-\lambda)y)-y|$$

$$\leqq \lambda|\lambda u(s)+(1-\lambda)y-u(s)|+(1-\lambda)|\lambda u(s)+(1-\lambda)y-y|$$

$$= 2\lambda(1-\lambda)|y-u(s)|.$$

So, we obtain the desired result.

Let x and y be elements of E. Then we denote by $[x, y]$ the set $\{\lambda x+(1-\lambda)y : 0 \leqq \lambda \leqq 1\}$.

Lemma 4 [14]. Let C be a closed convex subset of a uniformly convex Banach space E with a Frechet differentiable norm and $\{x_\alpha\}$ a bounded net in C. Let $z \in \bigcap_\beta \operatorname{co}\{x_\alpha : \alpha \geqq \beta\}$, $y \in C$ and $\{y_\alpha\}$ a net of elements in C with $y_\alpha \in [y, x_\alpha]$ and

$$\| y_\alpha - z \| = \min\{\| u-z \| : u \in [y, x_\alpha]\}.$$

If $y_\alpha \to y$, then $y = z$.

By using Lemmas 3 and 4, we can prove the following:

Lemma 5. Let E be a uniformly convex Banach space with a Fréchet differentiable norm, let $\{u(t): t \in G\}$ be an almost-orbit of $S = \{S(t): t \in G\}$ and $F(S) \neq \phi$. Let

$$z \in \bigcap_{s \in G} \overline{\operatorname{co}}\{u(t) : t \geqq s\} \cap F(S)$$

and $y \in F(S)$. Then, for any positive number ε, there is $t_0 \in G$ such that

$$\langle u(t)-y, J(y-z)\rangle \leqq \varepsilon \| y-z \|$$

for every $t \geqq t_0$.

Proof. Since $F(S) \neq \phi$, we may suppose that $\{u(t): t \in G\}$ is bounded. Let $z \in \bigcap_{s \in G} \overline{\operatorname{co}}\{u(t) : t \geqq s\} \cap F(S)$, $y \in F(S)$ and $\varepsilon > 0$. If $y = z$, Lemma 5 is obvious. So, let $y \neq z$. For any $t \in G$, define a unique element y_t such that $y_t \in [y, u(t)]$ and

$\|y_t - z\| = \min\{\|u - z\| : u \in [y, u(t)]\}$. Then, since $y \neq z$, by Lemma 4 we have $y_t \not\to y$. So, we obtain $c > 0$ such that for any $t \in G$, there is $t' \in G$ with $t' \geqq t$ and $\|y_{t'} - y\| \geqq c$. Setting

$$y_{t'} = a_{t'}u(t') + (1-a_{t'})y, \ 0 \leqq a_{t'} \leqq 1,$$

we also obtain $c_0 > 0$ so small that $a_{t'} \geqq c_0$ for every t'. In fact, since

$$c \leqq \|y_{t'}-y\| = a_{t'}\|u(t')-y\| \leqq a_{t'} \cdot \sup_{t \in G} \|u(t)-y\|,$$

we may put $c_0 = c/ \sup_{t \in G} \|u(t)-y\|$. Since the limit of $\|u(t)-y\|$ exists, putting $k = \lim_t \|u(t)-y\|$, we have $k > 0$. If not, we have $u(t) \to y$ and hence $y_t \to y$, which contradicts $y_t \not\to y$. Let r be a positive number such that $\varepsilon > r$ and $k > 2r$. And choose $a > 0$ so small that

$$(R + a)(1 - \delta(\frac{c_0 r}{R+a})) < R,$$

where δ is the modulus of convexity of the norm and $R = \|z - y\|$. Then, by Lemma 3, there exists $s_1 \in G$ such that

$$\|S(s)(c_0u(t)+(1-c_0)y) - (c_0S(s)u(t)+(1-c_0)y)\| < a \quad \ldots (6)$$

for all $s \in G$ and $t \geq s_1$. We also choose $s_2 \in G$ such that $|u(t)-y| \geq 2r$ for every $t \geq s_2$ and $|u(t+s)-S(t)u(s)| < r$ for every $t \in G$ and $s \geq s_2$. Fix t' with $t' \geq \max(s_1, s_2)$. Then since $a_{t'} \geq c_0$, we have

$$c_0 u(t')+(1-c_0)y \in [y, a_{t'}u(t')+(1-a_{t'})y] = [y, y_{t'}].$$

Hence

$$|c_0 u(t')+(1-c_0)y-z| \leq \max(|z-y|, |z-y_{t'}|)$$

$$= |z-y| = R.$$

By using (6), we obtain

$$|c_0 S(s)u(t')+(1-c_0)y-z| \leq |S(s)(c_0 u(t')+(1-c_0)y)-z| + a$$

$$\leq |c_0 u(t')+(1-c_0)y-z| + a \leq R+a$$

for every $s \in G$. On the other hand, since $|y-z| = R < R+a$ and

$$|c_0 S(s)u(t')+(1-c_0)y-y| = c_0|S(s)u(t')-y|$$

$$\geq c_0(|u(s+t')-y|-r) \geq c_0 r$$

for all $s \in G$, we have, by uniform convexity,

$$\left\| \frac{1}{2}((c_0 S(s)u(t') + (1-c_0)y - z) + (y-z)) \right\|$$

$$\leqq (R+a)(1 - \delta(\frac{c_0 r}{R+a})) < R$$

and hence

$$\left\| \frac{c_0}{2} S(s)u(t') + (1 - \frac{c_0}{2})y - z \right\| < R$$

for all $s \in G$. This implies that if $u_s = \frac{c_0}{2} S(s)u(t') + (1 - \frac{c_0}{2})y$, then $\| u_s + \alpha(y - u_s) - z \| \geqq \| y - z \|$ for all $\alpha \geqq 1$. By Theorem 2.5 in [8], we have

$$\langle u_s + \alpha(y - u_s) - y, J(y-z) \rangle \geqq 0$$

and hence $\langle u_s - y, J(y-z) \rangle \leqq 0$. Then $\langle S(s)u(t') - y, J(y-z) \rangle \leqq 0$. Therefore

$$\langle u(s+t') - y, J(y-z) \rangle$$

$$\leqq \| u(s+t') - S(s)u(t') \| \, \| y - z \| + \langle S(s)u(t') - y, J(y-z) \rangle$$

$$< \varepsilon \| y - z \|$$

for all $s \in G$. This completes the proof.

4. Asymptotic behaviour

In this section, we study the asymptotic behaviour of an almost-orbit $\{u(t) : t \in G\}$ of $S = \{S(t) : t \in G\}$.

Theorem 1. Let E be a uniformly convex Banach space with a Fréchet differential norm, let $\{u(t) : t \in G\}$ be an almost-orbit of $S = \{S(t) : t \in G\}$ and $F(S) \neq \phi$. Then, the set

$$\bigcap_{s \in G} \overline{co}\{u(t) : t \geqq s\} \cap F(S)$$

consists of at most one point.

Proof. Let $y, z \in \bigcap_{s \in G} \overline{co}\{u(t) : t \geqq s\} \cap F(S)$. Then, since $\frac{y+z}{2} \in F(S)$, it follows from Lemma 5 that for any $\varepsilon > 0$, there is $t_0 \in G$ such that

$$\langle u(t+t_0) - \frac{y+z}{2}, J(\frac{y+z}{2} - z)\rangle \leqq \varepsilon \left| \frac{y+z}{2} - z \right|$$

for every $t \in G$. Since $y \in \overline{co}\{u(t+t_0) : t \in G\}$, we have

$$\langle y - \frac{y+z}{2}, J(\frac{y+z}{2} - z)\rangle \leqq \varepsilon \left| \frac{y+z}{2} - z \right|$$

and hence $\langle y-z, J(y-z)\rangle \leqq 2\varepsilon |y-z|$. Then we have $|y-z| \leqq 2\varepsilon$. Since ε is arbitrary, we have $y = z$.

By using Theorem 1, we study the problem of the weak convergence of $\{u(t) : t \in G\}$. For a function $u:G \to C$, let $w(u)$ denote the set of all weak limit points of subnets of the net $\{u(t): t \in G\}$.

Theorem 2. Let E be a uniformly convex Banach space with a Fréchet differentiable norm, let $\{u(t): t \in G\}$ be an almost-orbit of $S = \{S(t): t \in G\}$, and $F(S) \neq \phi$. If $w(u) \subset F(S)$, then the net $\{u(t): t \in G\}$ converges weakly to some $z \in F(S)$.

Proof. Since $F(S) \neq \phi$, $\{u(t) : t \in G\}$ is bounded. So, the net $\{u(t): t \in G\}$ must contain a subnet $\{u(t_\alpha)\}$ of $\{u(t)\}$ which converges weakly to some $z \in C$. Since $w(u) \subset F(S)$ and $z \in \bigcap_{s \in G} \overline{co}\{u(t) : t \geq s\}$, we obtain

$$z \in \bigcap_{s \in G} \overline{co}\{u(t) : t \geq s\} \cap F(S) .$$

Therefore, it follows from Theorem 1 that $\{u(t) : t \in G\}$ converges weakly to $z \in F(S)$.

Corollary 1. Let E be a uniformly convex Banach space with a Fréchet differentiable norm, let $\{u(t): t \in G\}$ be an almost-orbit of $S = \{S(t): t \in G\}$, and $F(S) \neq \phi$. If $\lim_t \| u(t+h)-u(t) \| = 0$ for all $h \in G$, then the net $\{u(t) : t \in G\}$ converges weakly to some $y \in F(S)$.

Proof. By Theorem 2, it suffices to show that $w(u) \subset F(S)$. Let $\{u(t_\alpha)\}$ be a subnet converging weakly to some $y \in C$.

Let $\varepsilon > 0$. Then

$$|u(t+t_\alpha)-S(t)u(t_\alpha)| < \frac{\varepsilon}{2} \quad \text{and} \quad |u(t+t_\alpha)-u(t_\alpha)| < \frac{\varepsilon}{2}$$

for all large enough α and $t \in G$. Since

$$|S(t)u(t_\alpha)-u(t_\alpha)| \leqq |S(t)u(t_\alpha)-u(t+t_\alpha)| + |u(t+t_\alpha)-u(t_\alpha)|,$$

we obtain $|S(t)u(t_\alpha)-u(t_\alpha)| < \varepsilon$ for all large enough α and $t \in G$. Since $(I-S(t))$ is demiclosed [5], we have $S(t)y = y$. Thus $y \in F(S)$.

The following theorem is a generalization of Moroșanu's result [16], which has been prove in the case when E is a Hilbert space. The proof is similar to [12].

<u>Theorem</u> 3. Let E be a uniformly convex Banach space, let $\{u(t): t \in G\}$ be almost-orbit of $S = \{S(t): t \in G\}$, and $F(S) \neq \phi$. Let P be the metric projection of E onto $F(S)$. Then the strong $\lim_t Pu(t)$ exists and $\lim_t Pu(t) = z_0$, where z_0 is a unique element of $F(S)$ such that

$$\lim_t |u(t)-z_0| = \min\{\lim_t |u(t)-z| : z \in F(S)\}.$$

<u>Proof</u>. Since $F(S) \neq \phi$, we know that $\{u(t) : t \in G\}$ is bounded and $\lim_t |u(t)-z| = \rho(z)$ exists for each $z \in F(S)$. Let $R = \min\{\rho(z) : z \in F(S)\}$. Then, since ρ is convex and continuous on $F(S)$ and $\rho(z) \to \infty$ as $|z| \to \infty$, there exists

$z_0 \in F(S)$ such that $\rho(z_0) = R$; see [4:p. 79]. On the other hand, since $|u(t)-Pu(t)| \leqq |u(t)-y|$ for all $t \in G$ and $y \in F(S)$, we have

$$\inf_t \sup_{t \leqq s} |u(s) - Pu(s)| \leqq R.$$

Suppose that $\inf_t \sup_{t \leqq s} |u(s) - Pu(s)| < R$. Then we can choose $\varepsilon > 0$ and $t_0 \in G$ such that $|u(s) - Pu(s)| \leqq R - \varepsilon$ for all $s \geqq t_0$. Observe that as in the proof of Lemma 1

$$|u(t) - Pu(s)| - |u(s) - Pu(s)| \leqq \phi(s)$$

for all $t \geqq s$. Since $\lim_s \phi(s) = 0$, we can choose $s \geqq t_0$ such that

$$|u(t)-Pu(s)| \leqq |u(s)-Pu(s)| + \frac{\varepsilon}{2} \leqq R - \varepsilon + \frac{\varepsilon}{2} = R - \frac{\varepsilon}{2}$$

for all $t \geqq s$. Thus $\lim_t |u(t)-Pu(s)| \leqq R - \frac{\varepsilon}{2} < R$. This is a contradiction. So we conclude that

$$\inf_t \sup_{t \leqq s} |u(s) - Pu(s)| = R.$$

We claim that $\lim_t Pu(t) = z_0$. If not, then there exists $\varepsilon > 0$ such that for any $t \in G$, $|Pu(t') - z_0| \geqq \varepsilon$ for some $t' \geqq t$. Let δ denote the modulus of convexity of E. There is a positive a such that

$$(R + a)(1 - \delta(\frac{\varepsilon}{R+a})) = R_1 < R.$$

We also have $|u(t') - Pu(t')| \leqq R+a$ and $|u(t')-z_0| \leqq R+a$ for all large enough t'. Therefore we have

$$|u(t') - \frac{Pu(t')+z_0}{2}| \leqq (R+a)(1 - \delta(\frac{\varepsilon}{R+a})) = R_1.$$

Since the points $w_{t'} = (Pu(t')+z_0)/2$ belong to F(S), again as in the proof of Lemma 1,

$$|u(t) - w_{t'}| \leqq |u(t') - w_{t'}| + \phi(t')$$

for all $t \geqq t'$. Also, since $\lim_s \phi(s) = 0$, there is t' such that

$$|u(t) - w_{t'}| \leqq |u(t') - w_{t'}| + \frac{R-R_1}{2}$$

$$\leqq R_1 + \frac{R-R_1}{2} = \frac{R+R_1}{2} < R$$

for all $t \geqq t'$. Then we obtain $\rho(w_{t'}) < R$. This is a contradiction. Thus we have $\lim_t Pu(t) = z_0$. Consequently, it follows that an element $z_0 \in F(S)$ with $\rho(z_0) = \min\{\rho(z) : z \in F(S)\}$ is unique.

By Theorem 2 and Theorem 3, we have the following:

Corollary 2. Let H be a Hilbert space, let $\{u(t): t \in G\}$ be an almost-orbit of $S = \{S(t): t \in G\}$, and $F(S) \neq \phi$. Then $u(t) \rightharpoonup y$ if and only if $u(t+h) - u(t)) \rightharpoonup 0$ for all $h \in G$. In this case, $y \in F(S)$ and $Pu(t) \to y$.

Proof. We need only prove the "if" part. Since $\{u(t)\}$ is bounded, there exists a subnet $\{u(t_\alpha)\}$ with $u(t_\alpha) \rightharpoonup z$. If $u \in F(S)$, then we have

$$0 \leqq \| u(t_\alpha) - z \|^2 - \| S(t)u(t_\alpha) - S(t)z \|^2$$

$$= \| u(t_\alpha) - u \|^2 + 2\langle u(t_\alpha)-u,\ u-z\rangle + \| u-z \|^2 - \| S(t)u(t_\alpha)-u \|^2 - 2\langle S(t)u(t_\alpha)-u,\ u-S(t)z\rangle - \| u-S(t)z \|^2$$

$$= \| u(t_\alpha)-u \|^2 - \| S(t)u(t_\alpha)-u \|^2 + 2\langle u(t_\alpha)-u,\ S(t)z-z\rangle + 2\langle u(t_\alpha)-S(t)u(t_\alpha),\ u-S(t)z\rangle + \| u-z \|^2 - \| u-S(t)z \|^2$$

and hence by letting t_α tend to infinity,

$$0 \leqq 2\langle z-u, S(t)z-z\rangle + |u-z|^2 - |u-S(t)z|^2.$$

In fact, note that

$$\lim_\alpha |S(t)u(t_\alpha)-u|^2 = \lim_\alpha |u(t+t_\alpha)-u|^2 = \lim_\alpha |u(t_\alpha)-u|^2.$$

Consequently $z \in F(S)$ and hence $w(u) \subset F(S)$. By Theorem 2, the net $\{u(t) : t \in G\}$ converges weakly to some $y \in F(S)$. On the other hand, since P is the metric projection of H onto $F(S)$, we know that

$$\langle u(t)-Pu(t), Pu(t)-z\rangle \geqq 0$$

for all $z \in F(S)$. So, if $Pu(t) \to u$ by Theorem 3, we have $\langle y-u, u-z\rangle \geqq 0$ for all $z \in F(S)$. Putting $y = z$, we obtain $-|y-u|^2 \geqq 0$ and hence $y = u$.

References.

[1] S. Aizicovici, On the asymptotic behaviour of solutions of Volterra equations in Hilbert space, Nonlinear Analysis, 7 (1983), 271-278.

[2] J. B. Baillon and H. Brézis, Une remaque sur le comportement asymptotique des semigroupes non linéaires, Houston J. Math. 2 (1976), 5-7.

[3] V. Barbu, Nonlinear semigroups and differential equations in Banach spaces, Noordhoff, Leyden, 1976.

〔4〕 V. Barbu and Th. Precupanu, Convexity and optimization in Banach spaces, Editura Academiei R. S. T., Bucuresti, 1978.

〔5〕 F. E. Browder, Nonlinear operators and nonlinear equations of evolution in Banach spaces, Proc. Sympos. Pure Math., vol. 18, no. 2, Amer. Math. Soc., Providence, R. I. 1976.

〔6〕 R. E. Bruck, Asymptotic convergence of nonlinear contraction semigroups in Hilbert space, J. Funct. Analysis 8 (1975), 5-26.

〔7〕 R. E. Bruck, A simple proof of the mean ergodic theorem for nonlinear contractions in Banach spaces, Israel J. Math., 32 (1979), 107-116.

〔8〕 F. R. Deutsch and P. H. Maserick, Application of the Hahn-Banach theorem in approximation theory, SIAM Rev., 9 (1967), 516-530.

〔9〕 J. Diestel, Geometry of Banach spaces, selected topics. Lecture notes in mathematics 485 (1975), Springer Verlag, Berlin-Heidelberg, New York.

〔10〕 C. W. Groetsch, A note on segmenting Mann iterates, J. Math. Anal. Appl., 40 (1972), 369-372.

〔11〕 N. Hirano and W. Takahashi, Nonlinear ergodic theorems for an amenable semigroup of nonexpansive mappings in a Banach space, Pacific J. Math., 112 (1984), 333-346.

〔12〕 D. S. Hulbert and S. Reich, Asymptotic behavior of solutions to nonlinear Volterra integral equations, J. Math. Anal. Appl., 104 (1984), 155-172.

〔13〕 K. Kido and W. Takahashi, Means on commutative semigroups and nonlinear ergodic theorems, to appear in J. Math. Anal. Appl..

[14] A. T. Lau and W. Takahashi, Weak convergence and non-linear ergodic theorems for reversible semigroup of nonexpansive mappings, to appear.

[15] I. Miyadera and K. Kobayashi, On the asymptotic behaviour of almost-orbits of nonlinear contraction semigroups in Banach spaces, Nonlinear Analysis, 6 (1982), 349-365.

[16] G. Moroşanu, Asymptotic behaviour of solutions of differential equations associated to monotone operators, Nonlinear Analysis, 3 (1979), 873-883.

[17] A. Pazy, On the asymptotic behaviour of semigroups of nonlinear contractions in Hilbert space, J. Funct. Analysis 27 (1978), 292-307.

[18] W. Takahashi, A nonlinear ergodic theorem for an amenable semigroup of nonexpansive mappings in a Hilbert space, Proc. Amer. Math. Soc., 81 (1981), 253-256.

[19] W. Takahashi, Fixed point theorems for families of nonexpansive mappings on unbounded sets, J. Math. Soc. Japan, 36 (1984), 543-553.

[20] W. Takahashi, A nonlinear ergodic theorem for a reversible semigroup of nonexpansive mappings in a Hilbert space, to appear in Proc. Amer. Math. Soc..

Printed and bound by CPI Group (UK) Ltd, Croydon, CR0 4YY
22/10/2024
01777392-0001

On a Factorization of Operators Through a Subspace of c_0

YAU-CHUEN WONG Department of Mathematics, The Chinese University of Hong Kong, Hong Kong

The notion of quasi-Schwartz linear mappings between locally convex spaces, induced by Randtke [5], is a natural generalization of the notion of precompact linear mappings. Recently, H. Junek [3] has shown that the class of all quasi-Schwartz linear mappings is exactly the inferior extension of the class of all compact operators between Banach spaces. Terzioglu [7] and Randtke [5] have given very nice characterizations of compact operators between Banach spaces. These results have been generalized by Wong [9, 10] to the case of quasi-Schwartz linear mappings. This note is mainly devoted to give such information as well as some applications.

In order to deduce the notion of quasi-Schwartz linear mappings, we require the following interesting result:

Theorem 1 (Terzioglu [7], Randtke [5]). *Let* X *and* Y *be Banach spaces and* $T : X \to Y$ *a continuous linear mapping. Then the following statements are equivalent.*

(a) T *is compact.*

(b) *There exists an* $[\zeta_n] \in c_o$ *and an equicontinuous sequence* $\{f_n\}$ *in* X' *such that*

$$\| Tx \| \leq \sup_n |\zeta_n f_n(x)| \qquad \text{(for all } x \in X).$$

(c) *There exists a closed subspace* H *of* c_o , *a compact linear mapping* $R : E \to H$ *and an* $S \in \mathcal{L}(H,F)$ *such that* $T = SR$.

Proof. We only given an outlined proof here, while details can be found in Terzioglu [7], Randtke [5] or Köthe [4].

The implication (a) $\Rightarrow$ (b) follows from Schauder's duality theorem on compact operators and a result concerning the structure of compact sets in a Banach space. (A subset K of X' is compact if and only if it is contained in the closed absolutely convex hull of some null-sequence g_n in X'.) For the implication (b) $\Rightarrow$ (c), we first notice that the mapping R , defined by

$$Rx = [\zeta_n f_n(x)]_{n \in \mathbb{N}} \qquad (x \in X) ,$$

is a continuous linear mapping from X into c_o with $\| R \| = \sup_n \| \zeta_n f_n \|$.

The compactness of R follows from the fact that the set, defined by

$$K = \{[\eta_n] \in c_o : |\eta_n| \leq \| \zeta_n f_n \| \qquad \text{for all } n \geq 1\} ,$$

is compact (since $\|\eta_n f_n\| \to 0$).

There are many remarkable applications. We mention two as follows:

Corollary 1.1 (Grothendieck). The following two statements are equivalent.

(a) Every Banach space has the A.P. (Approximation Property).

(b) Every closed subspace of c_o has the A.P.

Consequently, there is a closed subspace of c_o which does not have the A.P.

Proof. The implication (a) $\Rightarrow$(b) is obvious. To prove the implication (b) $\Rightarrow$(a), let X and Y be any Banach spaces and let $T \in \mathcal{L}(X,Y)$ be a compact linear map. By Theorem 1, there exists a closed subspace H of c_o, a compact linear map $R \in \mathcal{L}(X,H)$ and an $S \in \mathcal{L}(H.Y)$ such that $T = SR$. As H assumed to have the A.P., there exists a sequence $\{R_n\}$ of finite rank of continuous linear mappings from X into H such that $\lim_n \| R-R_n \| = 0$. Clearly $L_n = SR_n \in \mathcal{L}(X,Y)$ are linear maps of finite rank with $\lim_n \| T-L_n \| = 0$. Hence X has the A.P.

We say that a continuous linear map $T \in \mathcal{L}(X,Y)$ admits a compact factorization through a Banach space (resp. a locally convex space Z) if there are compact linear maps $R \in \mathcal{L}(X,Z)$ and $S \in \mathcal{L}(Z,Y)$ such that

$$T = SR .$$

Corollary 1.2 (Figiel [2]). Let X and Y be Banach spaces. An $T\in\mathcal{L}(X,Y)$ admits a compact factorization through a Banach space if and only if it admits a compact factorization through a closed subspace of c_o.

Proof. The sufficiency is obvious, while the necessity following from Theorem 1.

From the equivalence of (a) and (c) of Theorem 1, it is natural to ask in what cases a compact linear map $T : X \to Y$ admits a compact factorization through the whole space c_o. The following result gives such a criterion.

Theorem 2 (Terzioglu [8]). Let X and Y be Banach spaces and $T \in\mathcal{L}(E,F)$ a compact linear map. Then T admits a compact factorization through the whole space c_o if and only if T is an ∞-nuclear operator in the sense that there exists an $[\zeta_n] \in c_o$, an equicontinuous sequence $\{f_n\}$ in X' and a summable sequence $\{y_n\}$ in Y such that

$$Tx = \Sigma_n \zeta_n f_n(x) y_n \qquad \text{(for all } x \in X).$$

For a proof, we refer to Terzioglu [8] or Köthe [4, p. 227].

Theorem 1 suggests the following:

Definition 1 (Randtke [5]). Let E and F be locally convex spaces. A seminorm p on E is said to be precompact if there exists an $[\zeta_n] \in c_o$ and an equicontinuous sequence $\{f_n\}$ in E' such that

$$p(x) \leq \sup_n |\zeta_n f_n(x)| \qquad \text{(for all } x \in E).$$

An $T \in \mathcal{L}(E,F)$ is called a quasi-Schwartz operator if there exists a precompact seminorm p on E such that $\{Tx \in F : p(x) \leq 1\}$ is bounded in F.

Theorem 3 (Wong [10]). Let E and F be locally convex spaces and $T \in \mathcal{L}(E,F)$. The following statements are equivalent.

(a) T *is a quasi-Schwartz*.

(b) *There exists a* 0-*neighbourhood* V *in* E *such that* $T(V)$ *is* b-*pre-compact in the sense that there is a closed, absolutely convex, bounded subset* B *of* F *such that* $T(V)$ *is precompact in the normed space* $(F(B), r_B)$, where $F(B) = \bigcup_n nB$ *and* r_B *is the gauge of* B *defined on* $F(B)$.

(c) T *admits a quasi-Schwartz factorization through a vector subspace* H *of* c_o.

Proof. We only give an outlined proof here, while details can be found in Wong [10, Theorems 2 and 5].

(a) $\Rightarrow$ (b): We say that an $T \in \mathcal{L}(E,F)$, satisfied (b), is an b-*precompact operator*. It is easily seen that the composition of two continuous linear maps, in which one of them is b-precompact, is b-precompact. On the other hand, by a result of Randtke [5, (2.11)], every quasi-Schwartz operator admits the following decomposition:

$$E \overset{Q}{\to} X \overset{\tilde{T}}{\to} Y \overset{J}{\to} F$$

where X, Y are normed spaces, $\tilde{T}$ is precompact operator and Q, J are continuous linear maps. Finally, in a metrizable locally convex space, every precompact set must be b-precompact. Therefore the implication follows.

(b) $\Rightarrow$ (c): $T : E \to (F(B), r_B)$ is a precompact linear map. Using an argument of Randtke [5], one can show that $T : E \to (F(B), r_B)$ admits a precompact factorization through a subspace H of c_o.

It is clear that every b-precompact subset of F is precompact; but the converse is, in general, not true (except for the metrizable case of F). Thus, the concept of quasi-Schwartz operators is a natural generalization of that of precompact linear maps; consequently, Theorem 3 is a generalization of Theorem 1.

Corollary 3.1. *Let* E *and* F *be locally convex spaces and* $T \in \mathcal{L}(E,F)$. *If* E *is infrabarrelled with a fundamental sequence of bounded sets and if* F *is metrizable then an* $T \in \mathcal{L}(E,F)$ *is precompact if and only if the adjoint map* $T' : (F', \beta(F',F)) \to (E', \beta(E',E))$ *is compact*.

Proof. It is clear that every precompact linear mapping from a locally convex space into a *metrizable* locally convex space is quasi-Schwartz. The result then follows from Theorem 3 and Köthe [4, §42, 1(10)].

Corollary 3.2 (Dazord [1]). *Let* E *be a locally convex space. An* $T \in \mathcal{L}(E,c_o)$ *is compact if and only if* T *admits a compact factorization through the whole space* c_o. *Consequently, if* X *is a Banach space then* $S \in \mathcal{L}(X,c_o)$ *is compact if and only if* S *is* ∞-*nuclear*.

Proof. The sufficiency is obvious. To prove the necessity, we notice that T is quasi-Schwartz. The completeness of c_o ensures that T has a quasi-Schwartz factorization through a *closed* subspace H of c_o : $E \overset{R}{\to} H \overset{L}{\to} c_o$. As c_o has the compact extension property, L has a compact extension $\tilde{L} : c_o \to c_o$, hence

$$T = LR = \tilde{L}\, J_H R = \tilde{L}\, (J_H R) ,$$

where $J_H : H \to c_o$ is the canonical embedding, and thus $J_H R : E \to c_o$ is compact.

From Theorem 3, it is natural to ask in what cases a quasi-Schwartz operator has a quasi-Schwartz factorization through the whole space c_o. To answer this question, we require the following:

Definition 2 (Randtke [5]). A sequence $\{y_n\}$ in a locally convex space F is said to be strongly summable if it satisfies the following two conditions:

(i) for any $[\lambda_n] \in \ell_\infty$, the series $\Sigma_n \lambda_n y_n$ converges in F;

(ii) $\{\Sigma_n \lambda_n y_n : \| [\lambda_n] \|_{\ell_\infty} \leq 1 , [\lambda_n] \in \ell_\infty\}$ is bounded in F.

An $T \in \mathcal{L}(E,F)$ is called a Schwartz operator if there exists $[\zeta_n] \in c_o$, an equicontinuous sequence $\{f_n\}$ in E' and a strongly summable sequence $\{y_n\}$ in F such that

$$Tx = \Sigma_n \zeta_n f_n(x) y_n \qquad \text{(for all } x \in E).$$

Remark. It can be shown (see Wong [9]) that if F is sequentially complete and if $\{z_n\}$ is a weakly summable sequence in F, then for any $[\lambda_n] \in c_o$, the sequence $\{\lambda_n z_n\}$ in F must be strongly summable. Thus the notion of Schwartz operators is an extension of that of ∞-nuclear operators.

Theorem 4 (Wong [9]). Let E and F be locally convex spaces. A quasi-Schwartz linear mapping $T : E \to F$ is Schwartz if and only if T admits a quasi-Schwartz factorization through the whole space c_o.

Combining Theorem 3, we see that the preceding result is a generalization of Theorem 2.

References

[1] Dazord, J., Factoring operators through c_o. Math. Ann. 220, 105-120,(1976).

[2] Figiel, T., Factorization of compact operators and application to the approximation property, Studia Math. 45, 241-252, (1973).

[3] Junek, H., Locally convex spaces and operator ideals (Teubner-Texte zur Mathematik, Band 56) (1983).

[4] Kōthe, G., Topological vector spaces II (Springer-Verlag) (1979).

[5] Randtke, D., Characterizations of precompact maps, Schwartz maps and nuclear spaces, Trans. Amer. Math. Soc. 165, 87-101, (1972).

[6] Randtke, D., Representation theorem for compact operators, Proc. Amer. Math. Soc. 37, 481-485, (1973).

[7] Terzioglu, T., A characterization of compact linear mappings, Arch. Math. 22, 76-78, (1971).

[8] Terzioglu. T., Remark on compact and infinite-nuclear mappings, Math. Balkanika 2, 251-255, (1972).

[9] Wong, Yau-chuen, A factorization theorem for Schwartz linear mappings, Math. Japonica 26 (6), 655-659, (1980).

[10] Wong, Yau-chuen, A characterization of quasi-Schwartz linear mappings, Indag. Math. 85 (3), 359-363, (1982).

Trace Formula for Almost Lie Algebra of Operators and Cyclic One-Cocycles

DAOXING XIA Department of Mathematics, Vanderbilt University, Nashville, Tennessee

1. In the previous paper [4], the author introduced the concept of almost Lie algebra of operators in Hilbert space and gave a trace formula for almost Lie algebra of operators. As the author mentioned at the end of [4], R. Douglas pointed out that the functional introduced in [4] is closely related to Connes' cyclic cohomology. This relation will be illustrated in §5.

The aim of the present paper is to determine the first cyclic cohomology $Z^1_\lambda(\mathcal{A})$ of the enveloping algebra of Heisenberg algebra which generalized a result in [4].

2. Let A be any algebra over C, and $Z^1_\lambda(\mathcal{A})$ be the space of all bilinear functionals ϕ on A satisfying

$$\phi(a_1, a_0) = -\phi(a_0, a_1), \quad \text{for} \quad a_0, a_1 \in \mathcal{A} \tag{1}$$

and

$$(b\phi)(a_0, a_1, a_2) = 0, \quad \text{for} \quad a_0, a_1, a_2 \in \mathcal{A}, \tag{2}$$

where b is the Hochshild coboundary:

$$(b\phi)(a_0, a_1, a_2) = \phi(a_0a_1, a_2) - \phi(a_0, a_1a_2) + \phi(a_2a_0, a_1). \tag{3}$$

The formula (2) may be rewritten as

$$\phi(a_0, a_1a_2) + \phi(a_1, a_2a_0) + \phi(a_2, a_0a_1) = 0, \tag{4}$$

since ϕ also satisfies (1).

From (4), it follows that

$$\sum_{j=0}^{n} \phi(a_j, a_{\nu_{j+1}} \dots a_{\nu_{j+n}}) = 0, \tag{5}$$

This work is partly supported by NSF grant no. DMS-8502359

where $\nu_j = j$ for $j \leq n$ and $\nu_j = j - n - 1$ for $j > n$. The proof of (5) is similar to the proof for Lemma 2 in [4]. We use the same symmetrization symbols as in [4]. From (5), it follows that

$$\phi(S(\xi_1^{m_1} \dots \xi_k^{m_k}), \eta) = \sum_{j=1}^{k} \phi(\xi_j, S([\eta]\frac{\partial}{\partial \xi_j} S(\xi_1^{m_1} \dots \xi_k^{m_k}))), \tag{6}$$

for $\xi_i, \dots, \xi_k, \eta \in \mathcal{A}$, where $m_1, \dots, m_k$ are non-negative integers.

The cocycles in $Z_\lambda^1(\mathcal{A})$ possess the collapsing property and the covariance property as indicated in [4].

Lemma 1. If $\phi \in Z_\lambda^1(\mathcal{A})$, then

$$\phi(f(a), g(a)) = 0, \tag{7}$$

for all $a \in \mathcal{A}$ and polynomials f and g, and

$$\phi(a\xi a^{-1}, a\eta a^{-1}) = \phi(\xi, \eta) \tag{8}$$

for all invertible $a \in \mathcal{A}$ and $\xi, \eta \in \mathcal{A}$ satisfying $[\xi, \eta] = 0$.

Proof. Put $m_1 = \dots = m_k = 1, \xi_1 = \dots = \xi_k = a$ and $\eta = a^n$ in (6). Then

$$\phi(a^k, a^n) = k\phi(a, a^{k+n-1}) = -k\phi(a^{k+n-1}, a), \tag{9}$$

since ϕ satisfies (1). If $n = 1$, then from (9) it follows

$$\phi(a^k, a) = 0.$$

Using (9) again, one has $\phi(a^k, a^n) = 0$ for all non-negative integers n and k, which implies (7).

By means of (5), it is obvious that

$$\begin{aligned}\phi((a\xi a^{-1}), a\eta a^{-1}) + \phi(a, \eta a^{-1}(a\xi a^{-1})) + \phi(\eta, a^{-1}(a\xi a^{-1})a) \\ + \phi(a^{-1}, (a\xi a^{-1})a\eta) = 0,\end{aligned} \tag{10}$$

and

$$\phi(a, \xi\eta a^{-1}) + \phi(\xi\eta, 1) + \phi(a^{-1}, a\xi\eta) = 0.$$

On the other hand $\phi(b, 1) = 0$ for $b \in \mathcal{A}$ by (7). Hence (10) implies (8).

It is easy to seen that in the proof of Theorem 1, 2 and 3 in [4], the four properties of the functional in Lemma 1 of [4] are basic. Now, from Lemma 1, the cocycle ϕ in $Z^1_\lambda(\mathcal{A})$ also possesses those properties. By means of the same proof in [4], we have the following theorem.

Theorem 1. Let A be an algebra over C with finite generators $\eta_1, \ldots, \eta_n$. If $\phi \in Z^1_\lambda(\mathcal{A})$, then there are linear functionals M_{jk} on $\mathcal{A}$, satisfying $M_{jk} = -M_{kj}, j, k = 1, \ldots, n$ such that

$$\phi(\xi, \eta) = \sum_{j,k=1}^{n} M_{jk} \left(\frac{\partial}{\partial \eta_k} S \left([\eta] \frac{\partial \xi}{\partial \eta_j} \right) \right).$$

3. Hitherto, suppose $\mathcal{A}_j, j = 1, 2$ are two subalgebras of the algebra $\mathcal{A}$ satisfying

$$[a_1, a_2] = 0 \quad \text{for} \quad a_j \in \mathcal{A}_j, \quad j = 1, 2, \tag{11}$$

where $[a_1, a_2]$ is the commutator $a_j a_2 - a_2 a_1$. For $\phi \in Z^1_\lambda(\mathcal{A})$, define

$$\phi_j(\xi, \eta; \varsigma) = \frac{1}{2}(\phi(\varsigma\xi, \eta) + \phi(\xi, \varsigma\eta)), \quad \xi, \eta \in \mathcal{A}_j, \varsigma \in \mathcal{A}_{j'}, \tag{12}$$

for $j = 1, 2$, where $j' = 2$ for $j = 1$ and $j' = 1$ for $j = 2$.

Lemma 2. The functionals ϕ_j, $j = 1, 2$ are tri-linear,

$$\phi_j(\xi, \eta; \varsigma) = -\phi_j(\eta, \xi; \varsigma), \tag{13}$$

$$(b\phi_j)(\xi_0, \xi_1, \xi_2; \varsigma) = \phi(\varsigma; \xi_0\xi_1\xi_2), \tag{14}$$

$$\phi_1(\xi_0, \xi_1; [\eta_0, \eta_1]) = \phi_2(\eta_0, \eta_1; [\xi_0, \xi_1]), \tag{15}$$

and

$$\phi(\xi_1\eta_1, \xi_2\eta_2) = \phi_1(\xi_1, \xi_2; \eta_1\eta_2) + \phi_2(\eta_1, \eta_2, \xi_1\xi_2)) \tag{16}$$

$$= \phi_1(\xi_1, \xi_2; S(\eta_1\eta_2)) + \phi_2(\eta_1, \eta_2; S(\xi_1\xi_2)). \tag{17}$$

Proof. It is obvious that ϕ_j are tri-linear. To prove (13), notice that

$$\begin{aligned}\phi_j(\xi, \eta; \varsigma) + \phi_j(\eta, \xi; \varsigma) &= \frac{1}{2}(\phi(\varsigma\xi, \eta) + \phi(\eta\varsigma, \xi) + \phi(\xi, \varsigma\eta) + \phi(\eta, \xi\varsigma)) \\ &= \frac{1}{2}(\phi(\varsigma, \xi\eta) - \phi(\varsigma, \eta\xi)) = \frac{1}{2}\phi(\varsigma, [\xi, \eta]),\end{aligned} \tag{18}$$

by (1) and (4). As functions of ξ and η, $\phi_j(\xi, \eta; \varsigma) + \phi_j(\eta, \xi; \varsigma)$ is symmetric and $\phi(\varsigma, [\xi, \eta])$ is skew-symmetric with respect to ξ and η. Thus both sides of (18) are zero which proves (13) and

$$\phi(\varsigma, [\xi, \eta]) = 0, \tag{19}$$

for $\varsigma \in \mathcal{A}_{j'}$, $\xi, \eta \in \mathcal{A}_j$.

From

$$\phi(\varsigma\xi_1, \xi_2\xi_0) + \phi(\xi_0, \varsigma\xi_1\xi_2) + \phi(\xi_2, \xi_0\varsigma\xi_1) = 0,$$
$$\phi(\varsigma\xi_0, \xi_1\xi_2) + \phi(\xi_1, \xi_2\varsigma\xi_0) + \phi(\xi_2, \varsigma\xi_0\xi_1) = 0$$

and

$$\phi(\varsigma\xi_0\xi_1, \xi_2) + \phi(\xi_2\varsigma, \xi_0\xi_1) + \phi(\xi_0\xi_1\xi_2, \varsigma) = 0,$$

(14) follows. To prove (15), observe that

$$\phi((\xi_1\eta_1)\xi_2, \eta_2) + \phi(\xi_2\eta_2, \xi_1\eta_1) + \phi(\eta_2(\xi_1\eta_1), \xi_2) = 0$$

and

$$\phi(\xi_2(\xi_1\eta_1), \eta_2) + \phi(\eta_2\xi_2, \xi_1\eta_1) + \phi((\xi_1\eta_1)\eta_2, \xi_2) = 0.$$

Thus, if $[\xi_i, \eta_j] = 0$ for $i, j = 1, 2$, then

$$\phi([\xi_1, \xi_2]\eta_1, \eta_2) = \phi([\eta_1, \eta_2]\xi_1, \xi_2). \tag{20}$$

and similarly

$$\phi(\eta_1, [\xi_1, \xi_2]\eta_2) = \phi(\xi_1, [\eta_1, \eta_2]\xi_2). \tag{21}$$

From (20) and (21) it follows (15) immediately.

By a simple calculation, it is easy to see (16) and (17).

Theorem 2. Let the algebra $\mathcal{A}$ over C be generated by $\mathcal{A}_1$ and $\mathcal{A}_2$ satisfying (11). Then $\phi \in Z^1_\lambda(\mathcal{A})$, if and only if $\phi(\cdot\,,\cdot)$ is a bilinear functional on $\mathcal{A} \times \mathcal{A}$ satisfying (17), where $\phi_j, j = 1, 2$ are tri-linear functionals satisfying (13), (14) and (15).

Proof. By Lemma 2, we only have to prove the "if" part of the theorem. Suppose $\phi(\cdot\,,\cdot)$ is a bilinear functional satisfying (13), (14), (15) and (17). It is obvious that $\phi(\cdot\,,\cdot)$ is skew-symmetric, by (13) and (17). From (15), there is a bilinear functional $\psi(\cdot\,,\cdot)$ on $\mathcal{A}_1 \times \mathcal{A}_2$ such that

$$\phi_1(\xi_1, \xi_2; [\eta_1\eta_2]) = \psi([\xi_1, \xi_2], [\eta_1, \eta_2]) = \phi_2(\eta_1, \eta_2; [\xi_1, \xi_2]). \tag{22}$$

To prove (4), we have to prove

$$\begin{aligned}&\phi_1(\xi_0, \xi_1\xi_2; S(\eta_0[\eta_1\eta_2])) + \phi_1(\xi_1, \xi_2\xi_0; S(\eta_1[\eta_2\eta_0]))\\ &+ \phi_1(\xi_2, \xi_0\xi_1; S(\eta_2[\eta_0\eta_1])) + \phi_2(\eta_0, \eta_1\eta_2; S(\xi_0[\xi_1\xi_2]))\\ &+ \phi_2(\eta_1, \eta_2\eta_0; S(\xi_1[\xi_2\xi_0])) + \phi_2(\eta_2, \eta_0\eta_1; S(\xi_2[\xi_0\xi_1])) = 0.\end{aligned} \tag{23}$$

By means of (14) and (22), we have

$$\begin{aligned}&\phi(\xi_0, \xi_1\xi_2; \eta_0\eta_1\eta_2) + \phi_1(\xi_1, \xi_2\xi_0; \eta_1\eta_2\eta_0) + \phi(\xi_2, \xi_0\xi_1; \eta_2\eta_0\eta_1)\\ &= \phi(\eta_0\eta_1\eta_2, \xi_0\xi_1\xi_2) + \psi([\xi_1, \xi_2\xi_0], [\eta_1\eta_2, \eta_0]) + \psi([\xi_2, \xi_0\xi_1], [\eta_2, \eta_0\eta_1]),\end{aligned}$$

and

$$\begin{aligned}&\phi_1(\xi_0, \xi_1\xi_2; \eta_1\eta_2\eta_0) + \phi_1(\xi_1, \xi_2\xi_0; \eta_2\eta_0\eta_1) + \phi(\xi_2, \xi_0\xi_1; \eta_0\eta_1\eta_2)\\ &= \phi(\eta_0\eta_1\eta_2, \xi_0\xi_1\xi_2), +\psi(\xi_0, \xi_1\xi_2], [\eta_1\eta_2, \eta_0]) + \psi([\xi_1, \xi_2\xi_0], [\eta_2, \eta_0\eta_1])\end{aligned}$$

and the other two similar identities for ϕ_2. Thus the left side of (23) equals

$$\begin{aligned}&\psi([\xi_1, \xi_2\xi_0] + [\xi_0, \xi_1\xi_2], [\eta_1\eta_2, \eta_0]) + \psi([\xi_2, \xi_0\xi_1], [\eta_2, \eta_0\eta_1] + [\eta_1, \eta_2\eta_0])\\ &+ \psi([\xi_0, \xi_1\xi_2], [\eta_1\eta_2, \eta_0] + [\eta_2\eta_0, \eta_1]) + \psi([\xi_1, \xi_2\xi_0] + [\xi_2, \xi_0\xi_1], [\eta_2, \eta_0\eta_1]) = 0,\end{aligned}$$

which proves (23). Thus $\phi \in Z^1_\lambda(\mathcal{A})$.

4. In this section the structure of $Z^1_\lambda(\mathcal{A})$ will be given when $\mathcal{A}$ is the enveloping algebra of the Heisenberg algebra [3].

Let n be a positive integer, and $\{x_1, \ldots, x_n, y_1, \ldots, y_n, z\}$ be a basis for a vector space $\mathcal{N}$. Let H_n be the Lie algebra on the space $\mathcal{N}$ defined by setting

$$[x_i, y_i] = -[y_i, x_i] = z$$

and the other brackets of basic elements being zero. This H_n is called the Heisenberg algebra. Let $\mathcal{A}$ be the enveloping algebra of H_n. In [4], the author proved that in the case of $n = 1$, for every $\phi \in Z^1_\lambda(\mathcal{A})$, there is a linear functional f on $[\mathcal{A}, \mathcal{A}]$ such that

$$\phi(\xi, \eta) = f([\xi, \eta]), \quad \text{for} \quad \xi, \eta \in \mathcal{A}. \tag{24}$$

If $n > 1$, then the structure of $Z^1_\lambda(\mathcal{A})$ becomes more complicate. Let $\mathcal{P}_n$ be the set of all polynomials of $2n$ variables $u_1, v_1, \ldots, u_n, v_n$, and U be the mapping from $\mathcal{A}$ onto $\mathcal{P}_n$ defined by

$$\begin{aligned}&U(\sum c_{m_1k_1\ldots m_nk_n\ell}x_1^{m_1}y_1^{k_1}\ldots x_n^{m_n}y_n^{k_n}z^\ell)\\ &\qquad = \sum c_{m_1k_1\ldots m_nk_n0}u_1^{m_1}v_1^{k_1}\ldots u_n^{m_n}v_n^{k_n}.\end{aligned}$$

Theorem 3. Let A be the enveloping algebra of the Heisenberg algebra H_n. Then for every $\phi \in Z^1_\lambda(\mathcal{A})$, there exists a linear functional f on $[\mathcal{A}, \mathcal{A}]$ and a linear functional g on the space of differentials on $\mathcal{P}_n$ such that

$$\phi(\xi, \eta) = f([\xi, \eta]) + g(dU\xi \wedge dU\eta), \xi, \eta \in \mathcal{A} \tag{25}$$

Proof. We shall prove this theorem by mathematical induction on the number n. This theorem is true for $n = 1$. Suppose it is true for $n = k - 1 \geq 1$. We have to prove that it is true for $n = k$. Let $\mathcal{A}_1$ be the subalgebra generated by x_1, y_1 and z, and $\mathcal{A}_2$ be the subalgebra of $\mathcal{A}$ generated by $x_2, y_2, \cdots, x_k, y_k$ and z. Then $\mathcal{A}_1$ and $\mathcal{A}_2$ satisfy the condition (11), and for any $\phi_j \in Z^1_\lambda(\mathcal{A}_j)$. There is a linear functional f_j on $[\mathcal{A}_j, \mathcal{A}_j]$, a linear functional g_2 on the space of differentials on $\mathcal{P}_{k-1}$ and $g_1 = 0$ such that

$$\phi_j(\xi, \eta) = f_j([\xi, \eta]) + g_j(dU\xi \wedge dU\eta), \quad j = 1, 2. \tag{26}$$

If $\phi \in Z^1_\lambda(\mathcal{A})$, then $\phi(\xi_1, \xi_2)$ depends on $U\xi_1$ and $U\xi_2$ only for $\xi_j \in \mathcal{A}_j$ by (19). Thus there is a bilinear functional $\tilde{\phi}$ on $\mathcal{P}_1 \times \mathcal{P}_{k-1}$ such that

$$\tilde{\phi}(U\xi_1, U\xi_2) = \phi(\xi_1, \xi_2) \qquad \text{for} \quad \xi_j \in \mathcal{A}_j.$$

Denote the degree of a polynomial p by $\delta(p)$. There is a tri-linear functional $\tilde{\phi}_1(\cdot, \cdot\,; \cdot)$ on $\mathcal{P}_1 \times \mathcal{P}_1 \times \mathcal{P}_{k-1}$ such that

$$\tilde{\phi}_1(\xi, \eta\,; \varsigma) = \frac{-\delta(\xi) + \delta(\eta)}{\delta(\xi) + \delta(\eta)} \tilde{\phi}(\xi\eta, \varsigma),$$

for monomials ξ and η satisfying $\delta(\xi) + \delta(\eta) > 0$. Similarly, there is also a tri-linear functional $\tilde{\phi}_2(\cdot, \cdot; \cdot)$ on $\mathcal{P}_{k-1} \times \mathcal{P}_{k-1} \times \mathcal{P}_1$ such that

$$\tilde{\phi}_2(\xi, \eta; \varsigma) = \frac{\delta(\xi) - \delta(\eta)}{\delta(\xi) + \delta(\eta)} \tilde{\phi}(\varsigma, \xi\eta).$$

for monomials ξ, η satisfying $\delta(\xi) + \delta(\eta) > 0$. Define $\tilde{\phi}_j(\xi, \eta; \varsigma) = 0$ for $\delta(\xi) = \delta(\eta) = 0$. It is easy to see that

$$(b\tilde{\phi}_1)(\xi_0, \xi_1, \xi_2; \varsigma) = -\tilde{\phi}(\xi_0\xi_1\xi_2, \varsigma) \tag{27}$$

and

$$(b\tilde{\phi}_2)(\xi_0, \xi_1, \xi_2; \varsigma) = \tilde{\phi}(\varsigma, \xi_0\xi_1\xi_2). \tag{28}$$

Construct ϕ_i, $j = 1, 2$ by (12) and define

$$\hat{\phi}_j(\xi, \eta; \varsigma) = \phi_j(\xi, \eta; \varsigma) - \tilde{\phi}_j(U\xi, U\eta; U\varsigma) \tag{29}$$

From (14), (27), (28) and (29), it follows that

$$b\hat{\phi}_j = 0,$$

and $\hat{\phi}_j \in Z^1_\lambda(\mathcal{A}_j)$ for $j = 1, 2$.

By means of (26), there are bilinear functionals $f_i(\cdot\,,\,\cdot)$ and $g_j(\cdot\,,\,\cdot)$ such that

$$\hat{\phi}_j(\xi,\eta;\varsigma) = f_j([\xi,\eta],\varsigma) + g_j(dU\xi \wedge dU\eta,\varsigma), \tag{30}$$

where $g_1(\cdot\,,\,\cdot) = 0$.

From (15) and (29), it follows that

$$\hat{\phi}_1(\xi_0,\xi_1;[\eta_0,\eta_1]) = \hat{\phi}_2(\eta_0,\eta_1;[\xi_0,\xi_1]).$$

Thus

$$f_1([\xi_0,\xi_1],[\eta_0,\eta_1]) - f_2([\eta_0,\eta_1],[\xi_0,\xi_1]) = g_2(dU\eta_0 \wedge dU\eta_1,[\xi_0,\xi_1]), \tag{31}$$

and $g_2(dU\eta_0 \wedge dU\eta_1,\varsigma)$ depends on $[\eta_0,\eta_1]$ only for $\varsigma \in [\mathcal{A}_1,\mathcal{A}_1]$. Hence, we may choose $f_2(\cdot\,,\,\cdot)$ and $g_2(\cdot,\,\cdot)$ such that $g_2(\cdot\,,\varsigma) = 0$ for $\varsigma \in [\mathcal{A}_1,\mathcal{A}_1]$, and (31) becomes

$$f_1([\xi_0,\xi_1],[\eta_0,\eta_1]) = f_2([\eta_0,\eta_1],[\eta_0,\xi_1]). \tag{32}$$

Besides, $g_2(\cdot\,,\varsigma)$ depends on $U\varsigma$ only. Thus $g_2(\cdot\,,\,\varsigma)$ may be rewritten as $g_2(\cdot\,,U\varsigma)$. Therefore (30) may be rewritten as

$$\hat{\phi}_j(\xi,\eta;\varsigma) = f_j([\xi,\eta],\varsigma) + g_j(dU\xi \wedge dU\eta, U\varsigma), \tag{33}$$

where $g_1(\cdot\,,\,\cdot) = 0$. Define

$$\begin{aligned} \psi(\xi,\eta) &= f_1(\xi,\eta), && \text{for} \quad \xi \in [\mathcal{A}_1,\mathcal{A}_1] \quad \text{and} \quad \eta \in \mathcal{A}_2, \\ &= f_2(\eta,\xi), && \text{for} \quad \xi \in \mathcal{A}_1 \quad \text{and} \quad \eta \in [\mathcal{A}_2,\mathcal{A}_2]. \end{aligned}$$

This function $\psi(\cdot\,,\,\cdot)$ is well-defined on $[\mathcal{A}_1,\mathcal{A}_1] \times \mathcal{A}_2 \cup \mathcal{A}_1 \times [\mathcal{A}_2,\mathcal{A}_2]$, since (32) holds.

We have to prove that

$$\psi(z\xi,\eta) = \psi(\xi,z\eta). \tag{34}$$

Notice that

$$\phi(z\xi,\eta) = \phi(\xi,z\eta),$$

since $\phi(\xi,\eta z) + \phi(\eta,z\xi) + \phi(z,\xi\eta) = 0$ and $\phi(z,\xi\eta) = 0$. Hence

$$\phi_j(z\xi,\eta;\varsigma) = \phi_j(\xi,z\eta;\varsigma) = \phi_j(\xi,\eta,z\varsigma), \tag{35}$$

by (12). On the other hand, $Uz\xi = 0$. Thus

$$\hat{\phi}_j(z\xi,\eta;\varsigma) = \phi_j(z\xi,\eta;\varsigma)$$
$$\hat{\phi}_j(\xi,z\eta;\varsigma) = \phi_j(\xi,z\eta;\varsigma)$$

and

$$\hat{\phi}_j(\xi, \eta; z\varsigma) = \phi_j(\xi, \eta; z\varsigma),$$

which proves (34) by (35).

Define a linear functional f on $\mathcal{A}$ as follows.

$$f(x_1^{m_1} y_1^{n_1} \cdots x_k^{m_k} y_k^{n_k} z^{m_0}) = \psi(x_1^{m_1} y_1^{n_1} z^n, x_2^{m_2} y_2^{n_2} \cdots x_k^{m_k} y_k^{n_k} z^{m_0 - n})$$

for $0 \le n \le m_0$. This functional f is well-defined by (34) and satisfies the equality

$$f(\xi\eta) = \psi(\xi, \eta) \qquad \text{for} \quad \xi \in \mathcal{A}_1, \eta \in \mathcal{A}_2.$$

Therefore

$$f([\xi_1\eta_1, \xi_2\eta_2]) = \psi([\xi_1, \xi_2], S(\eta_1\eta_2)) + \psi(S(\xi_1\xi_2), [\eta_1, \eta_2]) \tag{36}$$

for $\xi_j \in \mathcal{A}_1$ and $\eta_j \in \mathcal{A}_2$, since $[\xi_1\eta_1, \xi_2\eta_2] = [\xi_1, \xi_2]S(\eta_1, \eta_2) + S(\xi_1\xi_2)[\eta_1, \eta_2]$. Let $g(\xi, \eta) = \phi(\xi, \eta) - f([\xi, \eta])$. Then $g \in Z_\lambda^1(\mathcal{A})$ and

$$\begin{aligned} g(\xi_1\eta_1, \xi_2\eta_2) &= g_2(dU\eta_1 \wedge dU\eta_2, U\xi_1\xi_2) + \tilde{\phi}_1(U\xi_1, U\xi_2; U\eta_1\eta_2) \\ &\quad + \tilde{\phi}_2(U\eta_1, U\eta_2; U\xi_1\xi_2) \end{aligned}$$

for $\xi_j \in \mathcal{A}_1$ and $\eta_j \in \mathcal{A}_2$. Thus $g(\xi, \eta)$ depends on $U\xi$ and $U\eta$ only. From a formula in §3 of [4], it follows that there is a linear functional $g(\cdot)$ such that

$$g(\xi, \eta) = g(dU\xi \wedge dU\eta)$$

which proves (25).

Remark 1. It is obvious that $\phi \in Z_\lambda^1(\mathcal{A})$ is a coboundary of a cyclic 0-chain if and only if

$$\phi(\xi, \eta) = \psi([\xi, \eta])$$

where ψ is a linear functional on $[\mathcal{A}, \mathcal{A}]$. Thus the 1-cocycle (25) is a coboundary of a cyclic 0-cochain if and only if there is a linear functional h on $\mathcal{P}_n$ such that

$$g(dU\xi \wedge dU\eta) = h\left(\sum_{i=1}^{n} \frac{\partial(U\xi, U\eta)}{\partial(u_i, v_i)}\right)$$

Remark 2. If $n > 1$, the functional g in (25) may not be zero. For instance, let g be a linear functional such that $g(dUx_1 \wedge dUx_2) \neq 0$, and

$$\phi(\xi, \eta) = g(dU\xi \wedge dU\eta).$$

Then $\phi(\xi,\eta) \in Z_\lambda^1(\mathcal{A})$ may not be represented as $f([\xi,\eta])$, since $[x_1,x_2]=0$ and $\phi(x_1,x_2) \neq 0$.

5. Suppose ρ is a linear operator from an algebra $\mathcal{A}$ over C into the algebra $\mathcal{L}(\mathcal{H})$ of all bounded linear operators in $\mathcal{H}$ satisfying the condition

$$\rho(a_0a_1) - \rho(a_0)\rho(a_1) \in C_1(\mathcal{H}) \qquad \text{for} \quad a_0, a_1 \in \mathcal{A},$$

where $C_1(\mathcal{H})$ is the trace ideal of $\mathcal{L}(\mathcal{H})$. For odd n, define

$$\rho(a_0,a_1,\ldots,a_n) = tr\left(\sum_{j=0}^{n}(-1)^j\rho(a_{\sigma_1})\cdots\rho(a_{\sigma_{j+n}} - \rho\left(\sum_{j=0}^{n}(-1)^j a_{\sigma_{j+n}}\right)\right), \quad (37)$$

where $\sigma_j^{(n)} = j$ for $j \le n$ and $\sigma_j^{(n)} = j-n-1$ for $j > n$. This ϕ gives rise to an element of $Z_\lambda^n(\mathcal{A})$. It is obvious that

$$\rho(a_1,a_2,\cdots,a_n,a_0) = (-1)^n\rho(a_0,a_1,\cdots,a_n).$$

We have to prove

$$b\rho = 0, \quad (38)$$

where b is the Hochshild coboundary:

$$(b\rho)(a_0,a_1,\ldots,a_{n+1}) = \sum_{j=0}^{n}(-1)^j\rho(a_0,\cdots,a_ja_{j+1},\cdots a_{n+1}) + \rho(a_{n+1}a_0,\cdots a_n).$$

Notice that in the expression of $\rho(a_0a_1,a_2,\cdots,a_{n+1})$, if the operator $\rho(a_0a_1)$ is replaced by $\rho(a_0)\rho(a_1)$ the difference which is made by this replacement is

$$tr((\rho(a_0a_1)-\rho(a_0)\rho(a_1))\rho(a_2)\cdots\rho(a_{n+1}) - \rho(a_2)\cdots\rho(a_{n+1})(\rho(a_0a_1)-\rho(a_0)\rho(a_1)) \\ +\cdots - \rho(a_{n+1})(\rho(a_0a_1)-\rho(a_0)\rho(a_1))\rho(a_2)\cdots\rho(a_n)) = 0.$$

Hence

$$\rho(a_0a_1,a_2,\cdots,a_{n+1}) = \\ tr(\rho(a_0)\rho(a_1)\cdots\rho(a_{n+1}) - \sum_{j=2}^{n+1}(-1)^j\rho(a_{\tau_j})\cdots\rho(a_{\tau_{j+n+1}})) \\ -\rho(a_0a_1\cdots a_{n+1}) - \sum_{j=2}^{n+1}(-1)^j a_{\tau_j}\cdots a_{\tau_{j+n+1}})),$$

where $\tau_j = j$ for $j \leq n+1$ and $\tau_j = j - n - 1$ for $j > n+1$. There are similar expressions for $\rho(a_0 \cdots, a_j a_{j+1}, \cdots, a_{n+1})$ for $0 < j < n+1$ and $\rho(a_{n+1}a_0, \cdots, a_n)$, from which it is easy to deduce (38).

References

[1] A. Connes, Non commutative differential geometry. Ch. I, II (preprint I.H.E.S.).

[2] P. Cartier, Homologie cyclique: Rapport sur des travaux récents de Connes, Karoubi, Loday, Quillen...,(preprint).

[3] J. Dixmier, Enveloping Algebras, North-Holland Pub. Com. (1977) Amsterdam, New York, Oxford.

[4] D. Xia, Trace formula for almost Lie algebra of operators, Integral Equations and Operators Theory, **8**(1955), 854-881.

Index

9 780824 777777